NASA

HUBBLE SPACE TELESCOPE

1990 onwards (including all upgrades)

Owners' Workshop Manual

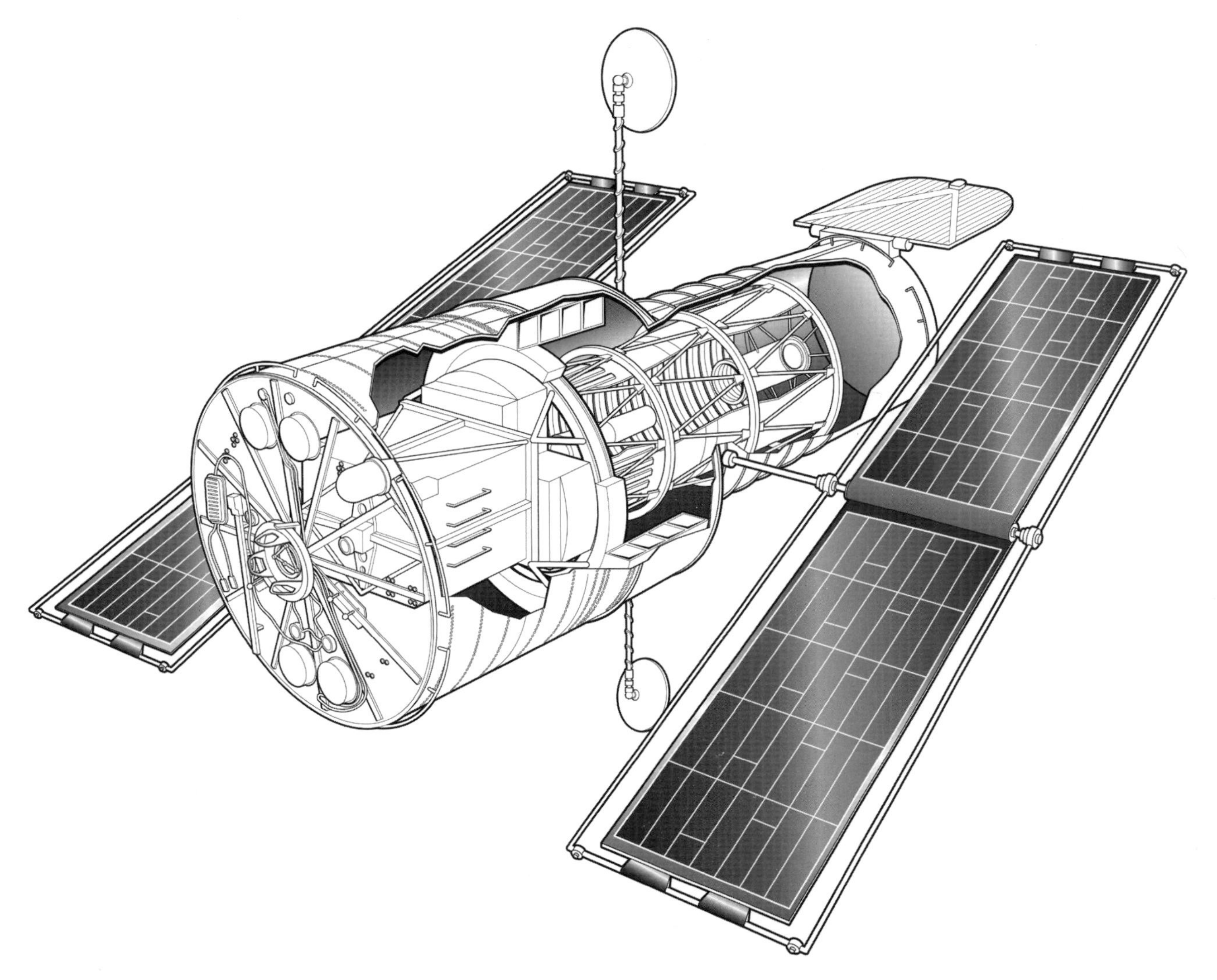

An insight into the history, development, collaboration, construction and role of the Earth-orbiting space telescope

David Baker

Contents

OPPOSITE The Hubble Space Telescope drifts away from the Orbiter Discovery as the Shuttle undertakes a gentle separation manoeuvre at the end of the third servicing mission. *(NASA)*

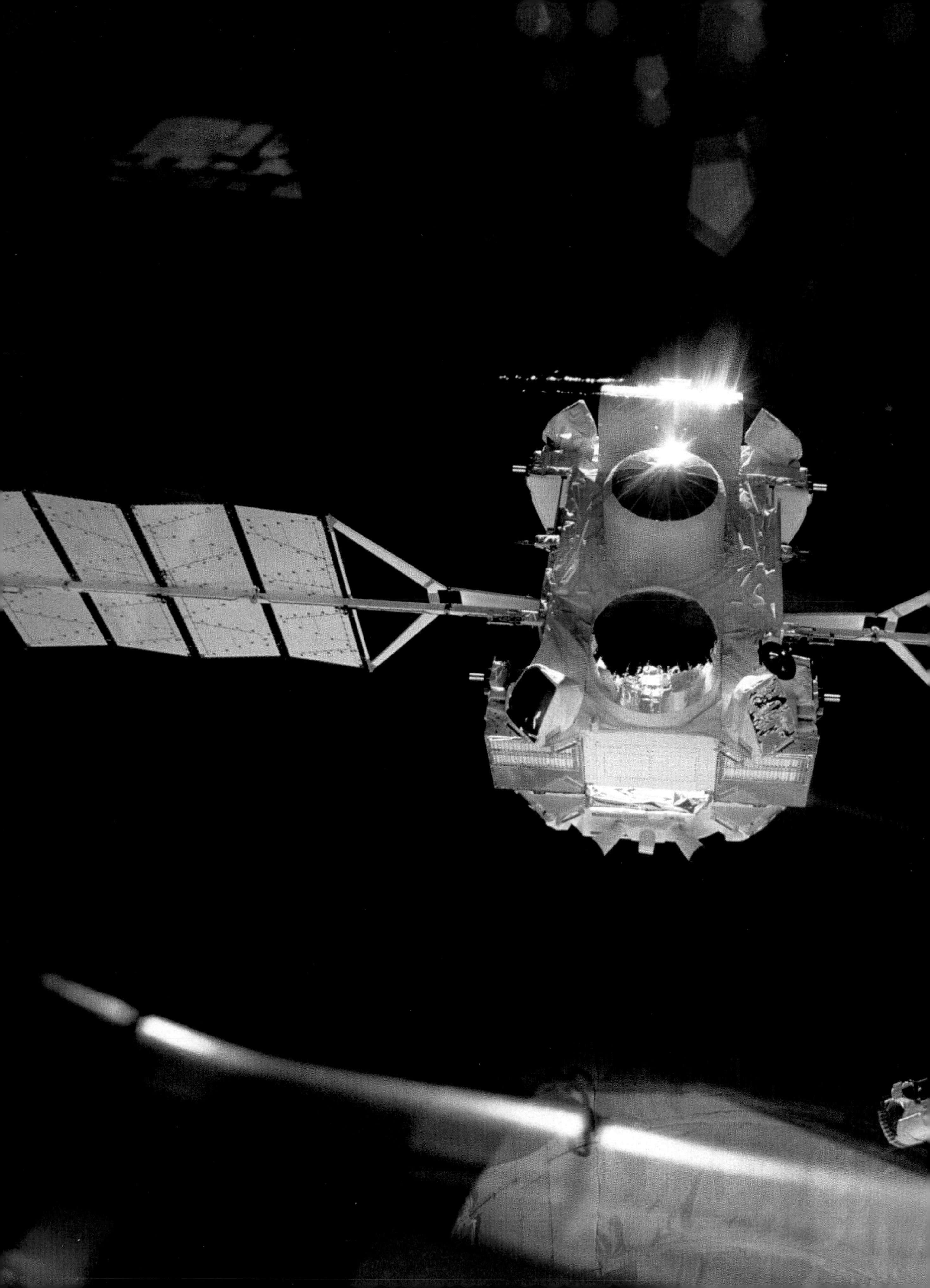

Chapter One

Origins and other observatories

The idea of placing an astronomical observatory in space to see the heavens with clarity, devoid of the obscuring effects of the Earth's atmosphere, is certainly as old as the space programme itself. In fact it was one of the justifications for putting people into space, at a time when electronics were new and the manual control of spacecraft systems was considered essential for even routine investigations of the universe from space.

OPPOSITE Launched on 5 April 1991, the Compton Gamma Ray Observatory was named after Dr Arthur Holly Compton, the American physicist who conducted studies of cosmic rays in the 1930s. With a weight of 34,450lb (15,623kg) it is one of the heaviest, and carried four science instruments for studying some of the most energetic events in the universe. Compton is seen here before it was deployed during the Shuttle STS-37 mission. *(NASA)*

RIGHT The Hubble Space Telescope represents one of four major astronomical observatories built during the 1980s and 1990s for advancing the science of cosmology, in addition to adding to knowledge supporting optical and non-optical astronomy. The first to be launched, Hubble would become perhaps the most famous telescope of all time, producing stunning images that inspire far beyond the realm of professional investigators. Designed to be visited several times by Shuttle astronauts, the HST is seen here during the final servicing mission in 2009. *(NASA)*

After Sputnik 1 ushered in the Space Age in October 1957, thoughts quickly turned to placing an observatory in orbit. In 1962, four years after NASA had been formed, a Space Studies Board summer study in the United States suggested that the time was right to plan for an orbiting telescope with a trained astronomer on board for limited but frequent visits.

NASA began studies to see how such a telescope might be designed and how it could be built, and to support such proposals arch supporter Aden Meinel set up a Space Division at the Kitt Peak National Observatory. But not all astronomers were enthusiastic. Ira Bowen, director of the Mount Wilson and Mount Palomar observatories, argued against such an idea, saying that such a telescope could never be made sufficiently stable in orbit to make it effective.

Nevertheless, with opinion strongly divided, by 1965 NASA had decided that it was time to start planning for such a mission, a move endorsed by another SSB summer study that year. A committee of the SSB under the chairmanship of Lyman Spitzer began a four-year study to define the appropriate instruments for what was soon to be known as the Large Space Telescope (LST). To support their plan, NASA mobilised action groups bridging the space agency with astronomers and specialist astronomical groups across the United States.

NASA set up a special LST Task Group in 1970 to define the engineering requirements and to make a list of scientific objectives so that a shortlist of appropriate instruments could be developed. During the next two years both the Goddard Space Flight Center and the Marshall Space Flight Center sent out commercial competitive bids for definition of the LST, and in 1972 overall management of the Telescope went to Marshall, with Goddard taking charge of the science instruments. But it was yet to be formally approved as a funded NASA programme and the agency was in a state of flux.

By 1972 the heyday of space was over. Man had landed on the Moon and the last expedition would visit the lunar surface in December that year; the budget was in serious decline and the Apollo/Saturn era was coming to an end. Only the Skylab space station of 1973/74 and the joint docking flight with the Russians in 1975 remained for Apollo-era hardware. NASA was looking to replace all expendable rockets with a reusable transportation system known as the Shuttle, and it had been formally approved by the Nixon administration in January 1972 with an anticipated operational debut long before the end of that decade.

The extraordinary development of electronics, computers and the operation of highly sophisticated and complex satellites and spacecraft meant that the LST need not be manned – in fact notions of sending astronomers up to make observations *in situ* had long disappeared from the list of possibilities. Neither was it desirable; even limited movement around the inside of a spacecraft would send it oscillating and bouncing like a trampoline, an effect hardly conducive to highly stable platforms with fine-pointing requirements.

But the Shuttle was absorbing a large portion of the budget and NASA was unable to get Congressional approval to fund the Telescope until 1977. By this time, as a way to reduce the cost, the newly formed European Space Agency had agreed to provide one of the five science instruments on board and to build the solar arrays, in return for 15% of observational time. Work on the Faint Object Camera began at ESA in 1975. Selection of the four US-built instruments was not so easy and an Instrument Science Team was formed to settle the matter. Bold decisions were made, such as the use of charge-couple devices (CCDs) for the first time in space, although as it turned out the Galileo spacecraft, launched ahead of the HST, made that breakthrough, despite it using a very different type of CCD.

As a compromise to gain approval, the preferred 9.8ft (3m) diameter primary mirror had been reduced to 7.9ft (2.4m), which allowed the use of existing manufacturing and fabrication techniques that in itself saved a lot of money. At the core of its design was an ability to periodically visit the Telescope using the Shuttle and replace ageing instruments with new ones. That in itself prompted a shift toward minimal testing and reduced tolerances, programme administrators deciding that since it could be readily accessed it would not be worth spending money on building a Telescope which could last for 10–15 years without attention.

Another aspect of the programme which some saw as causing concern was the paucity of NASA test and quality control engineers allowed into the facilities of both Lockheed Martin, the prime contractor, and Perkin-Elmer, who made the mirror. At the time, both companies were working on highly classified military surveillance and spy satellite programmes, and there was reluctance to grant passes to more than a few select individuals. Consequently, the number of personnel allowed in from outside these companies, especially the civilian space agency, was a tiny fraction of the number which would normally be expected to attend in residence for a project of this size.

Between 1980 and 1983 the programme ran into managerial and technical difficulties that some were convinced would bring cancellation. Conflict between the Marshall and Goddard facilities was toxic until a new programme director forged a better working relationship, and the way it was set up between NASA and Lockheed Martin brought hostile confrontations. When begun in 1977 the Space Telescope was expected to fly by 1983, but three years of delays rescheduled the launch for 1986.

In 1983 it was given its definitive name, after astronomer Edwin P. Hubble; as the Hubble Space Telescope it would encompass a new and much improved working relationship between NASA field centres and industry, much of which was catalysed by the robust efforts of NASA Administrator James E. Beggs. But this was a new era of scientific investigation using satellites and spacecraft. The 1960s had been all about establishing a role for humans in space, and the 1970s had been the great age of planetary exploration. The next decade would usher in the defining age of space astronomy that would endure long into the future.

By the 1980s there was a surge of anticipation that space-based observatories would advance the science of astronomy and propel it to new heights of knowledge and understanding about the most fundamental questions concerning the nature and structure of our universe. Through a series of four specialised spacecraft the broad electromagnetic spectrum would be observed from space above the obscuring effect of Earth's atmosphere in the infrared, optical, x-ray, and gamma-ray wavelengths. Thus was born the age of the 'Great Observatories' in space.

Leading the way was the Hubble Space Telescope launched in 1990, followed by the Compton Gamma Ray Observatory in 1991; the Chandra X-ray Observatory in 1999; and the Spitzer Space Infrared Telescope Facility, launched in 2003 and named after Lyman Spitzer, an enduring advocate of astronomy from space who had championed the optical telescope in orbit and had chaired the Space Science Board when it pushed for what became the Hubble Space Telescope.

LEFT Still the most powerful space-based X-ray telescope, the Chandra X-ray Observatory was launched on 23 July 1999 and continues to operate today. Unique in its capabilities, the telescope has made major contributions to the observation of x-radiation produced by the presence of black holes, and has provided independent evidence for the existence of dark energy, an as-yet undefined force which is accelerating the expansion of the universe. *(NASA)*

BELOW With a weight of 12,830lb (5,864kg), Chandra has a length of 45.3ft (13.8m) and a span of 64ft (19.5m) across its twin deployed solar arrays. Although designed for a life of five years, it continues to be supported by astronomical groups and by NASA's Astrophysics Division since a successor will not be available until at least the mid-2030s. *(NASA)*

Chapter Two

Edwin Hubble and the modern universe

Born on 20 November 1889 in Marshfield, Missouri, Edwin Powell Hubble is a giant among the greatest names in astronomy, and his discoveries gave birth to the idea that the universe is expanding and not static.

OPPOSITE The Hubble Space Telescope captures the breathtaking beauty of the Monkey Head nebula (NGC 2174). (*NASA*)

LEFT Edwin Hubble made some of the most important contributions to astronomy and reinforced new ideas in cosmology equal in scientific stature to Einstein's contribution to physics. Since one of his great discoveries was the presence of Cepheid variables in the M31 galaxy, which demonstrated that objects are receding at a rate proportional to their distance and thus gave birth to the 'Hubble constant', it is fitting that the Space Telescope, designed for deep-sky surveys, should be named after him. *(Franklin Institute)*

BELOW The 100in Hooker reflecting telescope at the Mount Wilson Observatory, used by Hubble for observing stars and nebulae in constructing the theory of an expanding universe. *(Mount Wilson Observatory)*

After completing undergraduate studies at the University of Chicago, he journeyed to England and attended Oxford as a Rhodes Scholar studying jurisprudence. Returning to Chicago to continue his studies he received his PhD after deciding that astronomy was far more appealing than law, and that it was to this that he should devote his time 'even if I were second rate or third rate; it was astronomy that mattered'. His dissertation was on the *Photographic Investigation of Faint Nebulae*. Between 1914 and 1917 Hubble carried out research at the Yerkes observatory before enlisting in the army after America entered the First World War in April 1917. Sailing to France, he was part of the American expeditionary force on the Western Front.

In 1919 Hubble got a position with the astronomy staff at the Carnegie Institution of Washington's Mount Wilson Observatory, and he remained there until his death on 28 September 1953 in San Marino, California. But his greatest work came early in his astronomical career. Using the 2.5m reflector at Mount Wilson, Hubble studied the Andromeda galaxy and other nebulae and went on to prove that they were stellar systems in their own right just like our own Milky Way galaxy. Hubble was looking for Cepheid variables, a type of pulsating star named after the constellation Cepheus in which they were first observed, their brightness increasing and decreasing in a highly precise and periodic manner. And Cepheids, thought Hubble, were the key to measuring the distance of very faint nebulae.

In 1912, when studying the Cepheids in the nearby Small Magellanic Cloud, Henrietta

Leavitt noted that there was a universal relationship between the mean brightness and the period: the period is related to the absolute magnitude, which is the brightness of a star when seen at a distance of 1 parsec (3.26 light years), determined as one parallax second of arc. In this way astronomers can measure the *apparent* magnitude as the luminosity they receive in their telescopes, and extrapolate from that the absolute magnitude by measuring the distance to that star as it drifts across the background of more distant objects as the Earth goes round the Sun.

This method is only good to a distance of approximately 100 parsecs (326 light years), because at that distance the motion of a star seen against a more distant background becomes so small as to be unmeasurable. What was needed was a means of determining the distance of stars that were so far away that they appeared motionless, even when measured across a baseline of approximately 186 million miles (315.4 million km) – the diameter of the Earth's orbit about the Sun. By relating the period of the variation in luminosity to its absolute magnitude, Leavitt gave astronomers the tool to calculate the distance, not by measurement but by noting the apparent magnitude and calculating the absolute magnitude from the period.

In 1923 Hubble discovered a Cepheid variable in Andromeda from which he calculated its distance to the Earth. This provided the first certain evidence that extragalactic nebulae were far beyond the boundaries of our own galaxy and this consolidated his view, and eventually became the proof, that they were stellar islands – galaxies. In the case of Andromeda we now know that this galaxy is more than two million light years from Earth. In the time this work was being conducted it was an astonishing discovery, and one that confounded many astronomers of the day.

Over the next several years Hubble observed that these nebulae were uniformly distributed across the sky and that their numbers increased with distance at a constant rate out to the limits of the observable universe. This enabled Hubble to take the next intellectual leap and use the studies of spectra of nebulae begun by V.M. Slipher in the United States, who in 1912 had showed that the H and K lines of ionised calcium appeared at longer wavelengths in these distant galaxies than in the spectra of light from the Sun and other stars in our galaxy. This change toward the red end of the spectrum

LEFT The Andromeda galaxy, M31 in the Messier catalogue of nebulae, in which Hubble discovered Cepheid variables. *(Mount Wilson Observatory)*

is the Doppler effect that arises from receding objects. From measurement of the precise displacement, the actual velocity of recession can be calculated.

In 1929 Hubble announced that the velocities of recession increase in a linear manner and that by using Cepheid variables to determine the distance of the nebulae (galaxy) the red shift can show the velocity. Just as the parallax method of geometric measurement and the observation of Cepheid variables enabled astronomers to calculate distance, calculations using the red shift would provide details of the motion of the universe over time. This fed directly into Albert Einstein's equation that $E = mc^2$ (energy equals mass times the square of the velocity of light [~299,000km/sec]), which caused the great scientist to revise his 1917 General Theory of relativity and make corrections to set the universe in motion rather than express it in a static frame of reference. In fact, Einstein had thought his work was taking him toward such a view but felt it so fantastic a notion that he backed away and settled for a static model.

What Hubble did was to set a constant that, if applied to any Cepheid variable and associated red shift, would tell not only the distance but the speed of recession. But he did more than that. Since light travels at approximately 186,000 miles/sec (299,000km/sec), these more distant objects were being observed not as they are now but as they were back at the evolution of the universe. In other words, the speed at which they are observed to move is not the speed they are moving at now, but the velocity at which they were travelling at that distance back in time.

Hubble provided the tools from which the speed at which the universe was expanding could be calculated, and this completely transformed the science of astronomy and provided a new and powerful science – cosmology. But this work was slow, and right up to the time the Hubble Space Telescope was being planned the most distant object defined by the red shift was at 41% of the speed of light, about four billion light years away, inferring the observation of an event four billion years ago. That is one-third the age of the universe known today.

Hubble assembled a vast body of work before his death, and because great priority was given in the 1960s and 1970s to very

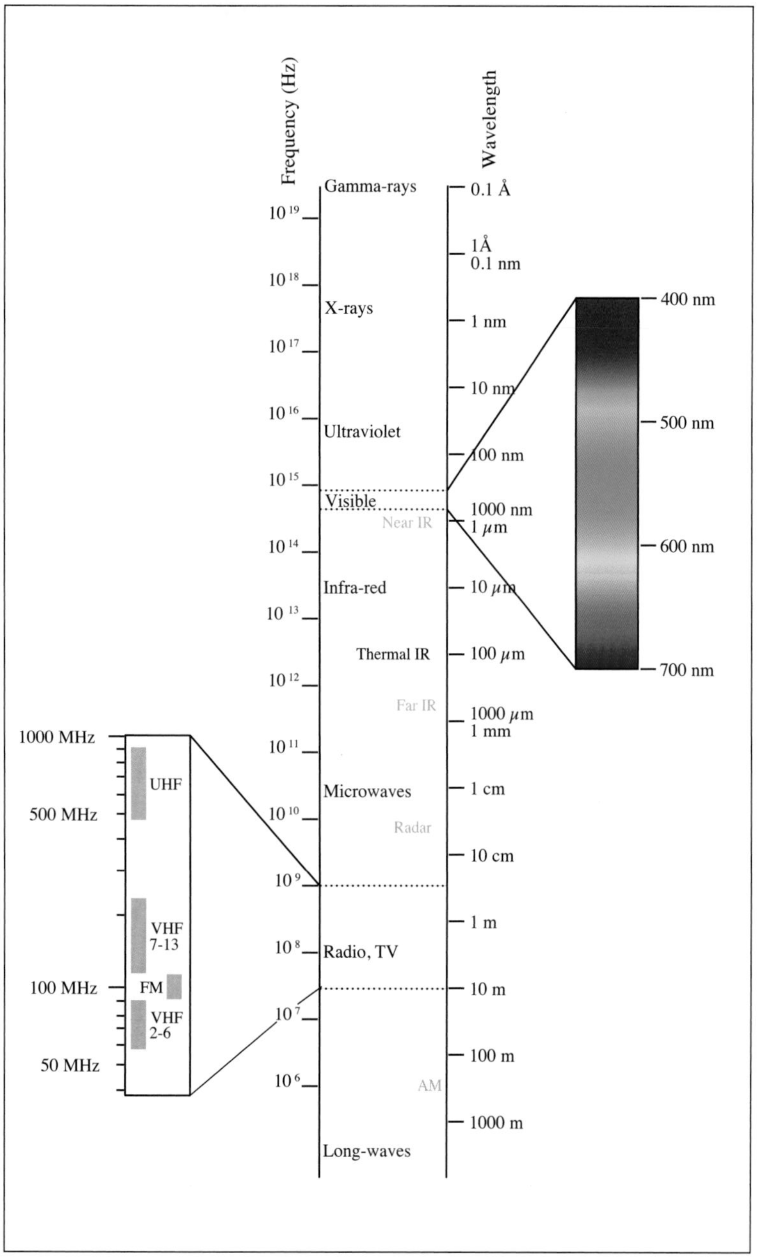

LEFT The electromagnetic spectrum displays the range of energies and values of wavelength and frequency from radio waves at one end to high-energy gamma rays at the other. Note the small window of visible light through which all astronomical observations were made until the advent of infrared and ultraviolet measurements. *(David Baker)*

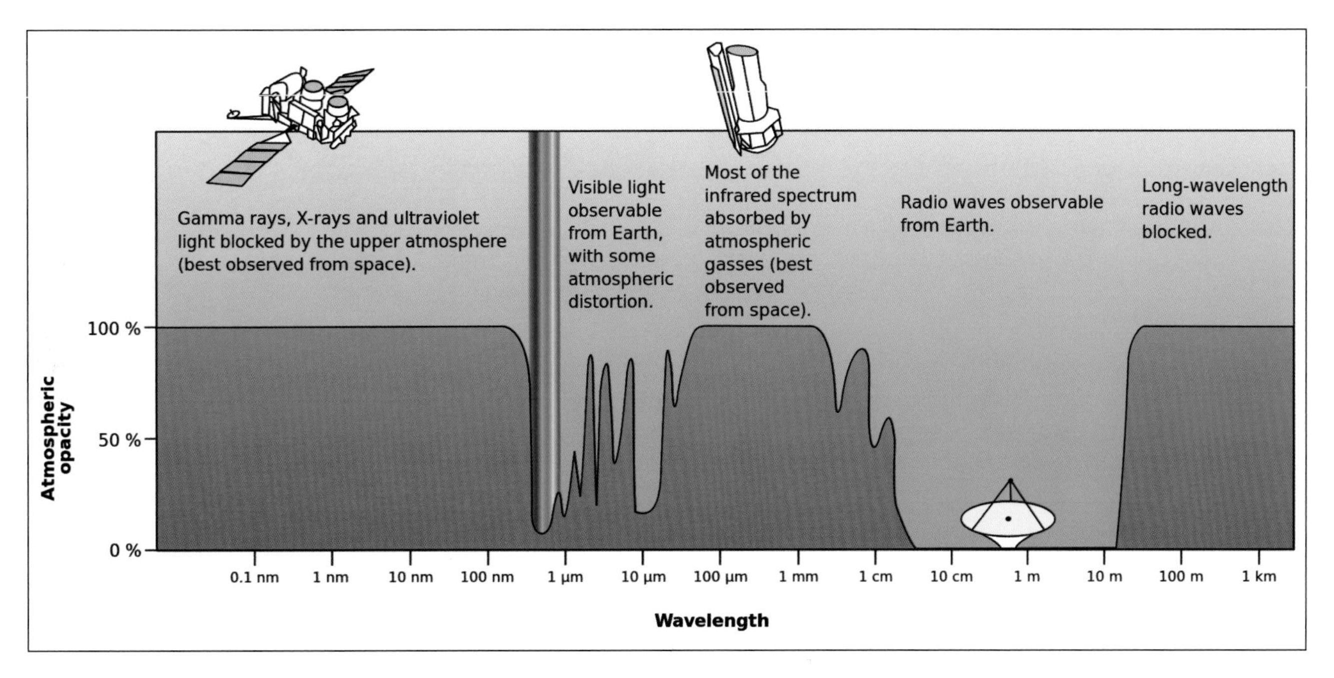

precise measurement of the distance of Cepheid variables with this new space-based observatory, it was for this reason that the Telescope was named after this giant among astronomers. The HST had as its primary task the very precise measurement of Cepheid variables, and because of that it is indeed fitting that his name should be forever associated with the first great optical telescope placed in space.

ABOVE The opacity of the atmosphere to various wavelengths on the electromagnetic spectrum. Note that wavelengths shorter than 100nm beyond the ultraviolet portion of the spectrum are blocked by the upper atmosphere, as are most of the infrared and radio waves longer than about 20m. The need for space-based observatories in these regions is self-evident. *(David Baker)*

BELOW This simple diagram shows how the apparent motion of a near-field star to a distant background of 'fixed' stars provides a trigonometric value that translates into distance. Here, positions A and B represent the Earth at six-month intervals in its annual orbit around the Sun. *(David Baker)*

PHOTO TAKEN FROM POSITION A

p

d

A

1 AU

SUN

EARTH'S ORBIT

B

PHOTO TAKEN FROM POSITION B

BELOW The division of one complete circle into 360° and from there into 60' (minutes) and 60" (seconds) of arc demonstrates the extreme accuracy of the Fixed Guidance Sensor on Hubble at 0.002". Accuracy such as this is vital for differentiating between separate stars for astrometry. *(David Baker)*

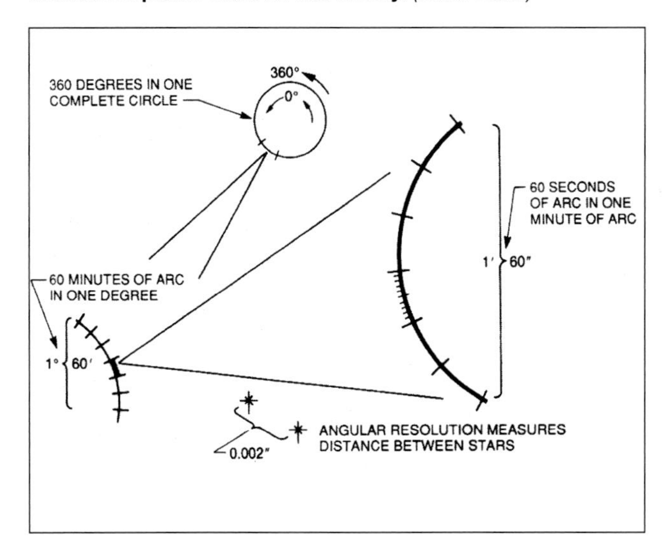

FLIGHT

Chapter Three

Design and development

As described in Chapter 1, development of the Hubble Space Telescope matured into a spacecraft that could be revisited by the Shuttle and refurbished, its instruments replaced as necessary for an evolving series of astronomical investigations, with its support systems upgraded as new technology became available.

OPPOSITE The Hubble Space Telescope displays the large doors at the base of the SSM Aft Shroud providing access to support systems and instruments. *(NASA)*

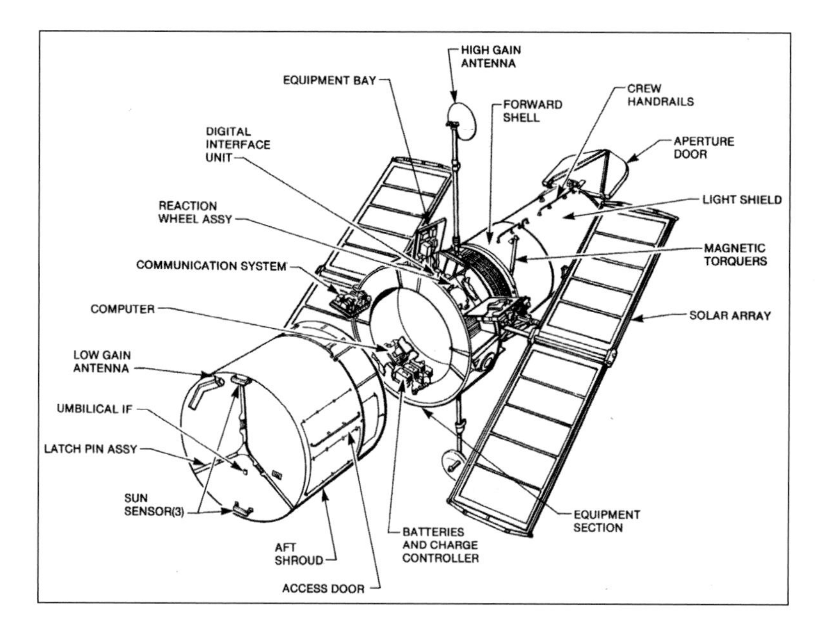

LEFT The essential elements of the Hubble Space Telescope are similar to those common in conventional Earth-based telescopes, except for the solar array panels for producing electrical power. The HST is structurally divided into a Support System Module, an Optical Telescope Assembly and the five science instruments. *(Lockheed)*

This in-orbit servicing was a clearly defined feature of a few satellites funded in the 1970s, along with development of the Shuttle with which they were compatible. Astronomers and scientists realised that engineering development would make available new and exciting possibilities unattainable when the Telescope was launched, and revisits by Shuttle would be a key feature of its projected longevity.

The Hubble Space Telescope is an international endeavour between NASA and its coterie of US-based contractors, subcontractors and suppliers, and the European Space Agency (ESA) with its headquarters in Paris. Structurally, the HST was manufactured in the US, with one of the five science instruments provided by the Europeans and the two solar arrays manufactured in the UK. The scientific use of the HST benefited from this international effort, which stimulated a broad global participation involving astronomers and scientists from around the world.

As launched in 1990 the basic configuration of the Telescope consisted of a Support Systems Module (SSM), an Optical Telescope Assembly (OTA) and, initially, five scientific

BELOW Integration of the Support System Module and the Optical Telescope Assembly allows a structurally rigid support for the primary and secondary mirrors and for the support systems for pointing and control. *(Lockheed)*

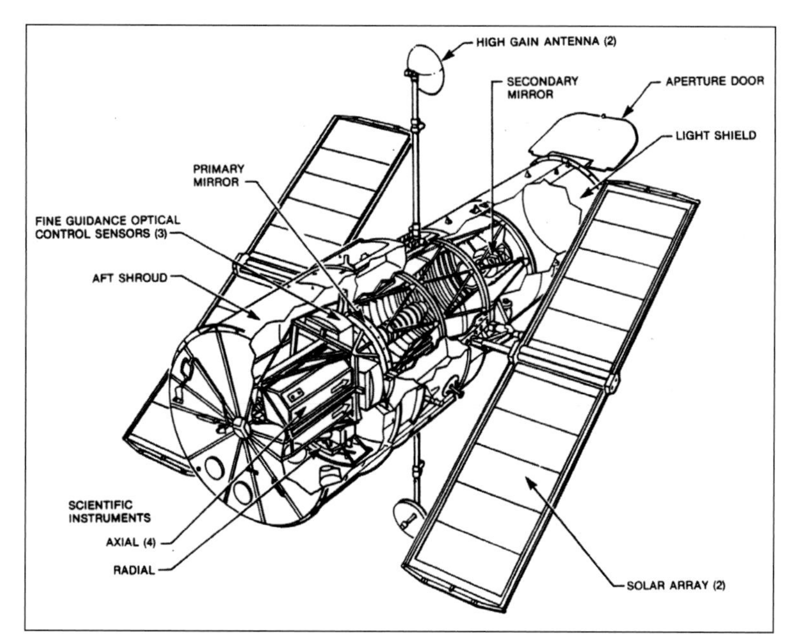

BELOW The axial reference frame for the HST serves as a coordinate matrix for locating equipment and appendages. In normal operations the Sun will be placed in the V1–V3 plane on the V3 side of the spacecraft with viewing positions outside a 50° zone on the Sun-line. *(Lockheed)*

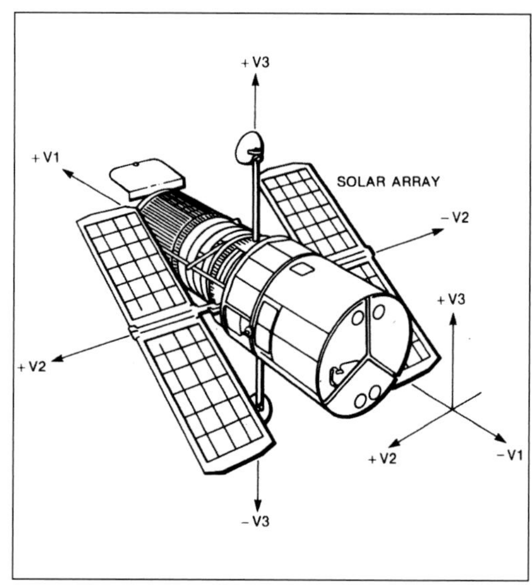

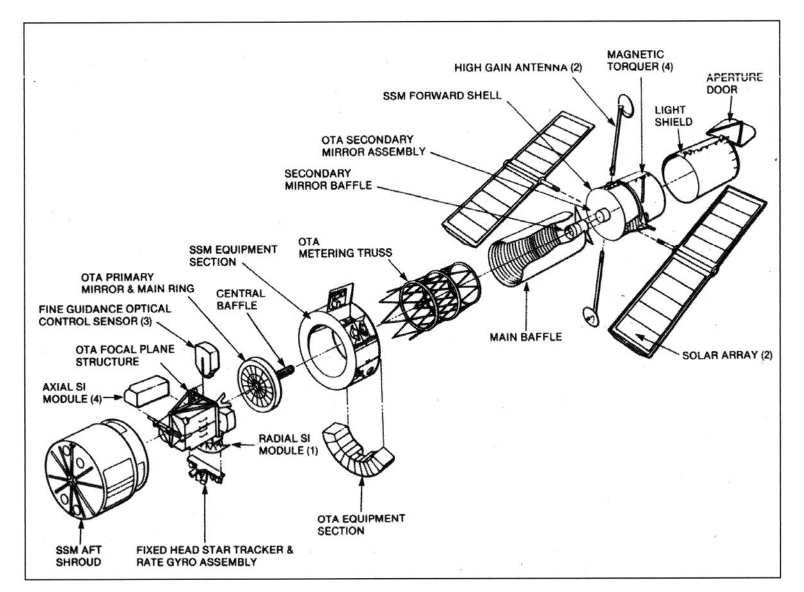

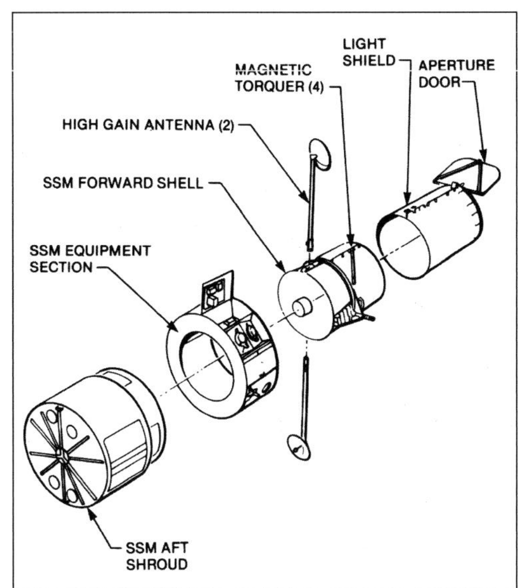

ABOVE LEFT The HST is broken down into separate elements, structures, systems and subsystems. *(Lockheed)*

ABOVE RIGHT The structural components of the Support System Module include the Aft Shroud, Equipment Section, Forward Shell and Light Shield with the aperture door to protect the primary mirror from direct sunlight. *(Lockheed)*

instruments. In one sense it can be considered to comprise a reflector telescope, the OTA, contained within a cylindrical satellite, the SSM. In another sense the SSM duplicates the function of an Earth-based observatory supporting the Telescope or OTA and its associated instruments. Either way, it is two precision-engineered structures manufactured by two separate companies and integrated as a single satellite-cum-spacecraft.

The then named Lockheed Missiles and Space Company received the contract to build the SSM and to integrate the two elements, while the then Perkin-Elmer Corporation was awarded a development and fabrication contract for the OTA. As with other science satellites and orbiting observatories, the experiments were designed and manufactured by the separate investigating groups, including non-government research establishments as well as NASA's Goddard Space Flight Center (GSFC). The HST has a total length of 42.5ft (12.9m) and a maximum diameter of 14ft (4.27m) with a launch mass constrained to 25,000lb (11,340kg), which is the uplift capability of the Shuttle to the operational orbit of 379 miles (610km) by direct ascent.

The SSM consists of an outer structure that provides core services and capabilities for operating the Telescope, such as the provision of electrical power, thermal control, data handling, communications equipment and attitude control. The OTA comprises the basic structural elements of the Telescope itself, gathering light and focusing it in the focal plane for the instruments.

The initial suite of scientific instruments was installed with a view to their being replaced with other instruments over the operational life of the

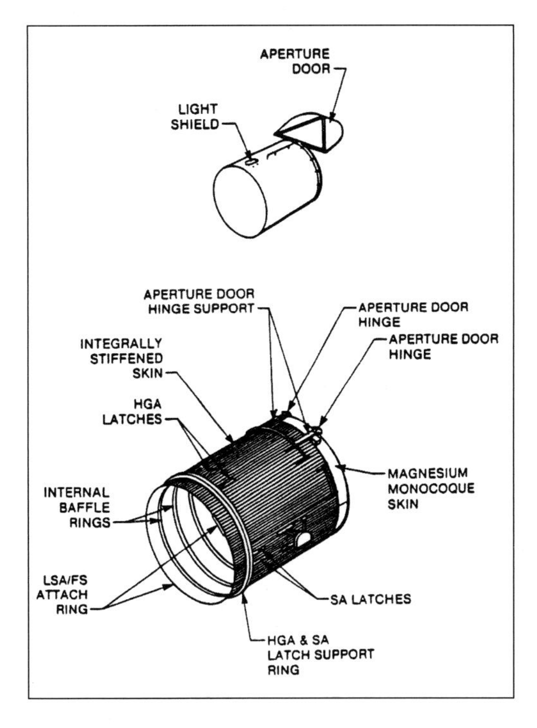

LEFT The magnesium monocoque Light Shield provides latches for the aperture door and hand rails for access during EVA operations. *(Lockheed)*

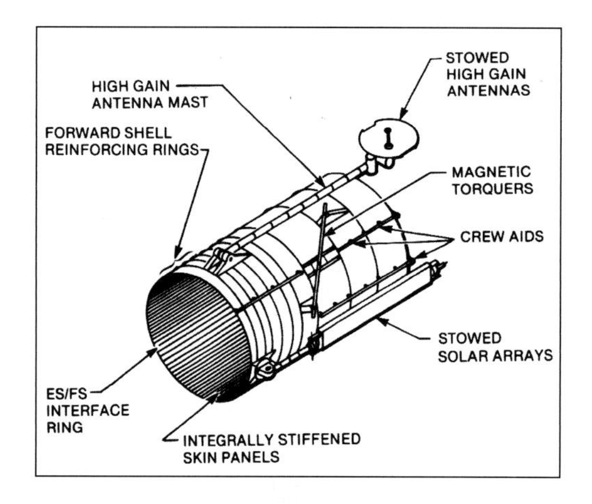

RIGHT Magnetic torque devices are located on the Forward Shell, and hand rails provide astronaut access to the region where the solar arrays are stowed for launch. *(Lockheed)*

Telescope, an important feature of its design and functional layout. Information gathered by the instruments is supplied to the on-board computers, where it is processed and either stored for later transmission or sent straight to the ground. In this way the SMM, the OTA and the instruments operate on a feedback loop.

The orientation reference for the Telescope was maintained through three axial planes: V1, V2 and V3. The primary V1 axis runs through the centre of the Telescope from one end to the other, the aft end referenced as -V1 and the forward (open) end of the Telescope as +V1, relative direction of the -/+ designation being aft or fore of the V2 axis. Motion around the V1 axis constitutes a roll in a clockwise or counter-clockwise rotation.

The V2 axis is parallel to the solar array wings and constitutes a line through their rotational hinge point, -V2 being to the right of the Telescope centreline and +V2 to the left. Movement around the V2 axis constitutes a pitching movement of the Telescope up or down. V3 is the vertical axis, perpendicular to the other two axes, and is the pivot around which the Telescope can be said to yaw to the right or to the left. For attitude reference, -V3 is in a downward direction, with +V3 upward. In this way the HST can be positioned to point in any desired direction by commanding it to move in any two or three axes. Uniquely among telescopes it is independent of the Earth's rotation. But more about that later on.

Support Systems Module (SSM)

The SSM consists of the exterior shell and containment structure into which is located the Optical Telescope Assembly. It comprises an outer structure of interlocking shells, rotating

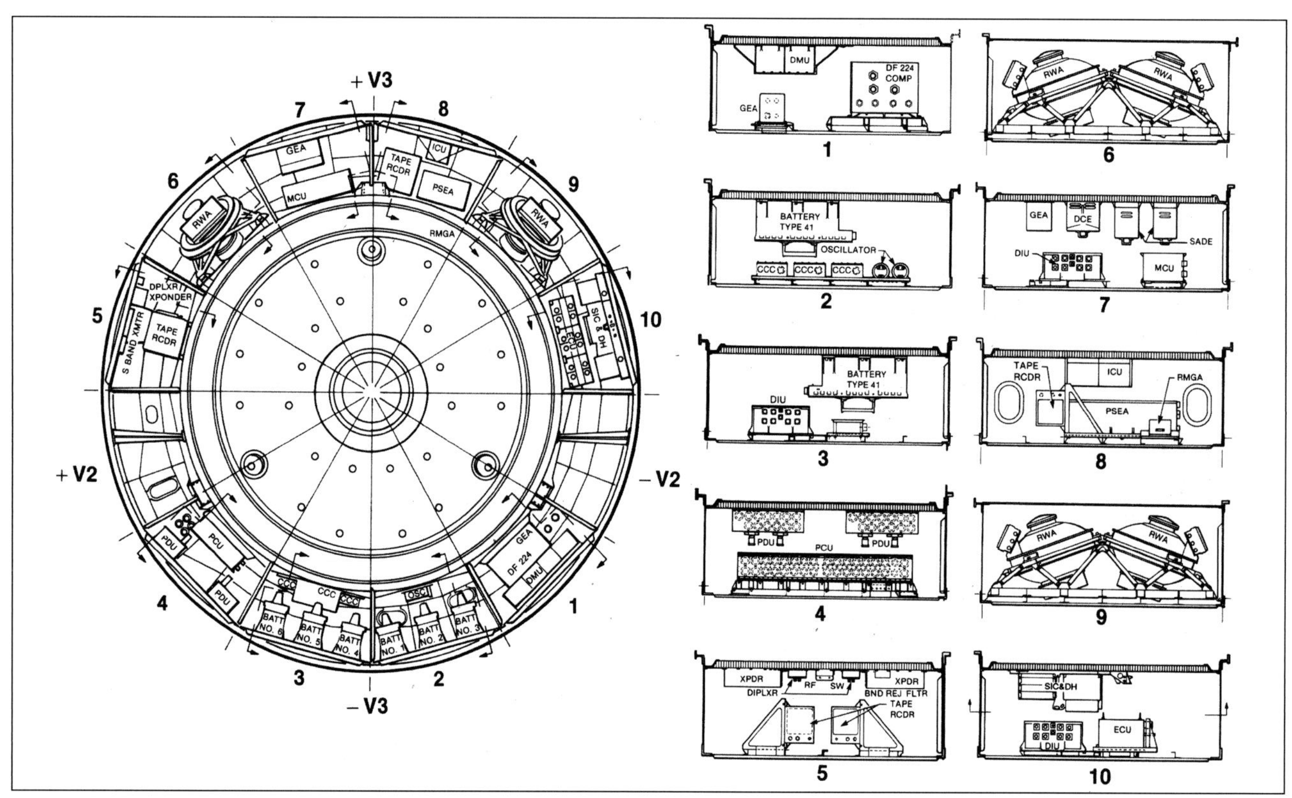

BELOW The SSM Equipment Section bays contain most of the systems and subsystems for operating the HST as well as looking after 'housekeeping' activity to maintain the Telescope as a working observatory. Use the list of abbreviations on page 178 to identify systems and subsystems by their acronyms on the diagram. *(Lockheed)*

reaction wheels with magnetic torque devices for attitude stabilisation, two solar arrays and their rotation and support structures, communications antennae, circumferential equipment bays, computers for operating spacecraft systems and for data handling, and thermal protection devices. The structure also supports three external doors, latches, rails and fasteners specifically designed to be accessed and used by astronauts conducting EVA, or spacewalking, with internal lighting to illuminate areas where astronauts could be called upon to work in dark or confined spaces.

The external structure of the SMM comprises (from the rear of the Telescope end to the front) the Aft Shroud & Bulkhead assembly, the Equipment Section, the Forward Shell and the Light Shield. The Aft Shroud & Bulkhead (ASB) supports the Focal Plane Structure (FPS) containing the four axial science instruments (the Wide Field/Planetary Camera, the Faint Object Camera, the Faint Object Spectrograph and the High-Resolution Spectrograph). Three Fine Guidance Sensors and the Wide Field/ Planetary Camera are installed radially near the attachment point for the Equipment Section.

The ASB is fabricated from aluminium with a stiffened skin containing internal panels and reinforcing rings together with 16 internal and external longerons for additional strength, rigidity and support. When assembled it is

11.5ft (3.5m) long and 14ft (4.27m) in diameter. The ASB is also used as the primary connection between the HST and the Shuttle Orbiter during launch into orbit and before deployment when the systems are checked out via umbilical lines prior to release for free flight. Physical attachment to the Flight Support Structure (FSS) includes three electrical umbilical points, which are located in the bulkhead to connect systems inside the ASB with the Shuttle.

The rear Low-Gain Antenna (LGA) is also attached to the aft face of the bulkhead, which is fabricated from 2in (5.1cm) honeycomb aluminium panels and three radially mounted aluminium support beams for structural

LEFT **The Aft Shroud & Bulkhead provides both protection for the internal working systems and access to the science instruments during EVA operations in orbit.** *(Lockheed)*

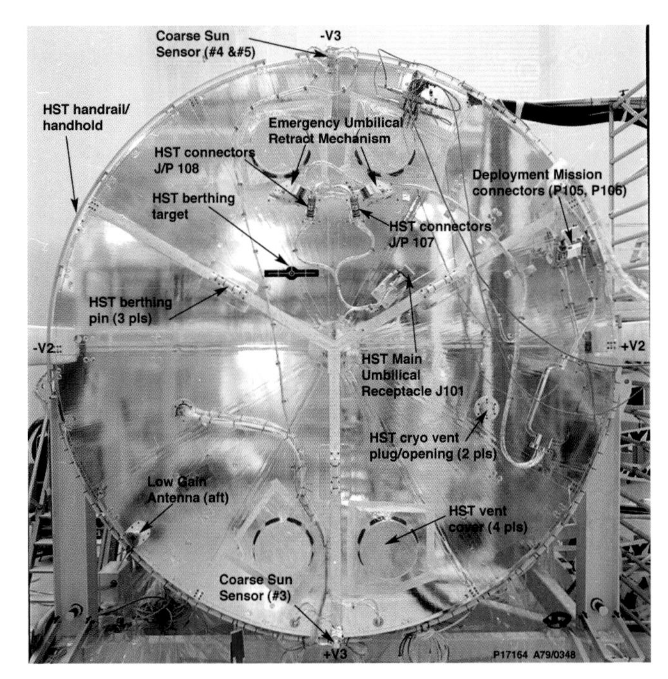

BELOW LEFT **The Aft Bulkhead provides a support structure for attaching the HST to the Flight Support Structure and also supports the course Sun sensors, berthing pins, low-gain antenna and umbilical connectors to the Shuttle Orbiter prior to release and during servicing missions.** *(Lockheed)*

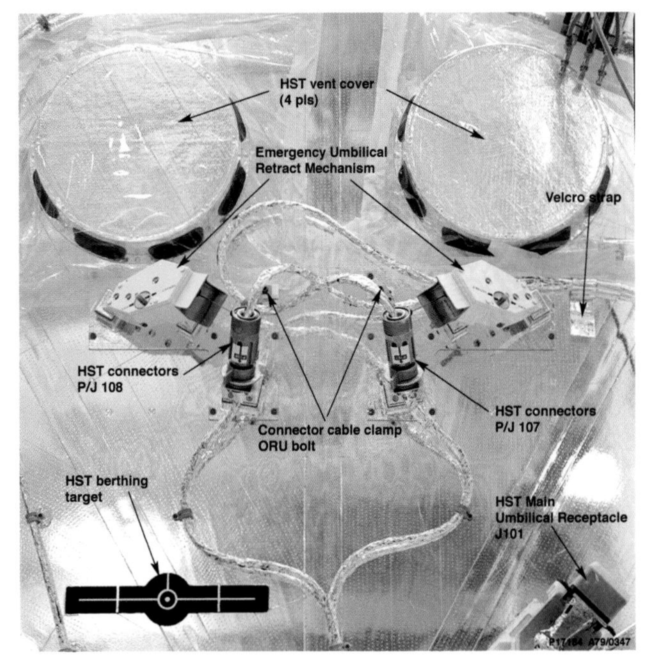

BELOW **Looking across the -V3 section of the aft bulkhead, the system's umbilical to the Orbiter can be seen along with the docking target.** *(Lockheed)*

RIGHT The High-Gain Antenna (HGA) has twin-axis gimbals for pointing at the TDRS or ground stations. *(Lockheed)*

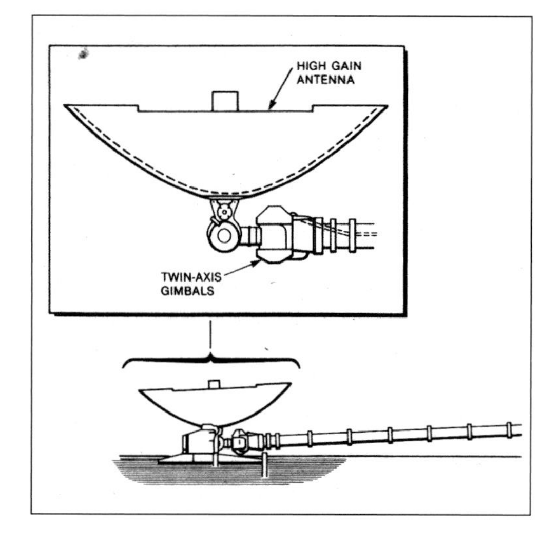

stiffening. A special gas purge system was fitted to prevent contamination of the delicate instruments before launch, vented through four specially designed, light-tight apertures on the back face of the bulkhead to prevent stray light waves from interfering with the wavelengths of light in the OTA.

All the HST operating equipment to run the observatory and handle data from the scientific instruments is carried within the Equipment Section (ES), 5.1ft (1.55m) high and 14ft (4.27m) in diameter. It consists of a toroidal set of 12 separate compartments, each with a depth of 4ft (1.22m), attached to an interface ring on top of the ASB. Ten compartments house equipment and two contain aft trunnion pins and scuff plates. Positioned between the ASB and the Aft Shroud, the ES is constructed from machined and stiffened aluminium frame panels attached to the inner aluminium barrel.

BELOW One of the two High-Gain Antenna dishes in its folded position against the Light Shield. Note the NASA 'worm' logo that temporarily replaced the traditional font, to which NASA reverted after a tide of complaints about the appearance of the new design. *(Lockheed)*

Eight bays support flat honeycombed aluminium doors packed with equipment; two carry thermally stiffened panels for protecting the reaction wheels (Bays 6 and 9), which are at approximately 90° spacing radially from each other. Forward and aft frame panels enclose the bays, which have an outer radial width of 3.6ft (1.1m) and an inner width of 2.6ft (0.79m). The two Power Distribution Units (PDUs) are mounted in Bay 4 together with the Power Control Unit (PCU). The six batteries and associated conversion units are divided equally between Bays 3 and 2. The Data Management Unit (DMU) is situated in Bay 1. The Science Instrument Control & Data Handling (SI C&DH) assembly is in Bay 10, with the Tape Recorder (TR) and the Pointing Safemode Electronics Assembly (PSEA) in Bay 8. The Mechanisms Control Unit (MCU) is in Bay 7, with the S-band transmitter and associated electronics in Bay 5.

The central section of the SMM consists of a Forward Shell (FS), 13ft (4m) long and 10ft (3m) in diameter, machined from aluminium plate and reinforced with external rings and internally stiffened panels. These rings are located on the exterior surface so as to provide clearance for the Optical Telescope Assembly inside. The FS constitutes a shield for the Telescope assembly main baffle and a structural support for the secondary mirror. Four magnetic torque devices are situated on the FS at 90° intervals on the exterior circumference.

Two grapple fixtures are mounted to the FS by which the Remote Manipulator System (RMS) on the Shuttle Orbiter can be controlled from inside the reusable spacecraft to reach out and be attached to the Hubble Space Telescope. In this case the HST can be secured to the Orbiter while the RMS manipulates the Telescope down on to a fixture in the payload bay (see Chapter 6). Attached to the ES by a circumferential interface ring, the FS also supports the two High-Gain Antenna (HGA) masts, located 180° apart and folded down against the sides until the HST is deployed in space, and the two stowed solar array booms which are also deployed – after the Shuttle has removed the HST from the payload bay – again by using the RMS on the Orbiter.

The front end of the Telescope consists of the Light Shield (LS), which is there to block stray light from influencing the measurement of light coming from the object observed. It consists of a tubular shell 13ft (4m) long and with an internal diameter of 10ft (3m). Machined from magnesium, it is structurally stiffened by a corrugated skin covered by a thermal blanket. Inside it has ten internal light baffles with a thin coating of flat black paint applied to suppress stray or incident light. The LS is attached to the Forward Shell by a circumferential ring and, like the FS, it carries a set of rails and hand-holds to allow EVA astronauts to work along its exterior surface to transition from one part of the structure to another.

The LS also carries scuff buffers – large metal plates mounted on struts extending 30in (76cm) from the surface of the HST – which, together with the trunnions, help support it while it is carried in the Shuttle Orbiter. The LS has attachment points for the forward element of the High-Gain Antennas, which are hinged at their attachment points on opposite sides of the lower section of the Forward Shell, and for latching the forward end of the stowed solar array wings as they lie parallel to the long axis of the Telescope. The extreme forward end of the Light Shield supports the Aperture Door (AD), about 10ft (3m) in diameter and hinged at the top of the LS where it is attached at two points reinforced with support beams running back some distance along the LS.

The Aperture Door is manufactured from honeycomb aluminium sheets, reinforced by a V-shaped cross-brace, with the external surface covered in a solar-reflecting material and painted black on the inner facing surface. The door is there to control the amount of sunlight falling on the OTA. It is electronically controlled using solar sensors on the door itself to provide warning when the Sun moves into an angular proximity to the mirror. The door can open a maximum of 105° from the stowed position and starts to close when the Sun approaches to within an angle of 35° to the V1 axis and finishes closing at an incidence angle of 20°.

The Telescope aperture allows a 50° field of view (FOV); but since direct sunlight would damage the mirror, complete closure can be achieved within 60 seconds of the initial alert signal. However, under certain circumstances the Aperture Door activating sensor system can be overridden by the Space Telescope Operations Control Center (STOCC) for observations within the 20° limit. This could

BELOW The Data Management Subsystem, which ties together the DF-224 computer, the data management unit, data interface units and three tape recorders. The functional input and output paths connect the Support System Module with the Optical Telescope Assembly. *(Lockheed)*

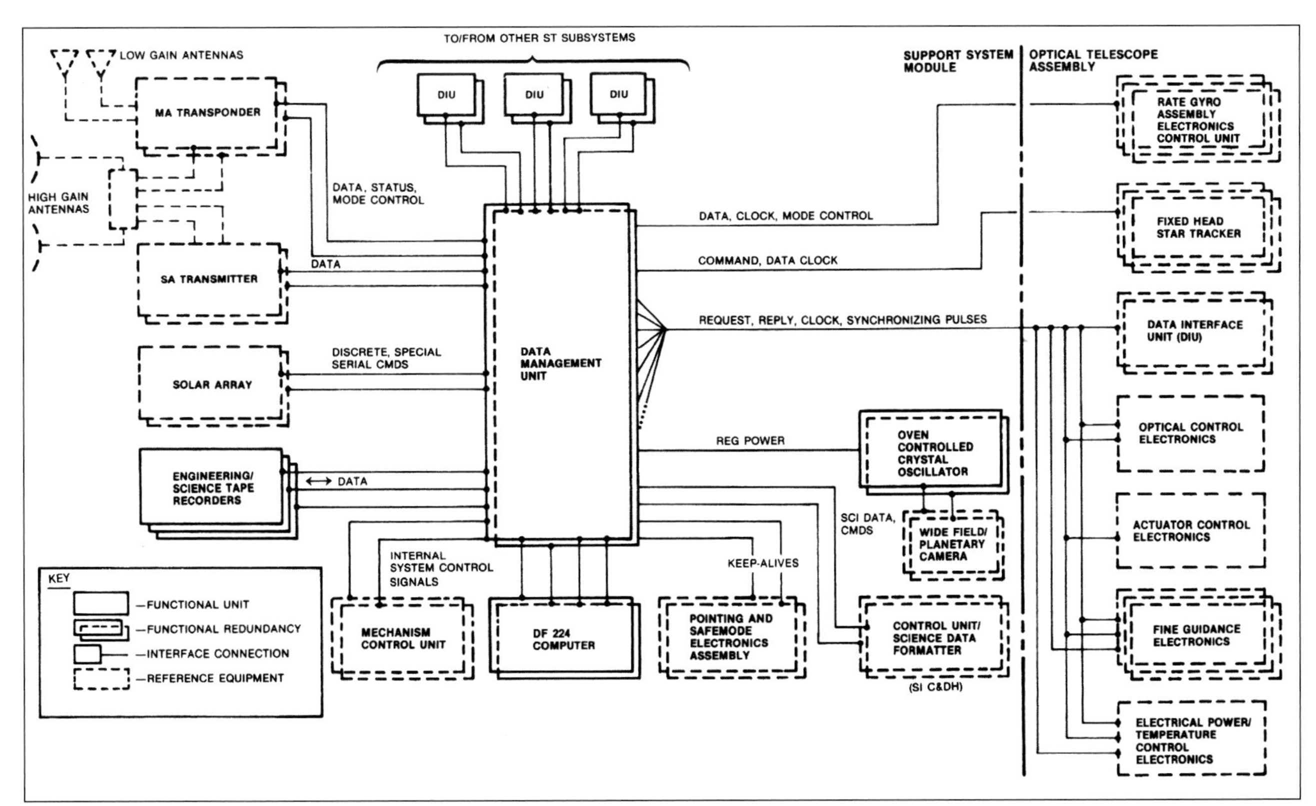

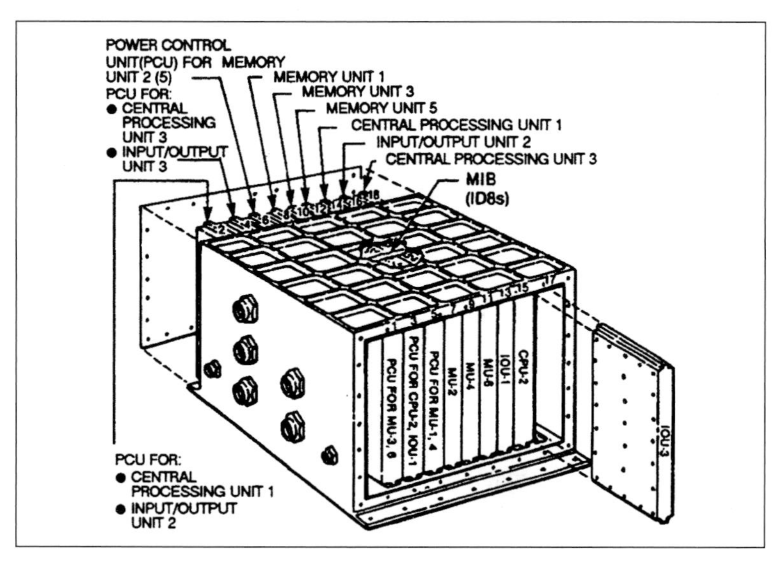

ABOVE The DF-224 computer installed in the HST during assembly was subsequently replaced with a considerably more powerful unit. It had three processing units with two working as backup and redundancy in the input/output units as well as the power converter units. *(Rockwell Autonetics)*

be, for instance, when the limb of the Moon is used to partially block the light when the HST is observing a particularly bright object, or it might be because the angle of the sunlight might approach, but not exceed, the design limit of the system, and not to prevent its closure would seriously interrupt an observation.

SSM systems and subsystems

The Support Systems Module accommodates a wide range of mechanical devices including electric motors, gimbal drives, hinges and latches to enable smooth and efficient maintenance, from remote commands sent from the ground to fully man-tended activity through EVA from the Shuttle. Functional design was based on this accessibility from the outset and the engineering of the HST incorporated these mechanisms as a priority. The SSM incorporates nine latch mechanisms, including four operating the antennas, four for the solar arrays and one for the Aperture Door at the top of the Light Shield. Each is designed to a common concept engaging a four-bar linkage and a rotary-drive actuator, or stepper-motor. The SSM carries three hinge drives: one for each of the two High-Gain Antenna arms and one for the Aperture Door. These also operate by rotary-drive actuator.

All 12 latch and hinge devices use hex-wrench fittings so that an EVA astronaut can manually operate the mechanism should a motor or actuator fail. The spacewalking astronaut would carry a toolkit during the spacewalk to allow on-site repair or activation of a stalled function. The requirement for on-orbit servicing was a very real possibility, as the life of the HST would depend to a great extent on the ability of trained spacewalkers conducting maintenance tasks. The ability to access the Telescope and conduct this kind of servicing was fundamental to the potential efficiency of maintaining the useful life of what constituted a major expenditure for NASA and the science community.

The collecting, collation and distribution of data from the Hubble Space Telescope provides engineers and scientists on the ground with detailed information about the condition of the spacecraft and the data from the instruments. Fundamental to this is the Instrumentation and Communication Subsystem (I&CS) which provides the communication link between the HST and the Tracking and Data Relay Satellites (TDRS) which send and receive messages, commands and other data through the High-Gain and Low-Gain Antennas and pass this along to the Data Management Subsystem (DMS).

The I&CS provides both multi-channel and single-channel systems for transmitting data and messages, the former providing a flow of commands or data on both channels simultaneously while the latter sends data only as it is gathered real-time. It can also send both science and engineering data from tapes

RIGHT The DF-224 computer occupies 2.25ft³ (0.06m³) and is considered an Orbital Replacement Unit (ORU). *(Rockwell Autonetics)*

as required under emergency conditions. Information sent through the High-Gain and Low-Gain Antennas goes directly to the TDRS satellites. Due to the TDRS being in geostationary orbit and within continuous line-of-sight of Earth receiving stations, this allows a much greater volume of information to be sent to the ground.

The HST is equipped with two High-Gain Antenna (HGA) assemblies, located on opposing sides of the SSM in the +V3 and the –V3 positions. They are stowed flat against the sides of the HST during launch and deployment, rotated 90° to the operational position on station. The system can send and receive data to and from the TDRS satellites for 90 minutes of each 97-minute orbit, communications therefore being maintained for almost 93% of the life of the Telescope. On occasions when the HGAs are not in the extended position, the Low-Gain Antenna (LGA) is used for sending and receiving information.

Each HGA consists of a parabolic reflector antenna attached to a mast on a two-axis gimbal with associated electronics to permit movement 100° in either direction. Prime contractor General Electric was responsible for the design and fabrication of these elements, manufactured from aluminium honeycomb with graphite-epoxy face sheets. Each antenna can be aligned to a fixed position with a 1° pointing error which is quite adequate for the less demanding requirement of the HGA compared to the $0.0000277_r°$ pointing requirement of the OTA. The accuracy is consistent with a beam width of more than 4°. The movement of the HGA is independent of the movement of the Telescope and its motion has no effect on the pointing accuracy or stability of the Telescope. The antennas transmit at 2255.5MHz or 2287.5MHz, +/-10MHz.

The two omnidirectional Low Gain Antennas are used for communicating with the HST for ground commands and the exchange of engineering data on a multi-channel access protocol. One antenna is fixed to the aft bulkhead of the Support Systems Module and another to the Light Shield. They take the form of spiral cones and are positioned so that they are 180° apart on opposing sides of the spacecraft, thus ensuring continuous communication whatever the orientation of the spacecraft.

ABOVE Manufactured by Lockheed, the Data Management Unit serves as the central node from which messages are received, interpreted and passed to systems and subsystems, and receives data from the science instruments. *(Lockheed)*

These antennas provide low data rate telemetry and are usually reserved for communications during deployment and while the HST is subject to servicing missions, but they also double as back-up in the event primary communication through the HGA is lost. Manufactured by Lockheed, the Low Gain Antennas operate on frequencies from 2100 to 2300MHz.

The HST's Data Management Subsystem receives information from the Space Telescope Operations Control Center (STOCC) and data from the SSM, in addition to data from the scientific instruments. It processes this information, which constitutes communications, subsystem calibrations and commands from the STOCC, and processes, stores and/or sends this information to the designated recipient subsystem or element. This includes the DF-224 computer, the Data Management Unit, the four data interface units, any one of three engineering/science tape recorders, or the two oscillators. All except the data interface unit, which is located in the Optical Telescope Assembly, are situated in the Equipment Section of the SSM.

Five types of signals are handled by the DMS: commands to the HST systems and subsystems; commands sent up from the ground; data in the form of command verification and response or status data; science data from the Science Instruments Communications and Data Handling (SI C&DH) subsystem; and outputs such as the clock signals, essential for timing algorithm updates and verification of calibrated convergence with the ground computers. Other inputs take

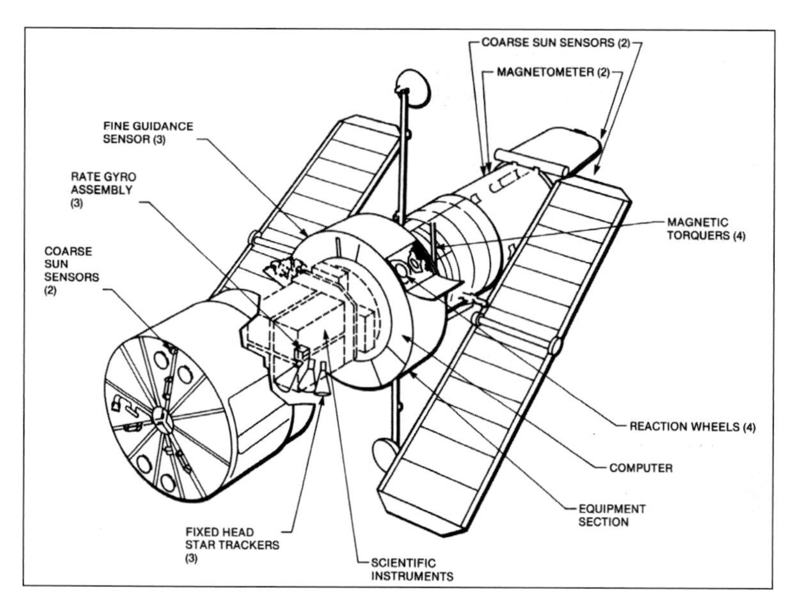

ABOVE The distribution of equipment for the Pointing and Control System, which includes four course Sun sensors (two on the Light Shield and two on the Aft Bulkhead), Fine Guidance Sensors, which also double for astrometry, and the rate gyro and reaction wheel assemblies. *(Lockheed)*

reporting-in signals from sensors and measuring devices, which can include the serial commands to the solar arrays, re-synchronisation pulses for the clocks, command-data clock calibration and data clock mode control status verification.

Produced by the then named Rockwell Autonetics (now Boeing), the DF-224 computer produced for the HST is a digital general-purpose device which is primarily used for on-board engineering computations. It is also required to execute stored commands, format data calculations into telemetry, point the solar arrays at the Sun, monitor and report the condition of the power system, and orientate the two High Gain Antennas. For these operations it utilises stored commands specifically written for these separate functions.

The computer was not unique to the HST and was developed as a general-purpose unit for a variety of potential science missions. It is biased toward reliability, with redundancy built in to extend the operating life which, while being an Orbital Replacement Unit (ORU), was nevertheless required to function for long periods between servicing missions. This unit would eventually be replaced by a much more powerful Intel 486 but for several years it would have to carry all the processing load built into the device installed for launch in 1990.

The DF-224 had three central processing units (CPUs) running on Intel 386 chips, with four active and two serving as redundant backups. There were six memory units with up to 48k word storage in total, three input/output (IOU) devices, one active as the primary unit and two IOUs for redundancy. The computer was equipped with six power converter units with overlapping functions, again for redundancy. It had a clock speed of 1.15MHz. The DF-224 weighed 110lb (50kg) and was contained within a box-structure 1.48ft x 1.48ft x 0.98ft (45cm x 45cm x 30cm) installed within Bay 1 of the SSM Equipment Section.

Designed and built by Lockheed, the Data Management Unit was directly linked to the DF-224 and encoded and sent messages to various systems and subsystems and all DMS units. It powered the oscillators and was the central timing system for the HST. It received and decoded all incoming commands and processed and passed each along for execution. The DMU also receives data from the Science Instrument Control & Data Handling subsystem as well as engineering data such as sensor readings on temperatures and voltages generated by the separate subsystems throughout the HST. These data can be stored in on-board tape recorders under the unusual circumstances in which the TDRS satellites are not within sight.

The DMU is also responsible for selecting from a menu of telemetry formats. These are divided into deployment format when the Low Gain Antenna is operational and diagnostic format for a faster transmission when a subsystem or system provides a readout that needs rapid evaluation. The DMU itself consists of a series of printed-circuit boards interconnected through a backplate and external connectors and is located on a door of Bay 1 of the Equipment Section. It weighs 83lb (37.7kg), contained within a box measuring 26in x 30in x 7in (60cm x 70cm x 17cm).

To provide a command and data link between the DMS and other HST systems and subsystems, four Data Interface Units (DIUs) are installed at various locations. Built by Lockheed, they receive instructions from the DMU to carry out a specified function, execute that function, and send back data or status information to the DMU. Each DIU is uniquely connected to the DMU to provide a fully integrated command and data link.

The Optical Telescope Assembly DIU is mounted in the Equipment Section of that element, while other units are located in Bays 3, 7 and 10 of the SSM Equipment Section. For redundancy each DIU is actually two units in one, with either male or female section carrying out all the data handling functions required. Each DIU is contained in a sealed compartment measuring 15in x 16in x 7in (38cm x 41cm x 18cm) which weighs 35lb (16kg).

Engineering information or science data that cannot be transmitted to the ground real-time can be stored on three tape recorders within the circuitry of the Data Management Subsystem. Each recorder has a capacity for up to one billion bits of information. Two are usually programmed for simultaneous use with a third held in reserve as a spare. This normally redundant, or standby, recorder can also be used for emergencies or for contingency operations. The selective difference between the two functions is that an emergency can trigger an immediate call to the ground for help at the nearest available point for a data-dump, while a contingency is a programmed response to a recognised malfunction for which there is a rectifying procedure. The three recorders are stored in Bays 1, 3 and 8 of the SSM Equipment Section and each weighs 20lb (9kg). They each measure 12in x 9in x 7in (30cm x 23cm x 18cm).

A highly stable timing source is vital for effective and successful management of operating procedures, HST 'housekeeping' operations, where standard procedures are carried out for the successful functioning of the support systems, and for the programming of scientific operations. This is carried out by the oscillator that is contained within a cylindrical housing measuring 4in (10cm) in diameter and 9in (23cm) in length and weighing 3lb (1.4kg). The primary, and a backup, oscillator are secured in Bay 2 of the Equipment Section.

Timing and highly accurate pinpoint attitude alignment are key to the successful operation of the science instruments when brought to bear on a target of choice. The Pointing and Control System (PCS) is responsible for ensuring attitude stability, pointing accuracy and designated target alignment with the selected object or field. The specification requires the Telescope to point to an accuracy of 0.01 arc-sec and to maintain that to an accuracy of 0.007 arc-sec.

ABOVE The three circular ports below the rectangular access to the science instruments are the apertures for the fixed-head star trackers. *(NASA)*

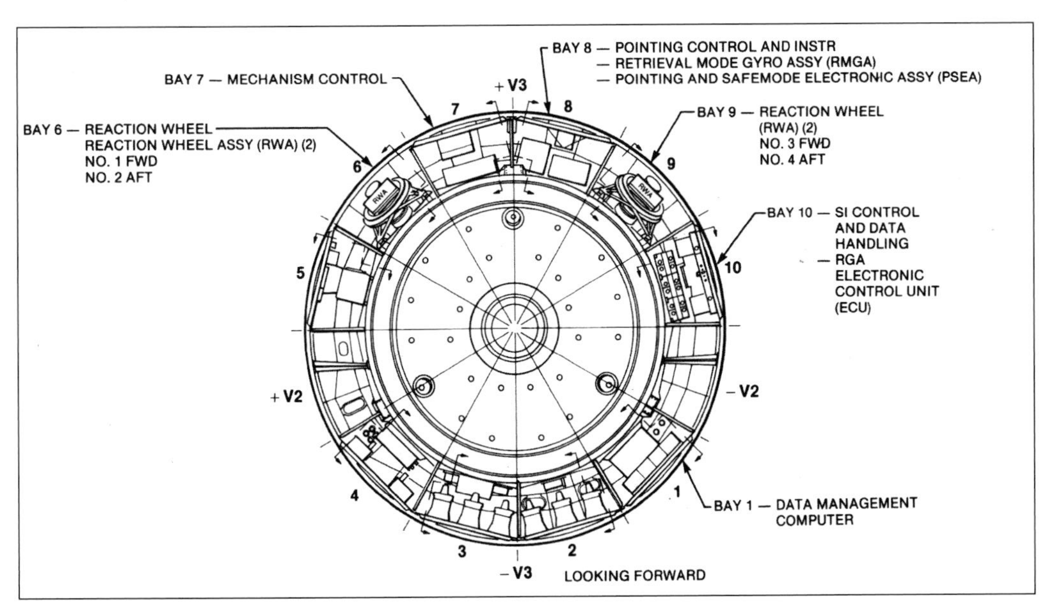

LEFT Looking forward across the Equipment Section of the SSM, this diagram displays the location of the primary subsystems employed in maintaining attitude control and fine-pointing accuracy. *(Lockheed)*

The Telescope was designed to achieve a resolution of 0.1 arc-sec, full width of the image at half the maximum intensity (0.06 arc-sec) and 70% of the total image energy encircled within 0.1 arc-sec radius, the calibration for performance being based on a light wavelength of 633nm. Because of the fine-pointing requirement and the longevity of the unattended segments of the overall life of the Telescope, a unique form of control was essential.

Satellites and spacecraft can be stabilised and controlled in attitude by several means, the earliest and most popular being the expulsion of a gas, usually produced by small rocket thrusters, to cause reaction until stopped by a thrust of equal magnitude in the opposite direction. These attitude control thrusters maintain a stable pointing angle for alignment with celestial objects, the limb of the Earth or the Moon or some distant star. The problem with this concept is that the ability of the system to maintain a fixed attitude, or rotate the satellite around one or other of its three axes, is limited by the quantity of propellant carried divided by the sum duration of periodic thruster firings.

The next most logical choice, where propellant consumption is a serious issue to the life of the satellite, or where the mass of the satellite is great and requires a large volume of propellant or a greater thrust output from the control motor, is the reaction-wheel assembly (RWA) or control moment gyroscope (CMG) concept. The RWA is akin to a flywheel to which an electric motor is attached so that when the rotation speed is changed the satellite or spacecraft begins to counter-rotate around its centre of mass in the relevant axis at a proportionate rate through the conservation of angular momentum. Reaction wheels can be made to rotate slower or faster to effect the desired change in axial orientation.

BELOW To change direction the Reaction Wheel Assemblies receive information from the magnetic torquers to change speed and use spin momentum to turn the HST to the required alignment. *(Lockheed)*

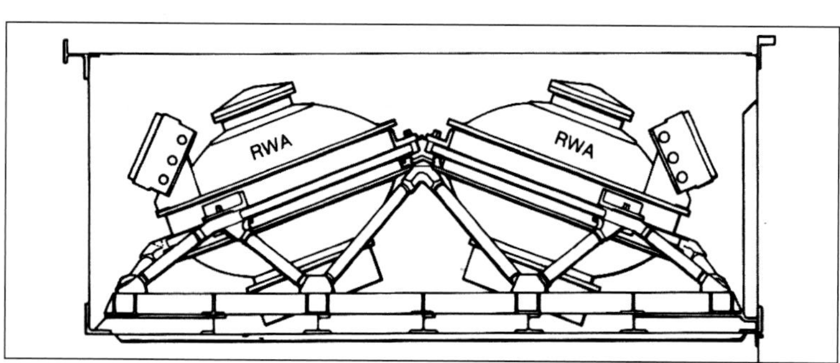

When used as momentum wheels they rotate at a constant speed to acquire a large quantity of angular momentum and this alters the rotation dynamic, enabling a torque to be applied perpendicular to one axis of the spacecraft, which is the axis parallel to the spin axis of the wheel. This does not result in angular motion about the same axis as the disturbance torque but rather a smaller angular motion, or precession, of that axis about a perpendicular axis. In this way the spacecraft can be stabilised to point in an almost-constant and fixed direction in an operating mode known as 'momentum bias'. The great advantage with the RWA is that it allows very fine motion and tiny angular variation in pointing angle, which is perfect for the Hubble Space Telescope.

An alternative method, the CMG operates on essentially the same principle but applies the energy for an axial change from a torque movement applied to the wheel itself. In most applications the momentum wheel is mounted in a one or two axis gimbal and when rigidly mounted to the structure of the spacecraft a constant torque is applied to the wheel by one of the motors, which builds momentum to produce an angular velocity about a perpendicular axis. In this way the spacecraft can be rotated around the relevant axis to achieve a desired pointing angle. They have the advantage over the RWA in that they are able to sustain larger torque energy with less electrical power. CMG was the attitude control system of choice for the Skylab laboratory launched by NASA in 1973 and for the International Space Station, assembled between 1998 and 2011.

The HST employs four reaction-wheel assemblies paired, two each in Bays 6 and 9 of the SSM Equipment Section. Designed and built by the then Honeywell Satellite Systems Operations (now Honeywell Aerospace), each wheel is 23in (59cm) in diameter and weighs approximately 100lb (45kg) with a nominal spin rate of 3,000rpm. The wheel axes are orientated so that the HST can run satisfactorily on three wheels if necessary. Using spin momentum to maintain the HST in position, braking or accelerating the wheels transfers momentum to the spacecraft itself.

The reaction wheels on the Telescope, which allow it to be pointed to any desired position in

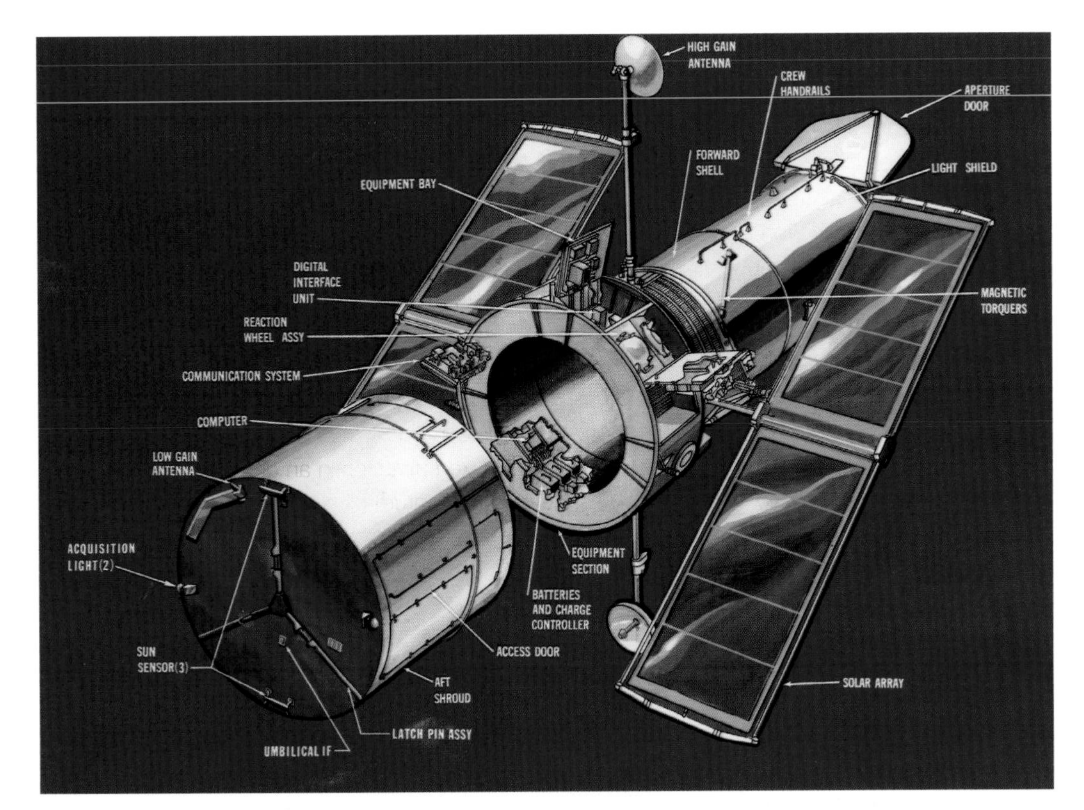

LEFT The general configuration of the pointing and control equipment together with the arrangement of the solar arrays that provide power for the HST. *(NASA)*

the sky, are unloaded using magnetic torque against the Earth's magnetic field. The torque reaction occurs in the direction that reduces the reaction-wheel speed by balancing the momentum. The torquers provide torque perpendicular to the Earth's magnetic field lines. The Pointing and Control System uses the DF-224 computer in the DMS to calculate the position updates and the torque values for attitude orientation and to translate the ground targeting commands into RWA operations. It also smoothes out the motion to reduce any adverse vibrations conducted through the structure to the science instruments in the OTA.

Error signals are provided by precision gyroscopes in all three axes. Gyro reference, in very low noise systems, is characterised by small drift rates which mean that the gyroscopes are sensitive to rates of angular motion over time. They are not, however, sufficiently accurate to determine the necessary exact pointing angle or to maintain it within the specified requirement for long periods of time.

To compensate, the Telescope carries fixed-head star trackers which have relatively wide fields of view that ensure the acquisition of bright catalogued stars but are capable of establishing direction to no better than a few arc-seconds. This degree of precision is adequate for course attitude determination but totally inadequate to place the small slits (0.1 arc-sec) of the ST spectrometers on the desired object or for extended exposures, and the integration at higher angular resolution, of which the detectors and the optics are capable of supporting.

Fine Guidance Sensors are carried to determine with high precision the position of reference stars relative to the Telescope's axes. They consist of a Koester's Prism interferometer/photomultiplier combination which can determine stellar positions to an accuracy of a few to several milliarc-seconds within a field of view of 15 milliarc-seconds overall. This provides a tracking signal. These Fine Guidance Sensors were developed by Perkin-Elmer. Only two Fine Guidance Sensors are necessary but a third is provided as backup. They are used to scan the 69 arc-min^2 field of view for stars of previously known coordinates, and these guide-stars are used as reference points to determine the difference between the actual and predicted positions so that the gyroscopes can be updated every second of time.

The required specification on pointing accuracy is achieved in this stepped fashion by use of the fixed-head star trackers and the Fine Guidance Sensors. The third Fine Guidance Sensor is employed on astrometric measurements of designated objects with respect to the reference objects, and considerable work was conducted by W. Jefferys of the University of Texas, Austin, in studying the capability which could be achieved on the astrometric measurement of objects as faint as m_V = 17 to a precision of 0.002 arc-sec.

It is essential for guide stars to be identified everywhere the Telescope is pointed within the relatively small angular field of view of the Fine Guidance Sensors. For this purpose it was necessary to measure such stars with precise positions of 0.3 arc-sec to m_V = 14.5. Prior to the launch of the HST no such catalogue existed, and one of the support efforts implemented during the 1980s was to compile a comprehensive catalogue. Moreover, it was necessary to determine the target position of guide stars brighter than 14.5 magnitude and visible within the field of view of the Fine Guidance Sensors to a precision finer than the desired maximum slit size on the spectrometers. For example, the 0.1 arc-sec slit on the Faint Object Spectrograph (FOS) required small area scans to locate the slit within the uncertainty of 0.33 arc-sec that is required for guide star position determinations.

The guide star catalogue was compiled from micro-densitometer scans of Schmidt telescope photographic plates taken by the 48in Palomar Schmidt telescope and the UK Schmidt telescope in Australia. The plates were measured with a precision accuracy of 0.25 seconds with the positions of reference stars taken from the old SAO catalogue published in 1966 and using much older observations. Nevertheless, the list of star positions and magnitudes of 20 million stars in the 9–15 magnitude range was compiled.

Across all the HST engineering challenges, the Pointing and Control System was the greatest technical challenge for design teams and technicians. Because it is used for every astronomical observation and by all research investigators, the engineers were concerned that its functions should be properly understood by the scientists involved in utilising the instruments throughout the life of the Telescope. Thus it was that the three-step approach to finessing the capability was made as robust and as simple as possible.

The three position indicators described above were crucial in achieving that: the fixed-head star trackers which are independent of the main telescope optics and can locate bright stars to within a precision of 1 arc-min; the rate gyroscopes which drift with an uncertainty of about 0.01 arc-sec/second; and the Fine Guidance Sensors which obtain their signals from stars near the edge of the Telescope's field of view.

The process of a typical acquisition would begin by slewing the Telescope to a predetermined spot, for which it has a maximum rate of travel of 15°/min. Because of the uncertainties of position due to gyroscope drift, and the inaccuracies in the initial direction (especially roll) which increase proportional to displacement, the slewing movement is kept in short periods. Tests show that these inaccuracies can be as much as several arc-minutes after a 90° slew. This is why the fixed-head star trackers are used first to determine the approximate position before the coordinates of the target star are fed into one of the two Fine Guidance Sensors.

One begins searching for the first guide star using a spiral search pattern with a 2 arc-sec aperture. When a star is found in the correct brightness range it stops and the other Fine Guidance Sensor searches for the guide star, and when that is found the relative positions provide the final confirmation of position. The search will resume if that initial confirmation is not forthcoming and this can be made over three arc-minutes. The sensitivity of the system is such that it will operate adequately at 14.5 magnitude, a brightness level that statistically provides two guide stars for 85% of the fields at the galactic poles. This acquisition phase normally takes about three minutes.

In addition to the magnetic sensing system, the rate gyro assembly, the fixed-head star trackers and the Fine Guidance Sensors, the PCS also has four course Sun sensors located on the Light Shield and the Aft Shroud. These send signals to the Pointing Safemode electronics located in the electronics assembly in Bay 8 of the SSM Equipment Section. These sensors provide information to the computer, which calculates the initial deployment position from the measured light incidence angle to determine when to command partial or full closure of the Aperture Door to prevent direct sunlight falling on the mirrors.

The magnetic sensing system consists of two magnetometers attached to the front end of the Light Shield and connected to the DF-224 computer providing data on the relative orientation of the HST with respect to the Earth's magnetic field. These sensors provide the information required by the magnetic torque system for working against the RWAs for pointing orientation and control. The rate gyro assembly comprises three sensing units attached to the underneath of the SSM Equipment Section and attached adjacent to the fixed-head star trackers. They send data to their electronic module situated in Bay 10. This suite of sensors and electronics senses and measures the Telescope's rate of rotation relative to the orbit plane, thus enabling the pointing system to control the orientation and line-of-sight of the HST and its optics.

The fixed-head star tracker is an electro-optical detector that locates and tracks a predetermined star within its field of view. The three devices are mounted below the Focal Plane Structure on the -V3 axis next to the rate sensor units. As described earlier, the STOCC uses the star tracker system as a calibration instrument when the HST is rotated into its final orientation for an observation. These trackers simultaneously calculate the position information before and after course positioning to help steer the attitude orientation within the field of view of the Fine Guidance Sensors and the rate gyroscopes, thus providing a target reference.

The Electrical Power Subsystem (EPS) provides the energy required to operate the two elements of the Hubble Space telescope: the Support Systems Module and the Optical Telescope Assembly. The EPS consists of two solar array wings, associated electronics, six batteries, six charge current controllers, a power control unit and four power distribution units. It also interfaces with the Shuttle Orbiter for electrical power when the HST is still in the payload bay and has yet to deploy its solar arrays and also when these are removed and replaced during routine servicing missions.

ABOVE The solar arrays operate on a system whereby the photo-voltaic cells are carried on thin sheets supported on bistem supports rolled up and held within a cassette for launch and unwound behind a spreader bar. In this view the deployment arm is at the top. *(Lockheed)*

With the exception of the solar arrays, all this equipment is located in separate bays in the Equipment Section.

The two solar arrays were a significant aspect of the international cooperation forged through the Telescope programme, these being developed by the then British Aerospace for the European Space Agency (ESA) which negotiated this participation. The design of the solar arrays had to be lightweight, folded up during ascent to orbit inside the Shuttle payload bay, rolled out and back in around a central drum like a roller-blind and stiff so as not to induce vibrations from thermal expansion and contraction such as might be occasioned by the very different temperature extremes as the

BELOW The solar array drive and deployment mechanism is located at the base of the deployment arm, where it is hinged to rotate 90° from the stowed position. *(Lockheed)*

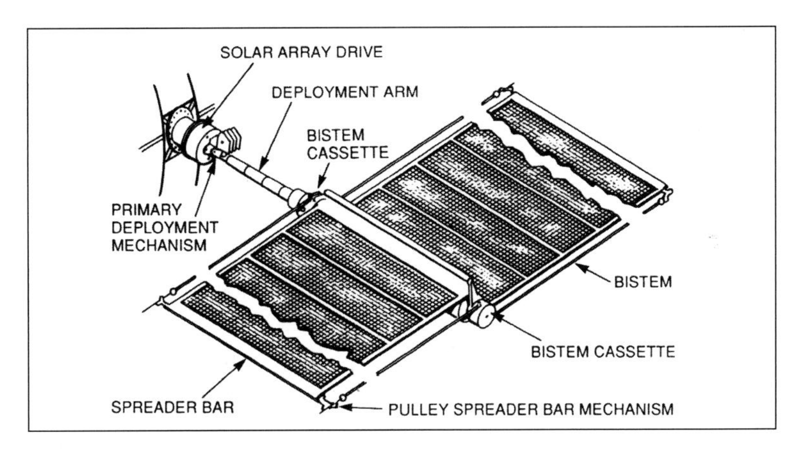

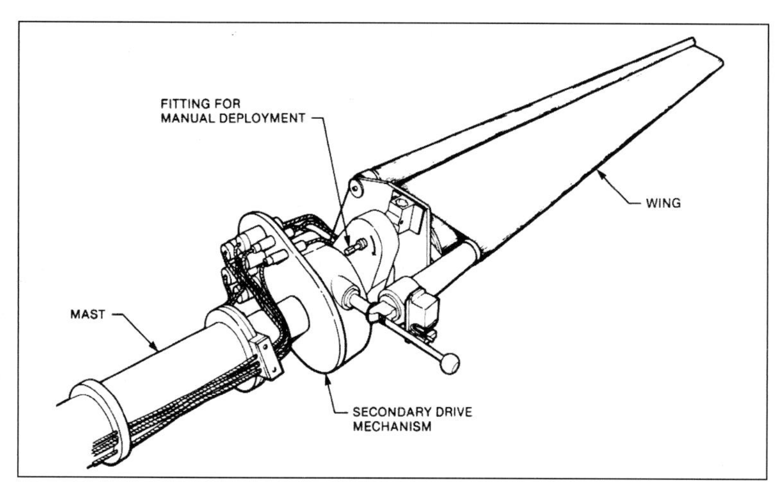

ABOVE In the event of a malfunction in the automatic deployment of the arm, an EVA astronaut can manually crank the solar arrays from their stowed location using the special fitting. *(Lockheed)*

BELOW A close-up of the bistem cassette, which clearly displays the boom deployment mechanism. A close look at the image shows a stray tension line creeping out from the panel. *(NASA)*

HST passed from day to night and back again. Moreover, there was no facility anywhere in the world where such large arrays could be tested for thermal and vacuum conditions encountered in space and all the projections regarding materials' response to this environment had to be theoretical or extrapolated from other solar arrays on different satellites and spacecraft.

There was no precedent for the sheer size and area the arrays on the HST would occupy. The closest arrays with this area were those attached to the Skylab space station launched in 1973, but arrays on that facility were more than twice the surface area of those attached to the HST and of a completely different design. Stiffness and rigidity were bought at a high price, with the requirement for strength and jitter-resistance much less than that required by a vehicle occupied by astronauts bounding off walls and causing 'floatation' in the structure. No observatory as large as the Telescope had

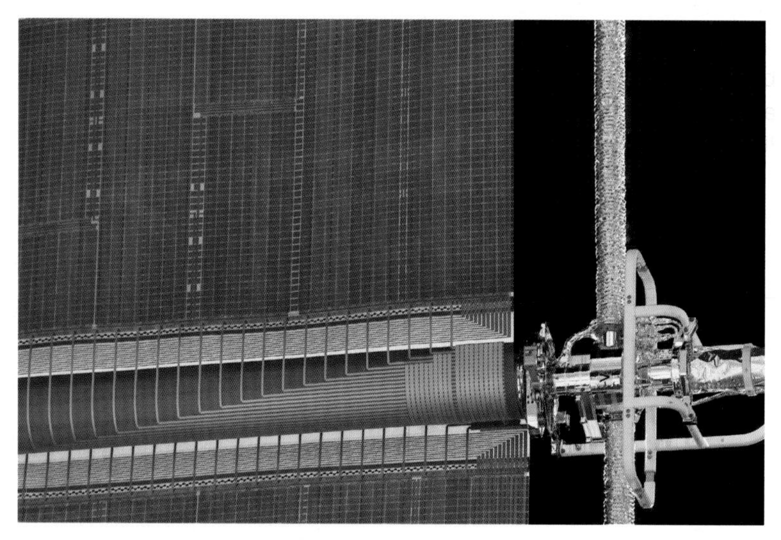

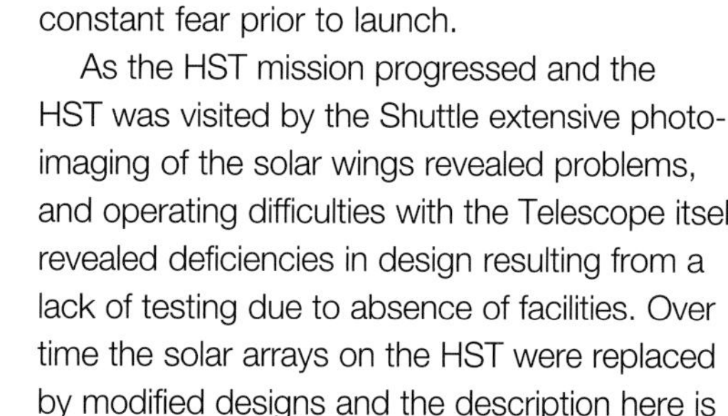

ever been deployed with solar cell arrays, and since astronomical observation requires an absence of attitude excursions the possible interference from flexing solar array wings was a constant fear prior to launch.

As the HST mission progressed and the HST was visited by the Shuttle extensive photo-imaging of the solar wings revealed problems, and operating difficulties with the Telescope itself revealed deficiencies in design resulting from a lack of testing due to absence of facilities. Over time the solar arrays on the HST were replaced by modified designs and the description here is for the initial configuration as launched in 1990. See Chapters 7 and 8 for additional information on the different arrays fitted to the Telescope on successive servicing missions.

From the outset, immediately the ESA received the go-ahead from NASA in 1977 for development of the solar arrays the challenges were known to require an exhaustive process of reducing a wide range of options to a single design concept. One of the most radical was the requirement for the arrays to be removed from the Hubble Telescope in orbit and returned to Earth by the NASA Shuttle, replaced with improved and updated arrays taking advantage of new technology. The initial requirement was for the Shuttle to visit Hubble every 2.5 years, but that was quickly revised upward and the design requirement focused on low mass and efficient arrays which could last for at least five years between replacement cycles.

The requirement to replace the solar arrays was one of the most difficult challenges. The arrays not only had to be unrolled on orbit but they had to be designed for up to 30,000 day/night cycles over a five-year life during which the temperature would move from -100°C to +100°C each orbit. While there was a requirement to keep the arrays taut for use they had to be sufficiently flexible to be rolled back up again after five years, their bistem cassette removed and returned to Earth in the Shuttle. Development of the solar arrays ran in parallel with the first flights of the Shuttle in the early 1980s, and when the first missions returned with severely degraded thermal blankets there was serious concern over the implications for the solar arrays.

Atomic oxygen in the near-Earth space

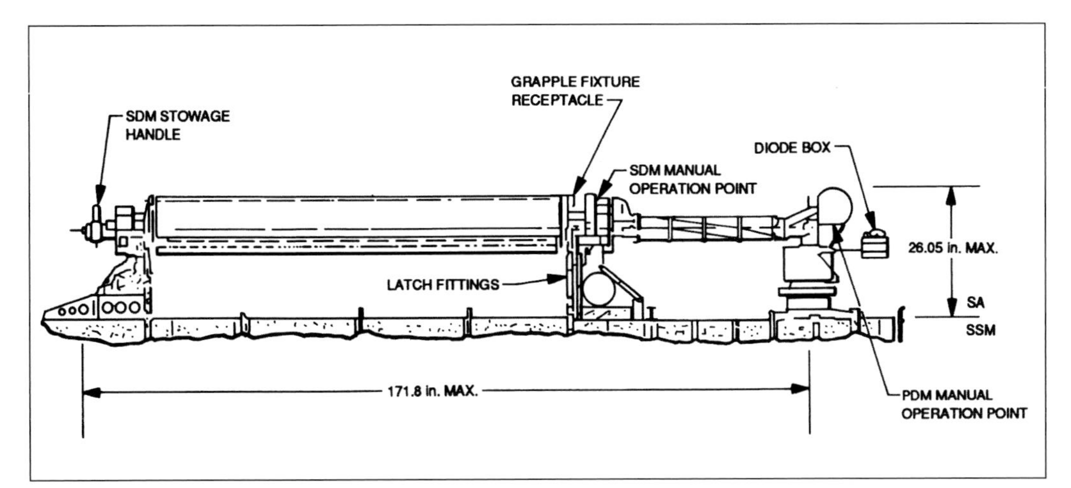

LEFT Manual operation can be used at the shown locations for rotating the arm from the stowed to the raised position and for unfurling the solar arrays from the storage cassette. *(Lockheed)*

environment to which the HST would be exposed was directly responsible for degradation to the thermal blankets and this effect would also have a reaction with material on the solar arrays. The highly reactive molecules could seriously impact the ability of the insulation material, which was similar to that used on areas of the Shuttle. To learn more about this phenomenon scientists took samples from the engineers working on the solar arrays and flew them on later Shuttle missions, discovering that the Kapton and silver linings would degrade away within a matter of weeks in space. The solution was to encapsulate the Kapton in a silicon coating and to replace exposed Kapton on the primary deployment arm with a polytetrafluoroethylene foil.

Each wing has ten panels, five on each half of the wing, rolled out from the cassette. These small panels support 2,438 solar cells attached to a glass-fibre/Kapton surface with silver mesh wiring underneath covered by another layer of Kapton; these blankets are less than 55 micrometres thick. The original cells had an efficiency of 12.7% but in the redesign of the materials used in the construction of the first flight arrays new cells with an efficiency of 14% were used instead. The cassette arm and drive unit is 15.5ft (4.8m) long and each wing (blanket) weighs 17lb (7.7kg). The total solar wing assemblies weighed 668lb (303kg). Fully extended it has a length of 40ft (12.1m) and a width of 8.2ft (2.5m). Together the two dual solar array wings provided 4kW of electrical power at 34v.

British Aerospace had received the contract to produce the solar arrays in a £13 million deal that produced the initial flight set in 1985.

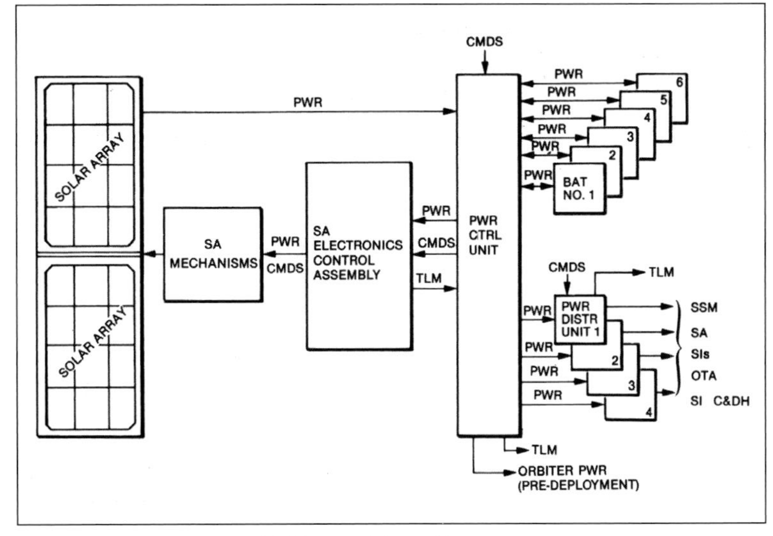

ABOVE Each solar array has a control mechanism and provides electrical current to the power control unit, which feeds into the six batteries and the four distribution units. *(Lockheed)*

When the launch of the HST was delayed by the *Challenger* accident and by other factors, British Aerospace received a further contract worth £2 million awarded for improvements to the design and replacement with a second flight set providing new blankets with BSFR back surface field reflectors, which improved output by 10%. As well as the protection against atomic erosion, which was required due to the launch coming later in the solar cycle, one new feature added was the silver interconnects between cells being replaced with molybdenum silver composite. The new arrays were delivered to Lockheed in April 1989. It was accepted that these were not the optimum design and British Aerospace received a further £11 million to provide another set of arrays incorporating improvements for the first servicing mission planned for 1993. These were delivered in May of that year.

The Solar Array Subsystems include the

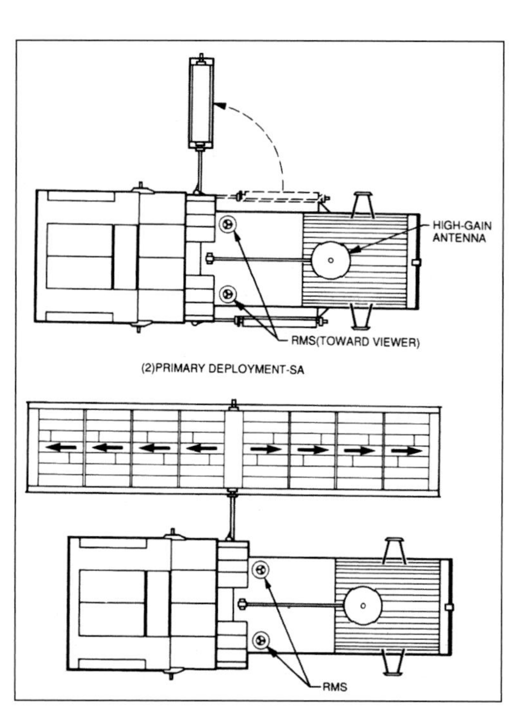

RIGHT For deployment of the solar arrays the HST is lifted up from the Shuttle Orbiter payload bay and rotated 180°, at which point the deployment boom is raised and the panel unrolled from the bistem. *(Lockheed)*

primary and secondary deployment mechanisms, the drive mechanisms and the electronic control assembly. The primary system raises the solar mast from the side of the Support System Module to an erect position 90° from its stowed alignment along the long axis of the Light Shield to a position perpendicular to the optical path of the Telescope. Each mechanism has motors to raise the mast and to support it in place with braces. The mast can be manipulated by an astronaut to raise it should a motor fail. Using a wrench fitting on the deployment drive a hand-crank allows the mast to be elevated after manually releasing the restraint latches.

After the solar array arm has been raised the secondary deployment mechanism unfurls the blanket of solar cells. Each wing has a secondary assembly including the cassette drum, to hold the panels, a cushion to protect the rolled array, and motors and their own subassemblies. After the assembly has rolled out the blanket, all the while maintaining tension, power is transferred along the wing assembly. The blanket can be rolled out fully or partially dependent upon the deployment sequence and whether or not any malfunction has been detected which would invalidate the array if fully unrolled. The secondary mechanism also has a manual override.

The drive mechanism for the solar array rotates the deployed wing toward the Sun, turning in either direction. This mechanism is situated at the base of each solar array wing and a brake can be applied to maintain the array in a fixed position if the HST is manoeuvred around in axial orientation. This also allows the array to be fixed and locked in one position for routine maintenance by astronauts. Each drive has a clamp ring that acts as a release if opened, and with this an astronaut can jettison a complete array if required by the STOCC. The electronic control assembly monitors all array functions and controls the primary and secondary deployment mechanisms and the drive system for the arrays.

The design of the arrays is driven by the functional requirement of the HST in orbit but other considerations played an important part in the design requirement. Not least, the need to fit the stowed arrays within the confines of the Shuttle Orbiter payload bay during the time the HST is carried into space. The array assembly extends over a length of 14.32ft (4.33m) and protrudes from the side of the Light Shield to a maximum 26in (65cm). After the STOCC is confident that the Telescope is deployed correctly in orbit the commands are sent to deploy the arm and its bistem extension so that the unfurled solar cell array is facing the Sun and drawing power down for the HST systems via the electrical power subsystem.

The EPS relies on six nickel-hydrogen batteries to provide power when the Telescope is in the shadow of the Earth and the solar arrays can no longer supply direct electrical power. Each 97-minute orbit of the Hubble Space Telescope is in sunlight for 61 minutes and darkness for 36 minutes, on average 37% of its lifetime without the solar arrays directly exposed to sunlight. When fully charged these batteries can provide power up to 68amp/hr, which is sufficient to power the SSM and the OTA for more than 3.5 hours. The limitation on amp/hr capacity is driven by the capacity of the thermal control system to maintain a stable thermal balance. The nickel-hydrogen batteries installed prior to launch were manufactured using the dry-sinter process.

Nickel-hydrogen batteries are preferable for space missions due to their reliability, long life, their suitability for deep-discharge and their tolerant life cycle. They are a rechargeable electrical power source based on nickel and hydrogen, which utilises gaseous hydrogen stored under pressure. They use potassium hydroxide as an electrolyte, and while the energy density is about 34% that of a lithium battery they are capable of working to extremely long life cycles, routinely enduring through more than 20,000 cycles with 85% energy efficiency. Nickel-hydrogen batteries were first used in space for the US Navy Navigation Technology Satellite-2 (NTS-2) launched in 1977 and were relatively new when specified for the HST. Since then they have been used in several planetary missions and are presently also in use aboard the International Space Station.

Power processed through the charge current controller is taken from the photo-voltaic cells in the arrays to charge the batteries as necessary and also provides a voltage-temperature control for the battery being charged. Each battery weighs 125lb (57kg) and consists of 22 cell plates in an aluminium case. The six batteries are attached to the doors of Bays 2 and 3 in groups of three. Each battery module weighs 460lb (207kg) and measures 36in x 32in x 11in (91.4cm x 81.3cm x 27.9cm). Designed to last five years, the batteries installed in the HST for launch lasted 19 years until replaced on Servicing Mission-4. The replacement batteries are described in Chapter 8 under subsection SM4.

The EPS has four power control and distribution units that handle the power between the current flowing from the solar arrays, the six batteries and the charge controllers. The Power Control Unit (PCU) provides the main power lines to the four distribution units and weighs approximately 120lb (55kg) in a structure measuring 43in x 12in x 8in (109cm x 30cm x 20cm). It is attached to the inside door to Bay 4 of the Equipment Section where the power lines switches, fuses, and electronic monitoring devices are contained. It is the nodal distribution hub taking power to all the HST systems and subsystems requiring energy.

The four distribution units are divided into functional roles. Two of the distribution units are dedicated to the OTA, the science instruments and the SI C&DH, and the other two support the Service Module. Each distribution unit is 10in x 5in x 18in (25cm x 12.5cm x 45cm) and weighs 25lb (11kg). A certain degree of redundancy is ensured by the design of the power lines and the arrangement of the four distribution boxes with the PCU. The CU receives commands from the DF-224 computer and from the ground sequence uplinks as well as being under the authority of the Shuttle Orbiter prior to deployment on orbit. The PCU is connected directly to the solar array electronics control assembly and also to the six batteries directly and independent of the four distribution units.

Maintaining a thermal balance inside a large structure such as the HST is a major part of the design emphasis and embraces the engineering of structures and materials as well as the disposition of systems and subsystems, those which each have a different degree of thermal emissivity and others which are active heat generators such as motors and electronics. The design was 'cold-biased' and would be insulated, with the interior heated to maintain a consistent temperature of about 70°F (21°C). Passive thermal control is essential for the HST and this is provided for with multi-layer insulation (MLI) covering 80% of the exterior surface. Together with electrical heaters, these provide a stable thermal environment.

The insulating blankets have 15 layers of aluminised Kapton, which, with an outer layer of

BELOW Passive thermal protection is an important part of overall temperature control, which also includes active and temperature-mitigation measures such as orientation into or out of sunlight during non-operational parts of the orbital cycle, and materials which control conduction and radiation (there is no convection in weightlessness). Chemglaze and multi-layer insulation (MLI) is applied in appropriate areas of the structure as shown, covering 80% of the exterior. *(Lockheed)*

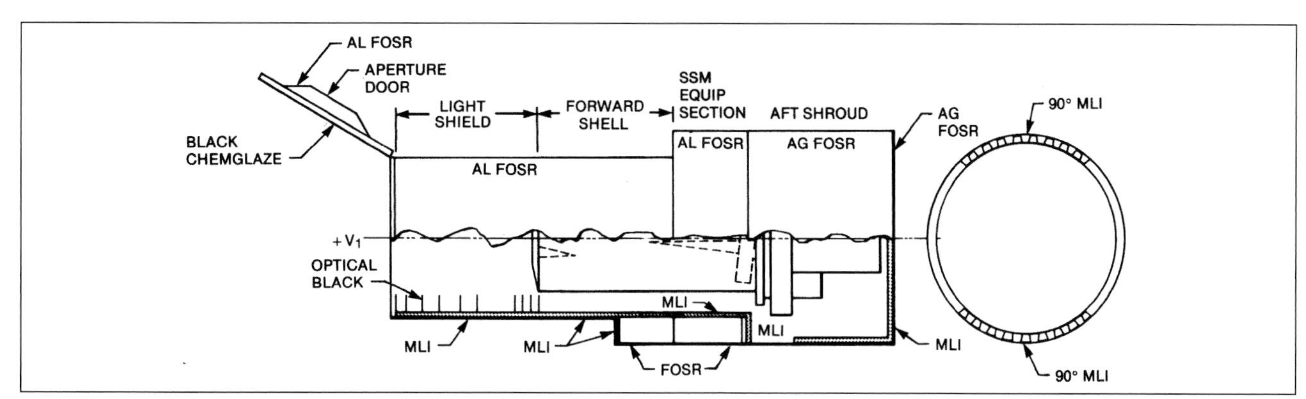

ABOVE A view of the Equipment Section of the SSM taken by an EVA astronaut on a servicing mission shows badly affected parts of the MLI covering under distress from thermal and particle erosion. *(NASA)*

FEP (see next column), is collectively referred to as the aluminised Teflon Flexible Optical Solar Reflector (FOSR) with aluminised or Teflon tape covering most of the remaining exterior surface. Special absorptive or reflective paints cover areas where harsher thermal environments are likely to build up, controlling the degree of thermal transfer across the surface and through into internal areas. An active thermal control system uses heaters on components and subsystems within the SSM Equipment Section and structural interface between the OTA and the science instruments.

Particular protective features of the SSM include MLI blankets for the Light Shield and the Forward Shell where thermal emissivity could pose structural expansion or contraction as well as providing a conductive path to other parts of the HST. Aluminium FOSR tape is applied to the Aperture Door side facing the Sun, with specifically designed shapes of FOSR and MLI blankets on the outer surfaces of the Equipment Section bay doors. Internal blankets of these materials are attached to the bulkheads to maintain thermal balance between bays. Silverised (AG) FOSR tape is added to the Aft Shroud and & Bulkhead exteriors, and radiation shields cover the inside of the Aft Shroud doors to protect the science instruments.

Outside the application of passive thermal control elements such as those described above, one of the most important aspects of thermal control and equilibrium was the layout of interior equipment taking account of the heat it might or might not generate and the degree of absorption and reflectivity it has as an engineered component. Consideration was also afforded to the interactive influence between closely grouped equipment items as to the degree of conducted thermal energy between elements. In a vacuum there is no convection, but conduction is a potential hazard for thermally sensitive items.

Bay spaces in the Equipment Section of the SSM were populated according to these criteria, with heat dissipating equipment on the side of the ES most likely to be in shadow for the majority of the time. Equipment was placed so as to take advantage of the natural thermal flow patterns throughout the structure, with the proviso that it would take several weeks for the Telescope to get a thermal balance from soakback effects replacing the thermal environment of the Orbiter payload bay with that of the diurnal cycle on orbit. To monitor the temperature profiles and for engineers to construct a thermal map, more than 200 temperature sensors were attached throughout the SSM, externally and internally, to monitor separate components and to control heater operations. These 'housekeeping' activities are crucial to maintain the SSM and the OTA in a suitably balanced equilibrium for the scientists to do their work through the instruments on board.

Portions of the MLI integral to the thermal control of the HST were returned 3.6 and 6.8 years after launch by two servicing missions (SM-1 and SM-2 respectively) and examined. On SM-2 astronauts retrieved materials which had torn and rolled up under their own stress and discovered potential long-term significance in the reaction of these materials to up to 40,000 solar cycles. Some degradation had been observed during the first servicing mission (SM-1), but just over three years later the condition of the MLI was much worse.

The outer layer consists of 0.127mm-thick aluminised fluorinated ethylene propylene (FEP) Teflon, which was found to have become severely embrittled. This is believed to have been caused by electromagnetic radiation across the entire solar spectrum, trapped radiation, atomic oxygen, and thermal cycling, although none could be replicated and produce the same effect. Only when an intense dose of unfiltered x-radiation was applied from a rotating anode generator in ground tests did the effect closely match the samples returned from space.

Assistance in tracing the source of the problem was obtained from scientists and engineers working on the Long Duration Exposure Facility (LDEF), which had been carried into orbit by the Shuttle *Challenger* on 6 April 1984 and eventually returned to Earth by *Columbia* on 20 January 1990. The LDEF was a large structure packed on external and internal surfaces with a wide range of materials, coatings, paints and coverings, optical fibres, crystals, electronics and optics which would provide detailed information on the effects from the space environment. LDEF was to have been retrieved by the Shuttle after a year in space, but delays to the flight sequence on which it was to have been recovered pushed that into 1986, and the loss of *Challenger* on 28 January 1986 put that off for almost another four years.

Scientists studying the results of ground tests on the returned samples from the HST, together with examining correlated analysis from LDEF samples, concluded that a combination of irradiation and annealing was the major cause of the embrittlement observed. Other scientists working on the industrial implications of this phenomenon for both space and ground-based materials narrowed the vulnerability to soft x-rays in 4–8Å range (1.55–3.2keV). This has helped determine the optimum materials for very long duration missions where such passive MLI is required – another way the re-serviceability and return to Earth of objects such as the solar arrays bistems has contributed to other space science projects and programmes.

Much of the information from the LDEF and from returned samples during the first and second HST servicing missions was applied to tests to inform a failure review board. The FRB was set up to examine the reasons for the degradation and to recommend suitable solutions including a replacement material in critical locations which could be carried up and installed on the HST on subsequent servicing missions. Ten candidate materials were chosen for detailed examination and exposed to ten-year equivalent doses of the simulated orbital environment.

Samples were exposed to low- and high-energy electrons and protons, atomic oxygen, x-rays, ultraviolet radiation and thermal cycling. Following these exposures, the mechanical integrity and optical qualities of the candidates were investigated using optical microscopy, scanning electron microscopy and a laboratory portable spectroreflectometer. Based on these analyses, two candidates were selected for possible use on future servicing flights (see Chapter 8).

Of equal importance to the routine operation of the Telescope systems, and to the acquisition of data on as near a continuous basis as possible, are the contingency and 'safing' procedures in the event of unexpected failure. The general

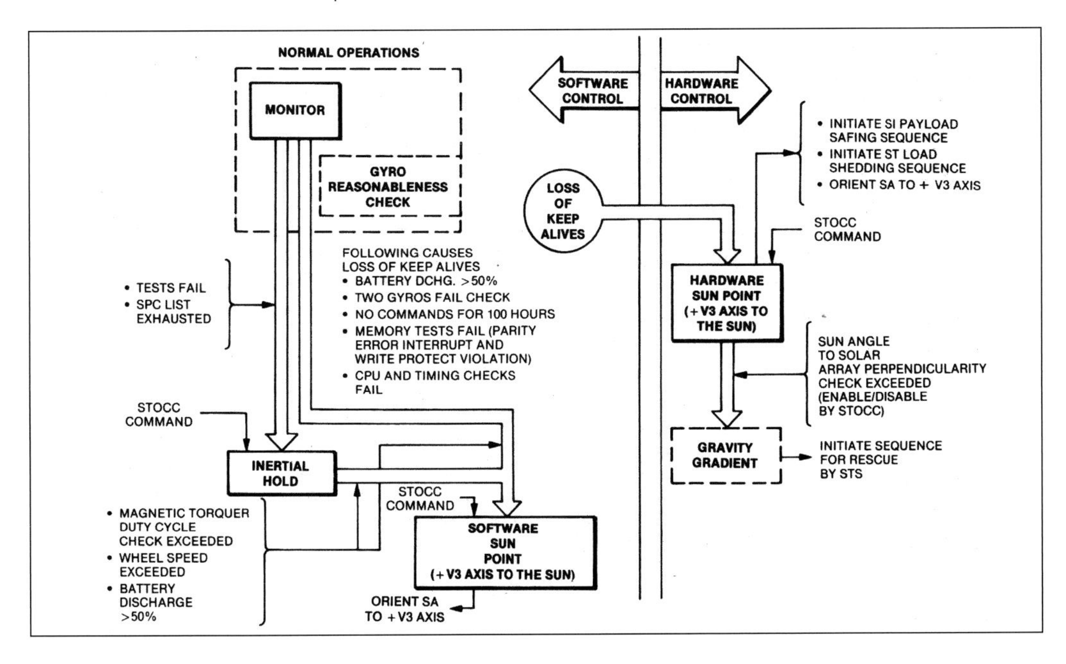

LEFT Safing and fault-mitigation flow patterns to prevent a cascade-effect from a single failure are built in to the Pointing Safemode Electronics Assembly (PSEA) – which weighs 86lb (39kg) and is installed in Bay 8 of the Equipment Section – and the Retrieval Mode Gyro Assembly (RMGA), which has three gyroscopes that are less accurate than the rate gyros. *(Lockheed)*

design philosophy of the HST was to design and build a robust and semi-autonomous vehicle to allow self-mitigation against equipment failure or system/subsystem breakdown. Considering the age of the Telescope and the original expectation that it should be periodically retrieved and returned to Earth by the Shuttle for upgrade and relaunch, the level of sophistication achieved is a credit to the Lockheed design teams as well as to the customer.

The basic design philosophy of the HST was laid down in the mid-1970s, more than 15 years before launch and now more than 40 years ago. Not only has the HST never been returned to Earth, it has performed considerably beyond its life expectancy and that is due in some considerable measure to the advanced thinking built into its design philosophy at a time when many of the rule books in use today were not even conceived, let alone written. It is a credit to Lockheed and to NASA that this spacecraft emerged in the vanguard of many safing and contingency planning procedures and yet has proven to be a robust exemplar of the genre.

Early in the concept phase, engineers decided they would use a fail-operational/failsafe approach and this went contiguous with the same philosophy in practice for the Shuttle. Wherever possible the spacecraft/vehicle would have a modest form of self-test and repair (STAR) by way of multiple redundancy and an inbuilt capacity for automatically switching to a safe mode until diagnostic analysis by ground controllers. Some spacecraft had gone further than this. Before the design of the Voyager spacecraft launched to the outer planets in 1977 had been finalised, this self-

1 A photo-survey of images stitched together showing the -V2 side of the HST with the solar array boom coming straight at the viewer and the two High-Gain Antennas at either side of the Light Shield in their stowed positions. *(NASA-JSC)*

2 Photo-survey montage of the -V3 side of the HST with the apertures for the fixed-head star trackers. Note the prolific labelling and the arrangement of EVA handholds and rails. *(NASA-JSC)*

3 The +V2 side of the Telescope reveals the folded solar array drive coming out of the picture with the two stowed High-Gain Antenna booms and two of the four magnetic torque booms at slant angles and covered with foil insulation. *(NASA-JSC)*

4 A good view of the High-Gain Antenna arm with two magnetic torque arms in this photomontage of the +V3 side. *(NASA-JSC)*

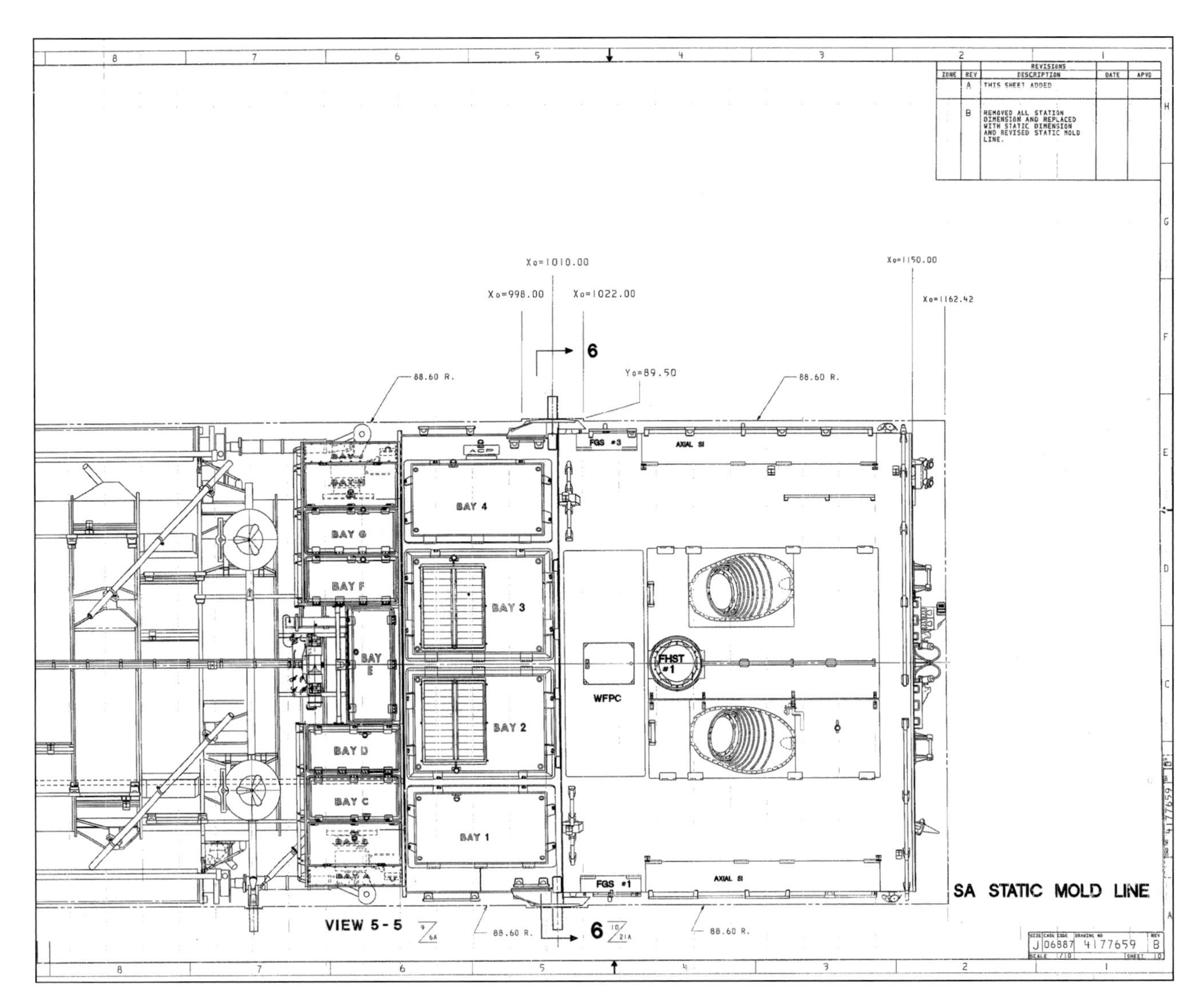

test and repair concept produced an advanced and highly sophisticated computer system for the so-called 'Grand Tour' of Jupiter, Saturn, Uranus, Neptune and Pluto on a sequence of separate flights.

The resulting methodologies for contingency operations with the HST grew out of that desire in the 1970s to produce a truly fully autonomous method of operating advanced space science satellites and spacecraft. However, for the HST the advanced computerisation required for distant planetary encounters, when communication times ran into several hours for a two-way exchange of data, the near Earth orbit of the Telescope rendered that unnecessary; but it is interesting to reflect on the status of conceptual autonomy at this early stage in a still-evolving generation of space science satellites and observatories.

In any event, the plans to routinely access the Telescope, whether to periodically return it to Earth for upgrade and relaunch or to conduct servicing activities in space, eliminated the rationale for full autonomy.

As designed, the Telescope's safing system uses existing elements, including pointing and data management components together with dedicated hardware in a network known as the Pointing Safemode Electronics Assembly (PSEA). This is a configuration of fail-paths and fault mitigation whereby the Telescope will maintain attitude lock, move the solar arrays toward the Sun for maximum energy, conserve electrical power by selectively shutting down unnecessary equipment, and balance energy consumption against power reserves. This system allows the HST to operate autonomously and await further instructions

ABOVE The general arrangement of equipment location in the SSM and Equipment Section. *(NASA)*

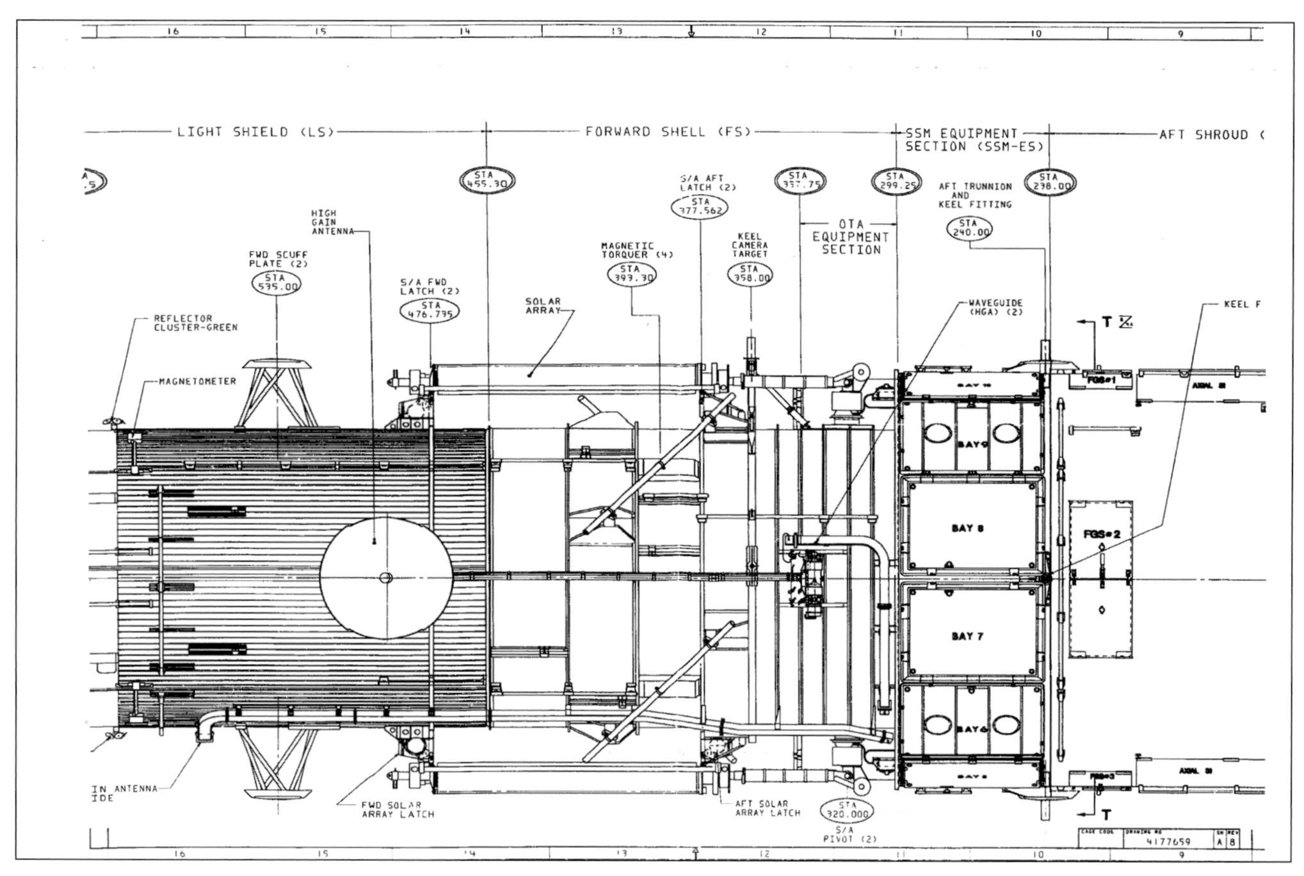
LIGHT SHIELD (LS)
FORWARD SHELL (FS)
SSM EQUIPMENT SECTION (SSM-ES)
AFT SHROUD
STA 455.30
S/A AFT LATCH (2)
STA 377.562
STA 337.75
STA 299.25
AFT TRUNNION AND KEEL FITTING
STA 238.00
HIGH GAIN ANTENNA
OTA EQUIPMENT SECTION
STA 240.00
MAGNETIC TORQUER (4)
KEEL CAMERA TARGET
FWD SCUFF PLATE (2)
STA 393.30
STA 358.00
STA 535.00
S/A FWD LATCH (2)
STA 476.735
SOLAR ARRAY
WAVEGUIDE (HGA) (2)
KEEL F
REFLECTOR CLUSTER-GREEN
MAGNETOMETER
BAY 9
BAY 8
BAY 7
FGS#2
FGS#1
FGS#3
AXIAL SI
FWD SOLAR ARRAY LATCH
AFT SOLAR ARRAY LATCH
STA 320.000
S/A PIVOT (2)
4177659

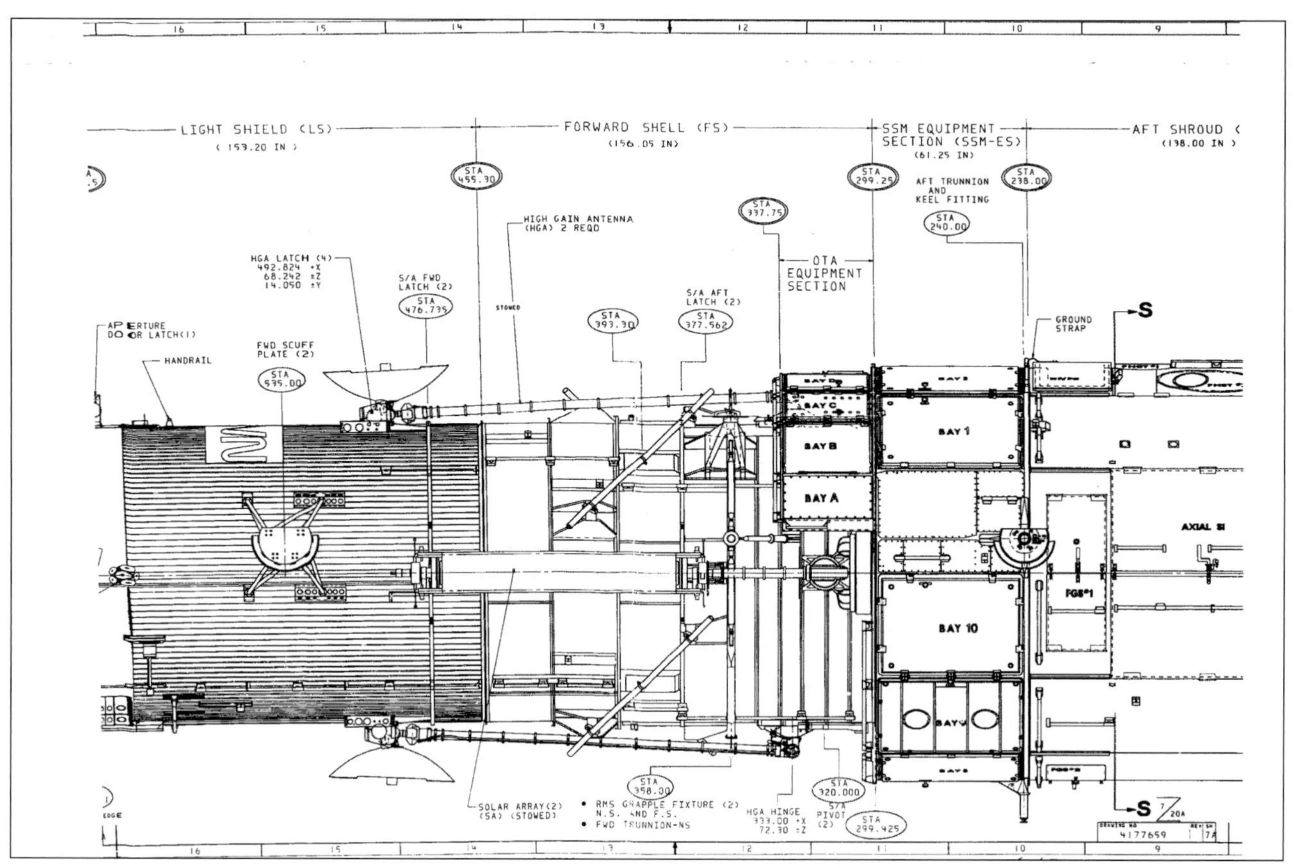
LIGHT SHIELD (LS)
(153.20 IN)
FORWARD SHELL (FS)
(156.05 IN)
SSM EQUIPMENT SECTION (SSM-ES)
(61.25 IN)
AFT SHROUD
(138.00 IN)
STA 455.30
STA 299.25
AFT TRUNNION AND KEEL FITTING
STA 238.00
HIGH GAIN ANTENNA (HGA) 2 REQD
STA 337.75
STA 240.00
OTA EQUIPMENT SECTION
HGA LATCH (4)
492.824 +X
68.242 ±Z
14.050 ±Y
S/A FWD LATCH (2)
STA 476.735
STOWED
S/A AFT LATCH (2)
STA 393.30
STA 377.562
GROUND STRAP
APERTURE DOOR LATCH(1)
HANDRAIL
FWD SCUFF PLATE (2)
STA 535.00
BAY 1
BAY B
BAY A
BAY C
BAY 10
AXIAL SI
FGS#1
STA 358.00
STA 320.000
SOLAR ARRAY(2) (SA) (STOWED)
RMS GRAPPLE FIXTURE (2) N.S. AND F.S.
FWD TRUNNION-NS
HGA HINGE
373.00 +X
72.30 ±Z
S/A PIVOT (2)
STA 299.425
4177659

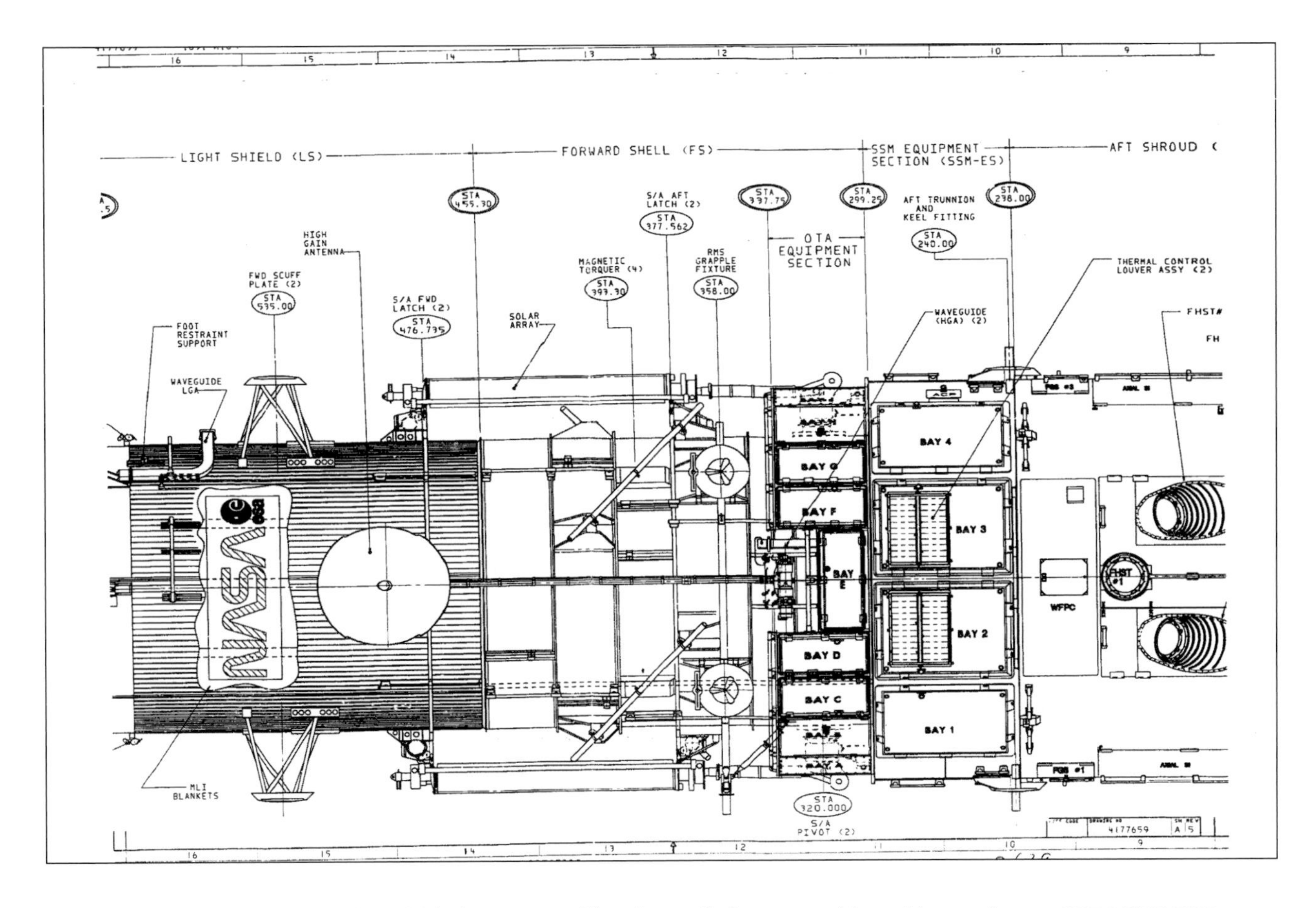

for up to 72 hours, a period during which the STOCC could be reasonably expected to reconnect through the communication system, retrieve telemetry from the tape recorders, analyse the problem and upload a new and corrective set of operating instructions.

During the emergency period the HST's safing system would self-monitor through its Monitor Mode, which can be turned on or off by the STOCC as required. This allows frequent confirmation through special tones sent to the ground that confirm that the Telescope is in a provisional holding pattern of internal activity and that it is in fact 'safed' and functioning. The STOCC has special standby teams of specialists, technicians, engineers and scientists to focus on fault-mitigation activities through recommendations to the flight controllers. These experts will identify whether the problem is a flight or data issue, the latter being indicated by inconsistencies or errors in downloading information to the ground. A flight problem involves the operation of the Telescope, usually its SSM equipment, and could range from a major subsystem failure to a minor electrical fault.

The diagnostic teams would provide a series of recommendations and compile instructions and protocols to be followed wherein the Telescope could be recovered to normal operations. They will advise on how best to operate the system or systems involved and present options as well as calculating the probabilities of success for each path chosen. There are a range of permutations involved in which the Telescope may move initially into Software Inertial Hold Mode (SIHM), in which case the Telescope will hold its last commanded position and report that fact to await updates. If a failure should occur while an attitude change or orientation is taking place, the safing system will complete the motion and hold the HST at that position.

If, however, the safing system encounters a small electrical power problem, or an internal Pointing and Control System safety check fails, the Telescope will automatically go into Sun Point Vehicle Mode (SPVM) in which the Telescope is manoeuvred to an orientation so that the solar arrays continuously point towards the Sun (on the sunlit side of the Earth) and the photo-

OPPOSITE TOP **Latch positions and dimensions for the Equipment Section and Forward Shell.** *(NASA)*

OPPOSITE BOTTOM **SSM, Forward Shell and Light Shield geometry with latch positions and rails.** *(NASA)*

ABOVE **Rotated 90° to the previous orientation, the fixed-head star trackers and thermal control louvres can be clearly seen.** *(NASA)*

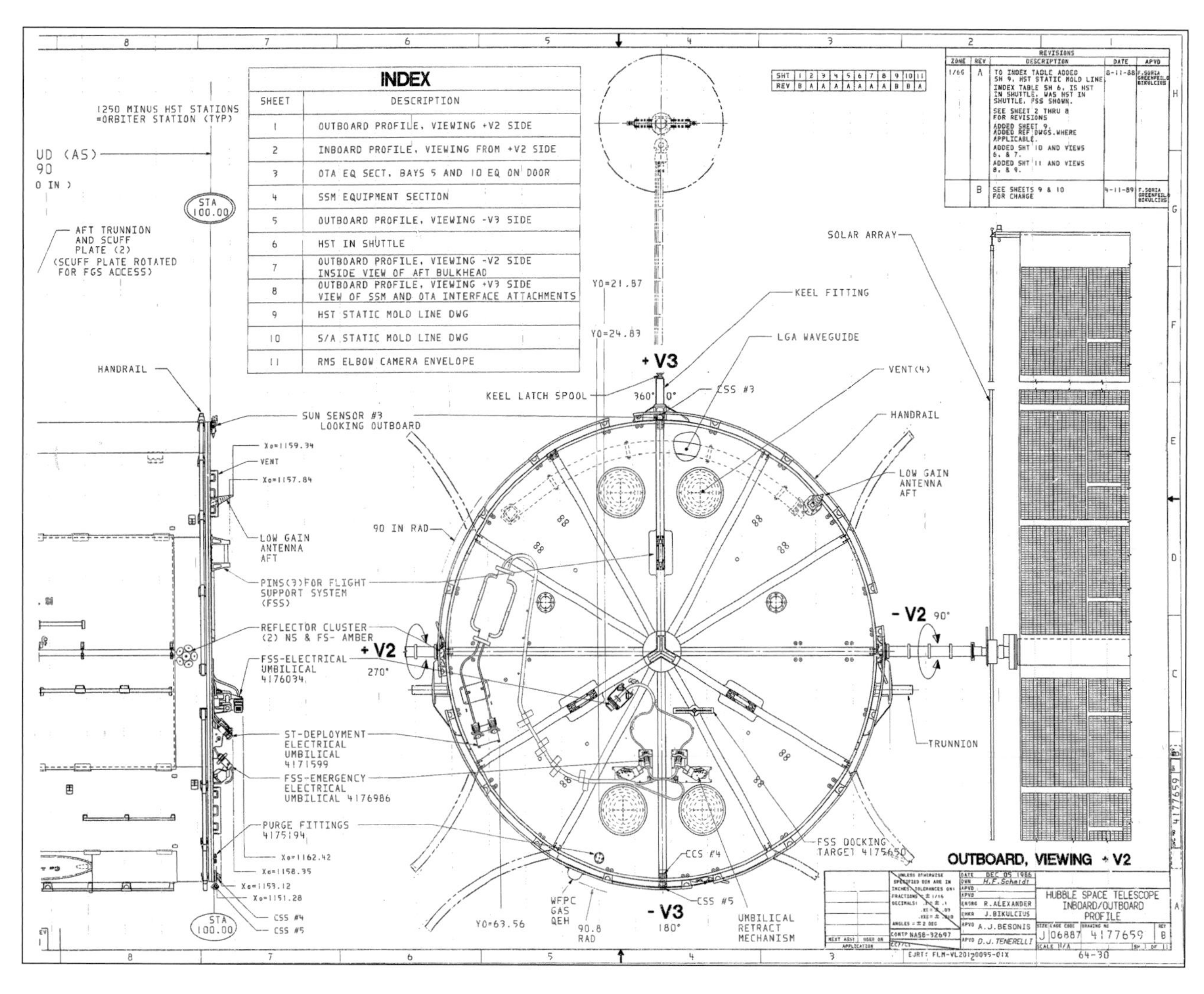

ABOVE The end elevation looking forward displays the servicing doors in the open position together with the relative position of the solar arrays. *(NASA)*

voltaic cells can generate a constant supply of electrical energy. The safing system will assume that further instructions from the STOCC will be forthcoming and will anticipate a return to normal operations by not compromising a return to standard activities. Equipment will be configured by the safing system so that thermal balance is maintained, which includes heater activation to contain any anomalies to the temperature profile. Normal operations would not return unless the STOCC intervened.

The two optional configurations identified above as the SAHM and SPVM sequences are controlled through the DF-224 software, but if the situation worsens the system will transfer authority to the Pointing Safemode Electronics Assembly (PSEA) and this will put the Telescope into Hardware Sun Point Mode (HSPM), a condition which can be promoted by several failures: a computer malfunction; batteries losing more than 50% of their charge; two of the three rate gyro assemblies failing; or the data management subsystem failing. In the HSPM routine, the 'keep-alive' signals will cease flowing to the ground.

The specific sequences triggered by the HSPM will command the PSEA computer to turn off the Telescope's standard equipment to conserve electrical power. This could result in shutting down the DF-224, because it may be the cause of the failures. The 'second-brain' in the Telescope would take over from the primary operating system. If the emergency condition continues for more than two hours the SI C&DH is also shut down, both to conserve electrical power and to isolate potential causes of the failure. After this a payload safing sequence will start, and if it has not already been done the PSEA computer will turn the HST to the Sun using the course Sun sensors. Unless the STOCC begins to rectify the situation quickly the 'brain' progresses through a series of shutdowns

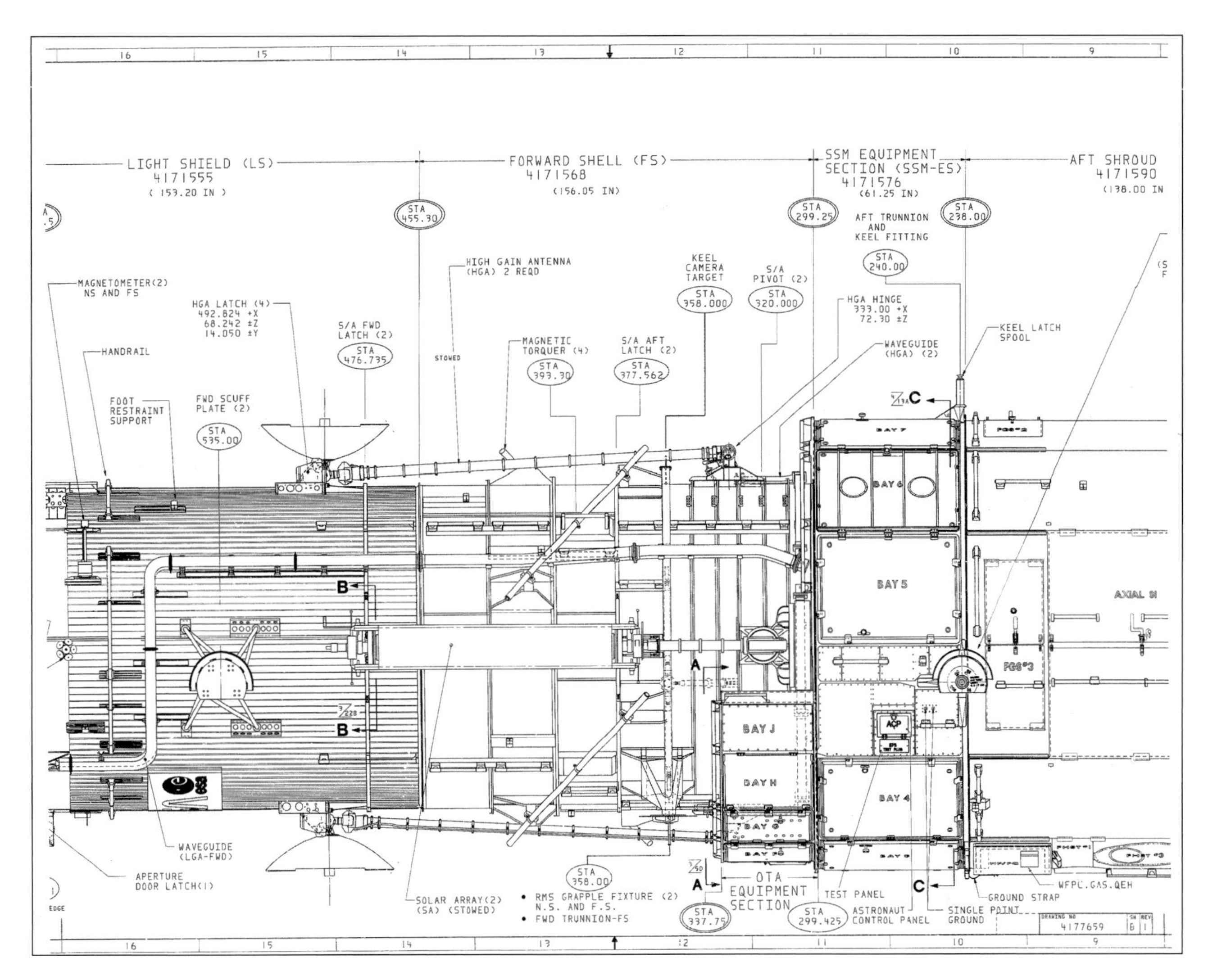

by removing power not essential for the survival of the Telescope.

If it gets to this stage the STOCC will have to go through an extensive and lengthy recovery procedure – but it can be done. However, this condition can only prevail for 72 hours without contact from the ground. If contact remains absent, or if more systems fail, the PSEA moves to Contingency Gravity Gradient Mode (CGGM), which will maintain the HST in a gravity-gradient stabilised attitude until it can be retrieved by the Shuttle, an option available only while the Shuttle was operational (to early 2011). The range of options available with the Shuttle allowed management to consider a flight to retrieve the Telescope and return it to Earth, but that eventuality was always a very low option. Nevertheless, the CGGM attitude-hold would prevent the HST from tumbling uncontrollably.

Gravity-gradient stabilisation is a relatively simple way of a large satellite of the Earth maintaining stable attitude without attitude propulsion or controlling gyroscopes. It operates on the physical law regarding gravity and inertia. Because gravity decreases on the inverse square law, a large object can be made to align with the centre of the Earth because the force of attraction will tend to favour the point closest to the centre of the Earth and stabilise the object so that its long axis is perpendicular to the orbital path. The minimum moment of inertia will seek to line up with the centre of the planet and in so doing maintain the satellite in that position, relatively stable and passive.

This technique was tried operationally during the NASA Gemini 11 mission in 1966 when a 100ft (30m) tether was attached to the Agena rocket stage, with which it had previously docked, by an astronaut on EVA. After the EVA the Gemini spacecraft undocked

ABOVE Side elevation showing the High-Gain Antenna positions and the interface between the Forward Shell and the Light Shield. *(NASA)*

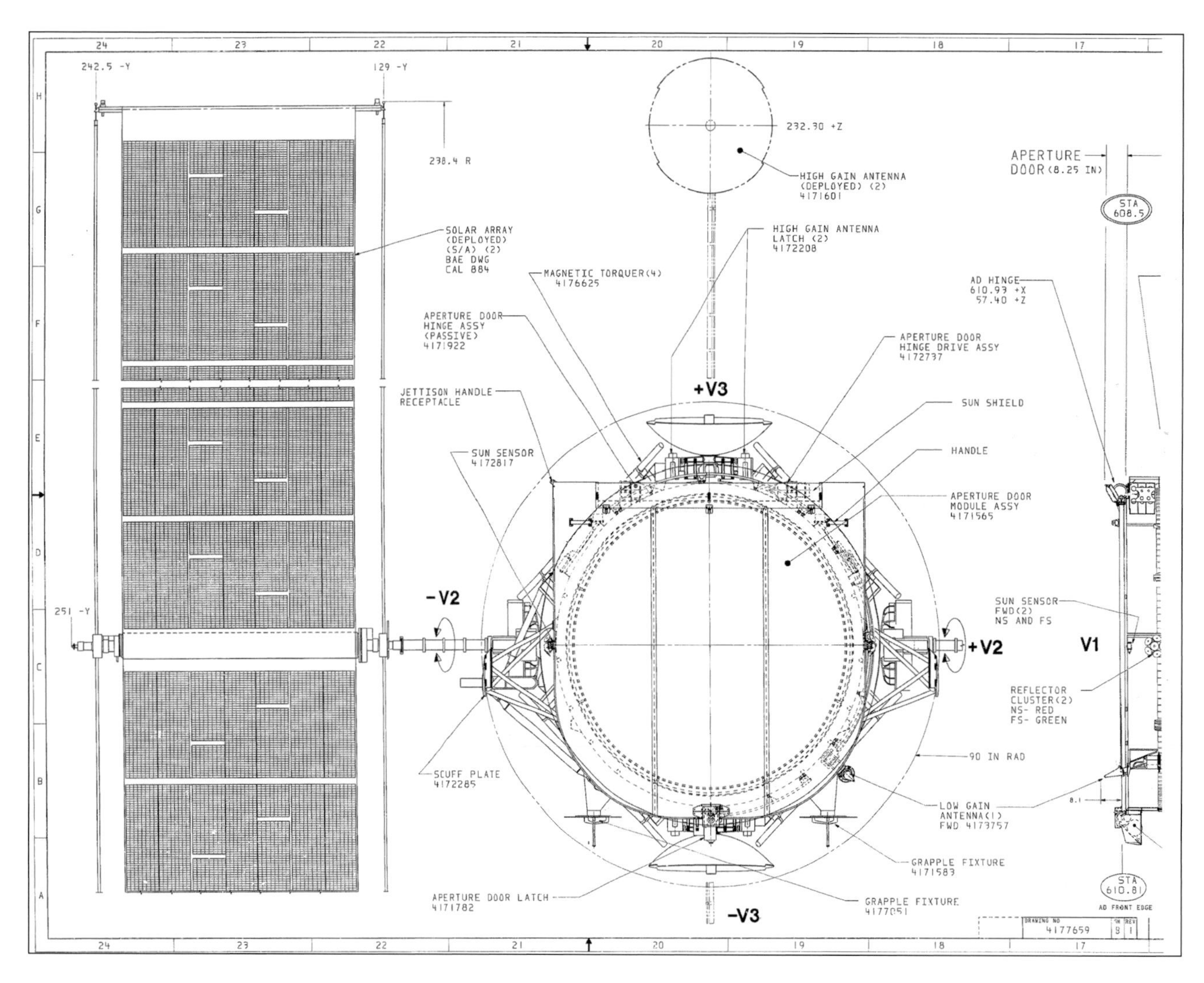

ABOVE Relative scale of the solar arrays as viewed from the front looking back, with the deployed position of the High-Gain Antenna indicated. *(NASA)*

and the two were aligned with the centre of the Earth in an attempt to demonstrate the principle. Insufficient gravity was sensed and the experiment was a failure due largely to the fact that forces within the tether caused a ripple motion that overwhelmed the minute difference between the Earth's gravitational attraction on the two structures. However, in 1967 the US launched a military experimental satellite that demonstrated the gravity-gradient effect from geosynchronous orbit. Between 1966 and 1969 several large US Applications Technology Satellites demonstrated the concept from low Earth orbit. But it was never tried with Hubble, simply because it was never needed.

The PSEA consists of 40 electronic printed-circuit boards with redundant electronics necessary to run the Telescope as described earlier. The PSEA weighs 86lb (39kg) and is located in Bay 8 of the Equipment Section. The Retrieval Mode Gyro Assembly (RMGA) is also in that bay and consists of three gyroscopes with less precision than the rate gyros. Without the Shuttle the neatly planned final option is not available and there is no vehicle currently in development, or planned, by any space agency anywhere that could rendezvous and grapple the HST for any further servicing operation or return it to Earth. In several respects, and seen most evidently in the safing options planning, the Telescope was made for the Shuttle and outlived it by a considerable margin.

Optical Telescope Assembly

The OTA is a highly complex piece of precision engineering that absorbed four-million man-hours over seven years of effort that took it from an idea on paper to a completed structure ready for launch. The contract to build the OTA was awarded to Perkin-Elmer Corporation in October 1977. Renowned for its optical devices utilised in a wide variety of applications, the company was taken over by Hughes Aircraft Company in 1989 and by the time of launch was renamed Hughes Danbury Optical Systems Inc. The OTA is not large as telescopes go but the precision with which it has been built and its operating position outside the Earth's atmosphere rendered it superior at launch to any existing ground-based observatory.

The justification for the Hubble Space Telescope was in its unobstructed view of the universe and the exacting tolerances built in to provide near-perfect images. As the early history of the HST would reveal, that was more difficult to achieve than at first believed, and some simple mistakes prevented the Telescope from realising its full potential until the first servicing mission in December 1993, more than three years after launch (see Chapter 7). In all respects the OTA is a facilitator, providing the fixed optics for an initial suite of scientific instruments that would change over time with successive revisits by Shuttle Orbiters, and the Fine Guidance Sensors essential to keeping the Telescope on target.

The engineering design of the Telescope itself was framed by requirements of a purely practical nature: it had to weigh as little as possible, gather the most amount of light as possible, be contained within a volume compatible with the Shuttle Orbiter payload bay and yet enable a large focal length to provide magnification. The obvious answer was a reflector of Cassegrain design, which uses a parabolic primary mirror designed to reflect incoming light to a convex, hyperbolic secondary mirror that lies within the focus of the primary and directs the light cone back through a central aperture in the primary mirror to the focal plane beyond. Because the light path is 'folded' back through the physical body of the telescope, a Cassegrain is short in length relative to the focal plane. The convex secondary mirror multiplies the focal length by the secondary magnification that is the focal length of the system divided by the focal length of the primary.

There are several different designs of Cassegrain. In the standard Cassegrain the primary mirror is parabolic, where all the rays of light are reflected to a single point at the axis of symmetry, the focal point, and the secondary mirror is hyperbolic. There are certain problems with this design, notably coma and spherical aberration (see also feature on 'The physics of the telescope' on page 170). Coma is inherent in parabolic mirrors and is a characteristic of incident rays that strike the primary mirror at an

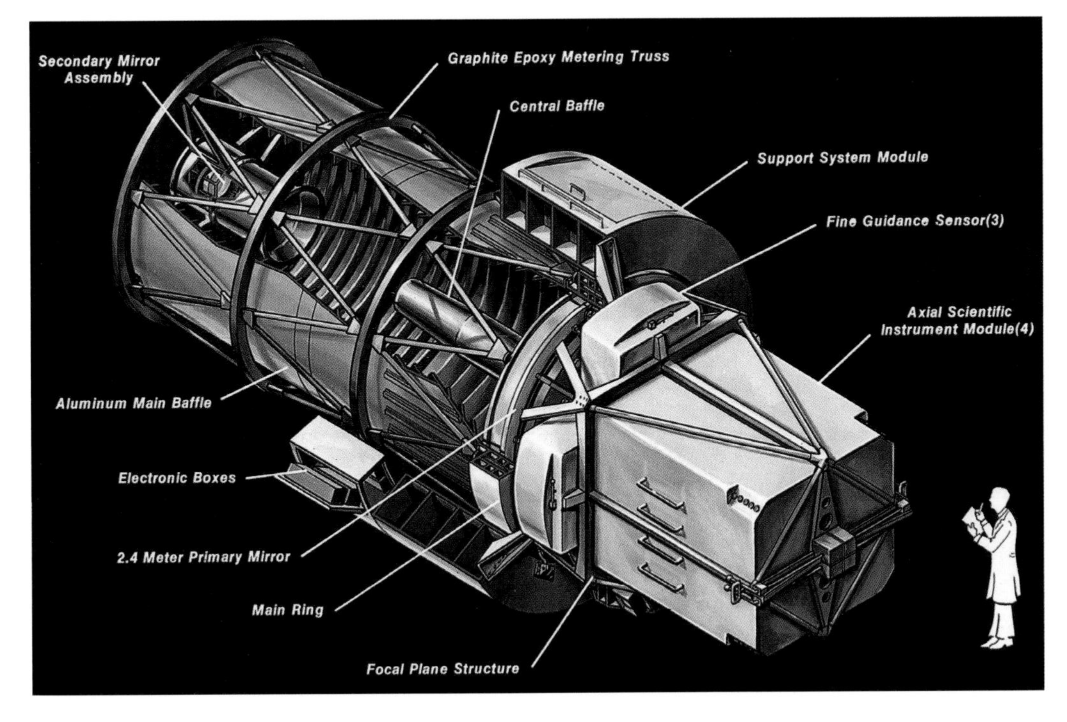

LEFT The second major element of the HST and the one for which the Telescope exists is the Optical Telescope Assembly. This fits inside the primary structural elements embraced by the Support System Module, the Forward Shield and the Light Shield, and comprises the primary and secondary mirrors and the four axial and one radial science instruments. *(NASA)*

angle and are not reflected to precisely the same point, causing the image to wander off-centre and appear to deform. Spherical aberration is a distortion produced when light rays are reflected from a mirror at its edge and cause a deviation resulting in an imperfect image. In short, they incur problems because light coming from the sides of the mirror focus to a slightly different point to light rays coming from the centre.

In the early part of the 20th century the American astronomer George Willis Ritchey and the French astronomer Henri Chrétien designed and built a new form of Cassegrain free from most coma effects and spherical aberration. In this design both the primary and the secondary mirror were hyperbolic, a design which brought its own anomalies and which requires expensive and prolonged preparation, rendering them suitable only for expensive telescopes used by professional astronomers. The hyperbolic shape is a conic section curved at the centre but with infinite divergence toward the outer edges. It requires exacting standards in manufacture and if the geometric shape of the conic section is not precise it will cause deformation in the image.

This Ritchey-Chrétien design of Cassegrain reflector, certainly not new to astronomical telescope design, achieves the objectives of the HST engineering constraints on size and mass while retaining a high standard of optical and operational flexibility for a wide range of science instruments. This 'folded' design packs a long focal length of 189ft (57.6m) into a small telescope length of 21ft (6.4m), with several smaller mirrors within some instruments capable of elongating the focal length even further. If engineered to almost exacting perfection, the image produced should be free of coma and aberration; but as explained in Chapter 3, this was not the case with the mirror prepared for the Hubble Space Telescope.

Even with a perfectly shaped hyperbolic primary mirror, the telescope would still produce some field curvature and astigmatism. Field curvature is where reflected light is focused to a curved plane which the eye can compensate for to some extent but which is difficult for an electronic observation system such as a CCD to handle; the plane of a CCD imaging detector is flat and will not coincide with the curved focal plane as the image is received from the mirror. Astigmatism is the same but with a symmetrical distortion that can inhibit astrometric readings. It is caused by a complex interaction between two light planes known as the tangential and the sagittal planes, which are perpendicular to each other. The tangential plane corresponds to the line of reflected light received at the fixed foci while the sagittal plane is at 90°, and each can produce a slightly different focal point. In the OTA optics both field curvature and astigmatism were designed to be zero at the centre and increased toward the outer reflected edge.

The light path of the HST begins by travelling down the tube past the baffles set there to attenuate light from unwanted sources (see section on SSM systems and subsystems), striking the 94.5in (2.4m) hyperbolic primary mirror and travelling back up the tube to the 12.2in (30cm) secondary mirror, where it is returned as reflected light down through the 23.5in (60cm) diameter opening at the centre of the primary mirror to the focus of the telescope, 3.3ft (1.5m) behind the mirror itself. The focal plane is shared by five science instruments and three Fine Guidance Sensors.

In the centre of the field of view is a small folding mirror that directs light into the Wide Field/Planetary Camera (WF/PC), mounted perpendicular to the long axis of the Telescope, with the remaining field divided between four axial science instruments, described later in this section. Around the circumference of the optical field is what is known as the guidance-field, divided up into the three Fine Guidance Sensors which each have their own folding mirrors, also described further on in this section.

The OTA is a facilitating host to a changeable suite of science instruments and maintains the

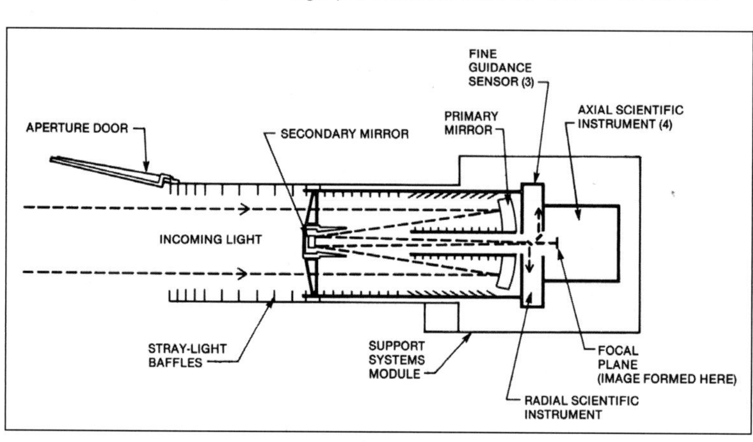

BELOW The light path follows a conventional Ritchey-Chrétien layout with the primary mirror reflecting back to a secondary mirror, which focuses the image through a hole in the centre of the primary to the focal planes formed within the axial science instrument package. *(Lockheed)*

structural support for the optical devices and instruments. The chief components are the Primary Mirror Assembly (PMA), the Secondary Mirror Assembly (SMA), the Focal Plane Assembly (FPA), and the OTA Equipment Section (ES). From contract award in October 1977, Lockheed was primarily responsible for the design and assembly of the OTA, with major subcontractors such as Boeing and Hughes-Danbury providing significant elements of the structure, optics and electronic control assemblies.

The Primary Mirror Assembly comprises the mirror itself supported inside the Main Ring, which serves as the structural backbone to the Telescope and supports the main and central baffles. It provides the central coupling to the rest of the spacecraft through a set of kinematic attachment brackets linking the Main Ring to the Support Systems Module. The assembly provides support to the primary mirror and the OTA baffles and has several integrated elements: the primary mirror, Main Ring structure, reaction plate and actuators and the main and central baffles.

Unlike ground-based telescope primary mirrors, the one manufactured for the HST is not formed from a solid piece of glass, which if fabricated at the required size would have weighed around 8,000lb (3,630kg). Instead, the contractor Corning Glass Works used a product identified by the company as 7971 ultra-low expansion silica glass, which was selected for its low expansion coefficient to ensure minimum sensitivity to temperature changes as well as for its very low weight. The mirror is composed of a core filling of glass honeycomb ribs, 10in (25.4cm) deep, between two lightweight facing sheets each 2in (5cm) thick. The result is a mirror assembly that weighs a relatively light 1,827lb (828kg).

The Corning mirror blank was ground to the required shape by Perkin-Elmer in its large optics fabrication facility in Danbury, and when finished to its hyperbolic profile it was moved to a computer-controlled polishing facility. The mirror blank was placed on a bed of 138 titanium rods to simulate the even loading of a gravity-free environment and polished by an abrasive pad attached to a swivelling arm. Several separate computerised runs, each lasting from six to seventy hours, imposed

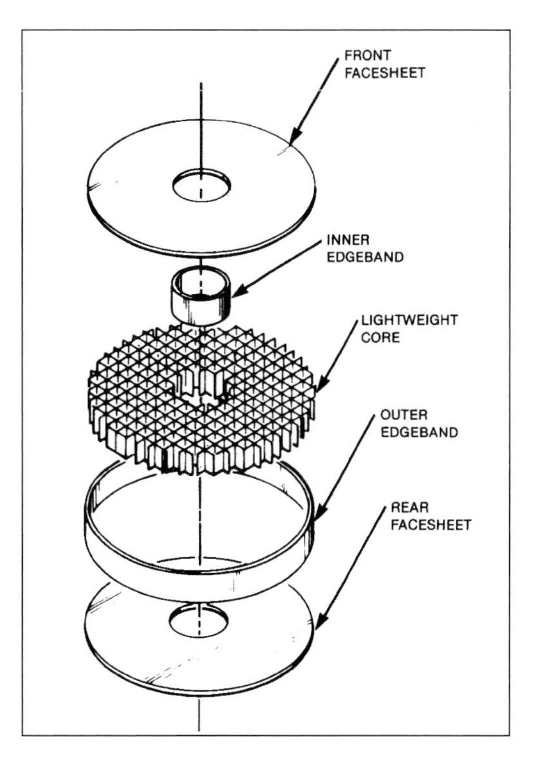

LEFT The separate component parts of the primary mirror are shown, which is manufactured from Corning 7971 ultra-low expansion silica glass. *(Lockheed)*

LEFT Measurements made on the primary mirror to determine its precise shape reveal the lightweight honeycomb structure. *(Perkin-Elmer)*

BELOW Mirror polishing with a numerically controlled machine. *(Perkin-Elmer)*

ABOVE The primary mirror is inspected prior to final polishing and weight balance measurements, hiding an imperfection that would threaten the ability of the Telescope to perform to specification. *(Perkin-Elmer)*

pre-programmed polishing loads, speed and rotation rates.

When the hyperbolic shape was believed to be at the correct size and profile across all parts of the 94.5in diameter surface, a null corrector and a special camera were brought in to give final judgement. Null correctors had been used before and Perkin-Elmer had simplified the technology to speed up the process. The device itself consisted of two small mirrors and a small lens. Measurements could be made which would detect surface smoothness to a fraction of a wavelength of light equal to a few millionths of an inch. To operate the system, light from a laser would be sent through the corrector and bounced off the mirror to pass back through the corrector and create an image of black and white lines. The pattern would be photographed and analysed, with any inconsistencies showing up as deformations in what should be a perfect set of geometrical lines.

This operation was so precise and sensitive to vibration that the tests were conducted in the middle of the night so that no vehicles were moving around the plant, and with the air conditioning turned off so that the running machinery would not cause vibrations. After the mirror shaping was passed the various component mirror elements were joined in a furnace at a temperature of 3,600°F (1,982°C). However, the outer edge-band designed to hold the separate components of the mirror together had fused to the interior slats, a possible cause of harm through uneven stresses. A special material known as Invar was used as a spacer to separate the null corrector from the surface of the mirror. The material was impervious to changes in temperature so once it had been calibrated by the supplier at a specified length it would not change under thermal stress.

In an attempt to cut costs the calibration of the null corrector was not performed as efficiently as it could have been, and several attempts were made to obtain a proper calibration of the instrument and then to apply it to the mirror being shaped as the flight article. The interferograms – photographs of the line patterns made by passing the laser light through the corrector mirrors – appeared to show alarming imperfections in the contour of the mirror itself. Because the computer-controlled polishing operation was reliant on the returned information from the null corrector, and because any corrections to the grinding commands relied on sound photographs, several tiny flaws in the shaping of the mirror crept in which would not be discovered until the HST had been launched and could be evaluated in orbit from the images it transmitted.

The primary mirror blank had been shipped to Perkin-Elmer in November 1978 for grinding and was to have been delivered to the Danbury plant for polishing in September 1979, but delays set that back to May 1980 and prevented the polishing from being completed before October 1981.This was followed by application of a reflective layer of aluminium of 65nm and coated with a further protective layer, of magnesium fluoride (MgF_2), which varied in thickness from 10nm to 25nm. The fluoride layer protects the aluminium from oxidation and enhances reflectivity at the hydrogen-emission line, also known as the Lyman-Alpha line, with

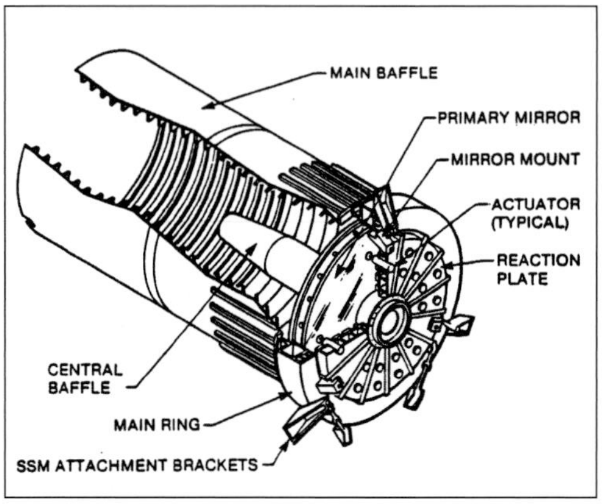

MAIN BAFFLE
PRIMARY MIRROR
MIRROR MOUNT
ACTUATOR (TYPICAL)
REACTION PLATE
CENTRAL BAFFLE
MAIN RING
SSM ATTACHMENT BRACKETS

RIGHT The primary mirror assembly sits within a reaction plate with central and main baffles all but totally eliminating reflective light rays bouncing off internal structures. *(Lockheed)*

a wavelength of 1,215.67Å. The surface finish of the primary mirror is to within a tolerance of 2nm with a reflectivity of 89% at 121.6nm and 78% at 632.8nm.

The primary mirror is attached to the Main Ring through a set of kinematic linkages that attach to the mirror with three rods. These penetrate right through the primary mirror glass for axial constraint and the ring is supported by three pads bonded to the back of the mirror for lateral support. Because the MR encircles the primary mirror it also supports the main baffle and the metering truss and provides a structural point of integration between the OTA and the SSM. Machined from titanium for strength and light weight, the Main Ring takes the form of a hollow box beam 15in (38cm) thick with an outside diameter of 9.8ft (2.9m), and is suspended inside the SSM by a kinematic support structure. The ring weighs 1,200lb (545.5kg).

The Reaction Plate (RP) structure consists of a wheel of I-beams which form a bulkhead behind the Main Ring and which spans its diameter. It radiates from a central ring that supports the central baffle, a long cone-shaped protrusion. The principle function of the RP is to support an array of 792 heaters which radiate warmth to the back face of the primary mirror to maintain its temperature at a consistent 70°F (21°C). The structure is machined from beryllium for stiffness and light weight and supports the 24 figure-control actuators attached to the primary mirror and arranged around the plate in two concentric circles. The actuators can be commanded from the ground to make minor adjustments to the shape of the mirror.

The OTA supports an outer and a central baffle for preventing stray light, not only from the Sun but also the Moon and reflected light from the Earth bouncing off the light tube and contaminating the optical image. Two of the three baffles on the HST are within the primary mirror location. The main baffle, an aluminium cylinder 9ft (2.7m) in diameter and 15.7ft (4.8m) long is attached to the front face of the Main Ring and carries a series of internal fins to help break up any stray light. Attached to the Reaction Plate through a hole in the centre of the primary mirror, the central baffle is in the form of a cone. With a length of 10ft (3m) it extends down the centreline of the Telescope tube and helps attenuate any light rays which fall across the centre of the light shaft. Both baffles are coated with flat black paint to further minimise reflections.

The Secondary Mirror Assembly is cantilevered off the front face of the Main Ring. It supports the mirror at the correct position in front of the primary mirror, which must be maintained to within an accuracy of 0.0001in (0.00025cm). The assembly consists of the mirror assembly, a light baffle and an outer truss support structure. The secondary mirror is attached to the truss structure by three pairs of actuators that together control its position and alignment. They are all enclosed within

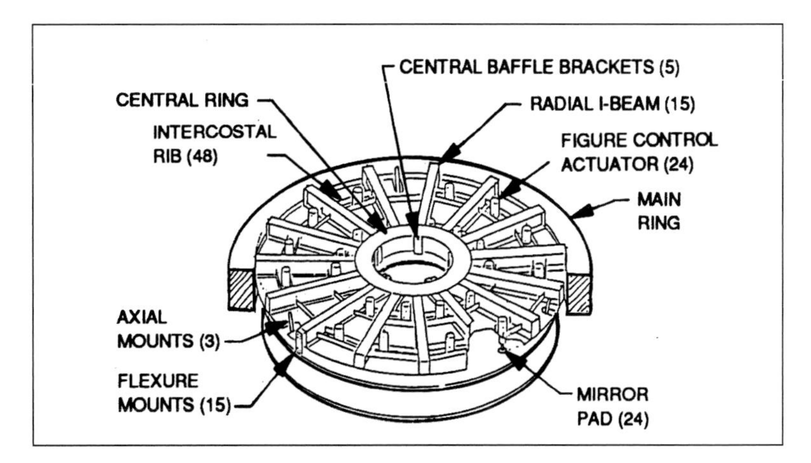

ABOVE The mirror sits on 24 pads with 48 intercostal ribs and is set on a reaction plate that can be adjusted by commands that adjust the position of the mirror using 24 figure-control actuators for small corrections to the shape of the mirror. *(Lockheed)*

BELOW The secondary mirror has its own light baffle and three pairs of alignment actuators to control the precise pointing of the reflected light rays back through the central orifice in the primary mirror. *(Lockheed)*

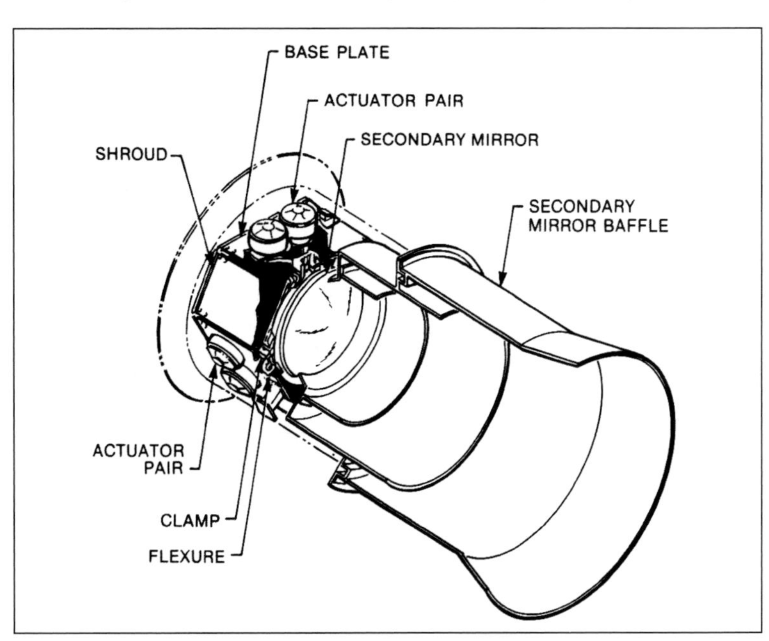

RIGHT The main mirror metering truss structure that sits inside the Light Shield, a part of the SSM, and which supports the secondary mirror. *(Lockheed)*

the central hub at the forward end of the truss structure. The enclosure and spider-beam mounts are fabricated from graphite epoxy and the hub covering the forward end was given a razor-sharp edge to reduce light scattering.

The secondary mirror has a magnification of 10.4x, converting the primary mirror converging rays from f/02.35 to a focal ratio system prime focus of f/24 and returning the beam back through the centre of the primary mirror where it passes through the central baffle to the focal point. The mirror itself is a convex hyperboloid, 12in (30cm) in diameter, made from Zerodur glass which is then coated with aluminium and magnesium fluoride in the same way that coatings are applied to the primary mirror (see above). The mirror has a reflectivity of 89% at 121.6nm and 73% at 632.8nm. The secondary mirror is steely convex and the surface accuracy exceeds that of the primary mirror, a technical achievement made possible by its much smaller radius.

The adjustable paired actuators can be commanded to align the secondary mirror to improve image quality, the magnitude of the adjustments being calculated on the ground using data collected by the optical control system's tiny sensors located in the Fine Guidance Sensors (see later). These minor adjustments may be required to compensate for asymmetric thermal loadings. Temperature variations as great as 110°F (61°C) can be experienced between opposing ends of the HST, and while materials were selected to inhibit any flexing, contracting or expanding under thermal loads the opportunity to carry out adjustments is retained throughout the life of the spacecraft.

The prime structural element of the Secondary Mirror Assembly is the metering truss. This consists of a cage with 48 latticed struts in a semi-geodetic form which uses the same load-path and force distribution maps as those first defined by Dr Barnes Wallis, the British inventor who designed such a structure for use in the Vickers Wellesley and later the Wellington bomber of World War Two. Dr Wallis went on to invent and develop a unique form of 'bouncing bomb' used to disable three German dams in the industrial area of the Ruhr in an epic raid carried out in 1943. The geodetic structure incorporated in the metering truss has been developed for several satellite and spacecraft configurations but its origin is frequently overlooked.

The truss is 16ft (4.8m) long and 9ft (2.7m) in diameter and, like the truss, the Focal Plane Structure and the Fine Guidance Sensors, it was fabricated by Boeing. The truss consists of a pattern of three bays arranged in a continuous 'W' pattern and fabricated from graphite epoxy, chosen due to its superb qualities of light weight, strength and rigidity plus a long life without deformation or degradation. This material almost totally lacks expensiveness to a level unattainable by any other material that could be practically used in the HST for orbital operations. However, graphite epoxy is hygroscopic and will absorb water from the atmosphere, so to prevent this the metering truss and the Focal Plane Structure were baked out at 190°F (88°C) for three days in a special vacuum chamber.

Some additional moisture was absorbed during the period between this baking procedure and launch but the moisture would

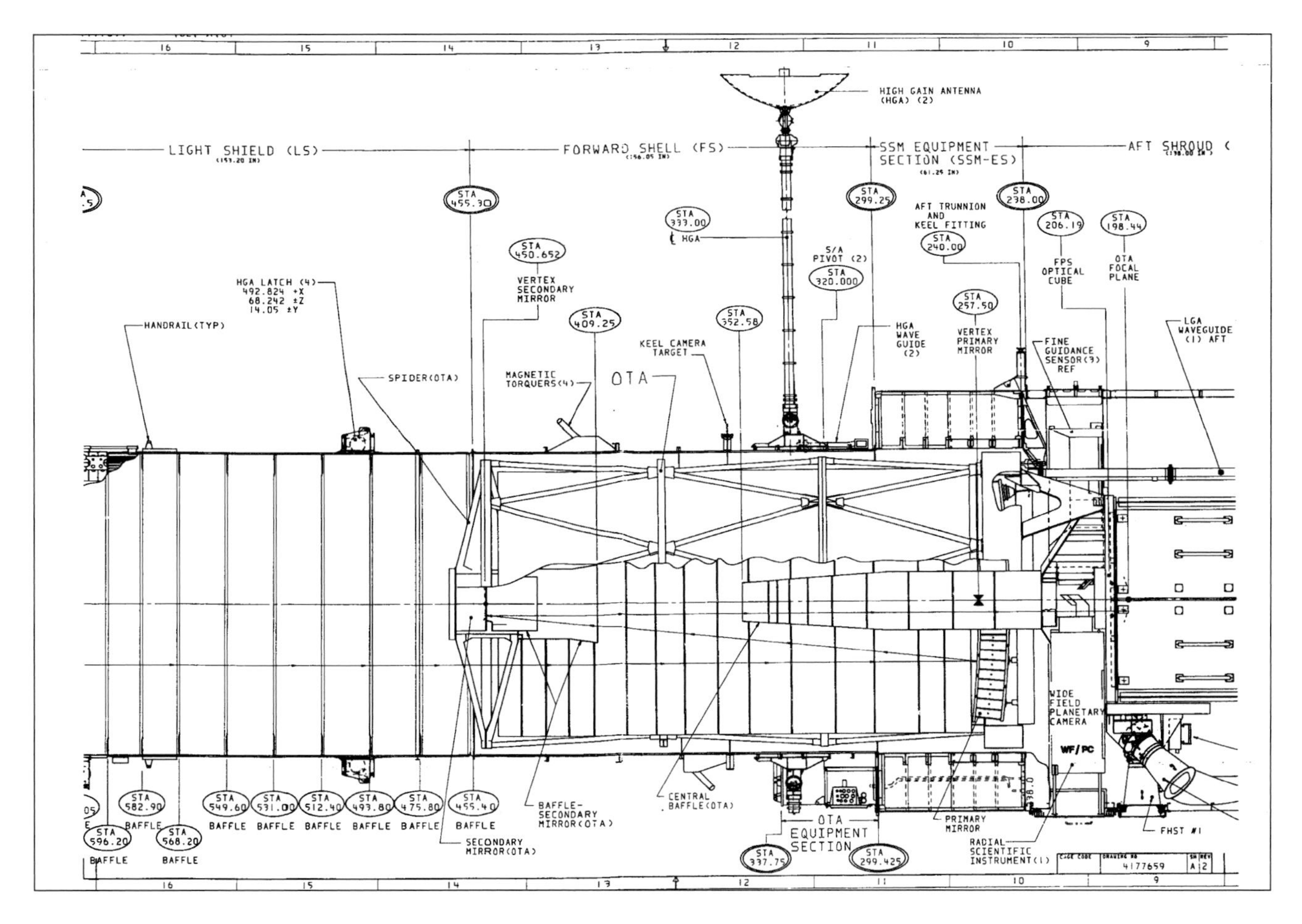

outgas in orbit and the telescope would contract slightly. During the first 60 days the HST was monitored closely as its entire thermal and hygroscopic condition stabilised in the cyclical range of day/night temperature cycles that would dominate its operational life. After about 200 days the structure would be stable and no further deformations would occur, but the presence of these minute changes made the actuators a vital part of the remote control of the Telescope.

The truss is attached at one end to the front face of the Main Ring of the Primary Mirror Assembly. At the other end there is a central hub that houses not only the secondary mirror but the baffle along the optical axis. Aluminised multi-layer insulation in the truss assembly compensates for routine temperature variations of up to 30°F (17°C) when the Telescope is in the shadow of the Earth which enable the primary and secondary mirrors to remain aligned. The conical secondary mirror subassembly light baffle extends almost down to the primary mirror and further reduces the risk of stray light bouncing around in the optical path.

The Focal Plane Structure (FPS) is situated behind the primary mirror Main Ring and exists to align the focal plane of the Telescope with the scientific instruments and the Fine Guidance Sensors. It is also the physical support structure for this equipment. The FPS is in the form of a square box 10ft (3.04m) long and is fabricated

ABOVE This sectional view shows the relative location of the primary and secondary mirrors and the internal layout of the Light Shield and metering truss structure. *(Lockheed)*

BELOW The Equipment Section for the Optical Telescope Assembly takes the form of a semi-circular set of compartments that fits on the forward shell of the SSM outside the spacecraft. See the list of abbreviations (page 178) for explanations of the equipment identified in this illustration. *(Lockheed)*

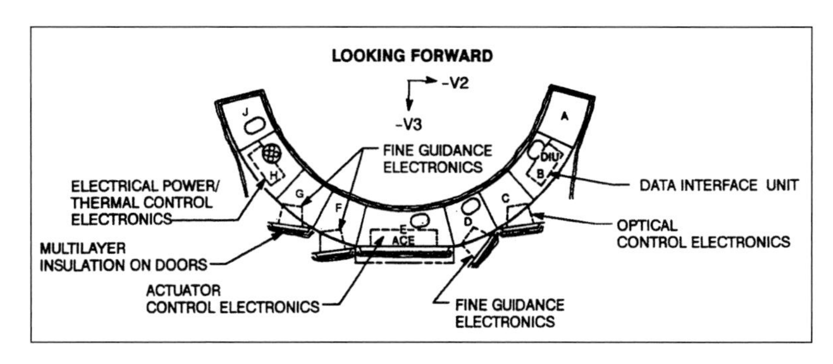

RIGHT Three Fine Guidance Sensors are installed in the HST, one of which is used for astrometry based on position fixes by the other two. Each FGS is a finely engineered arrangement of mirrors and prisms that provide fine-pointing and angular discrimination to maximise the value of a view free of atmospheric distortion and aberration. *(Lockheed)*

from graphite epoxy augmented with metal fasteners and metallic joints at locations where strength is critical and where thermal stability requires high levels of stiffness and resistance to expansion or contraction. The FPS proper weighs 860lb (391kg), with an additional 440lb (199kg) in the braces, fixtures and fasteners. It is required to carry almost three tonnes of science instruments, the guidance sensors and associated equipment.

The FPS cantilevers off the back of the Main Ring extending back into the Equipment Section (ES) and is secured by eight flexible attachment points that can be adjusted by remote command to eliminate any thermal distortions that do creep in. Nevertheless, the FPS is capable of holding alignment to within 0.0018 arc-sec. The FPS also has EVA-support equipment that includes hand rails and handles so that astronauts can access and replace the science instruments it contains. As with all elements of the HST, astronaut access was a driving criterion for the way the Telescope was designed and put together, an essential feature which was the only way the Telescope could be corrected and then kept up to date with successively more advanced instruments.

BELOW The optical path for the Fine Guidance Sensor is shown in this simplified diagram, where the light is passed through a collimator and on to star selectors and a beam splitter. *(Lockheed)*

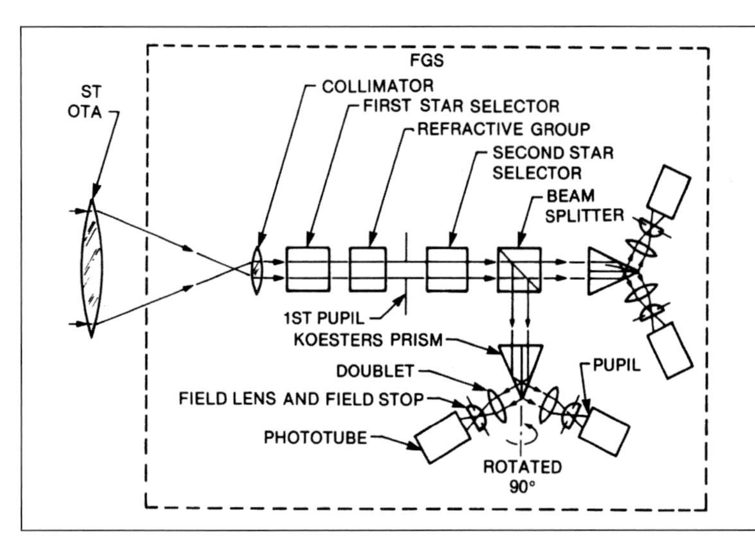

The Equipment Section of the OTA consists of a large semi-circular set of nine compartments mounted to the exterior of the spacecraft on the Forward Shell of the SSM. This contains the OTA electrical system, thermal control electronics system, the fine guidance electronics, actuator control electronics, optical control electronics and the fourth Data Management Subsystem data interface unit. Seven bays are used for equipment storage with two held in reserve for support. Each bay has an outward-opening door for ease of access and contains associated cabling and connectors for the electronics and a heater and insulation for thermal control.

The Actuator Control Electronics (ACE) provides the command and telemetry interface to the 24 actuators attached to the primary mirror and the six actuators attached to the secondary mirror. It selects which actuator to move and monitors its response to commands. Positioning commands are sent from the ground to the electronics through the data interface unit.

The Electrical Power/Thermal Control Electronics (EP/TCE) system takes power from the SSM and distributes it to the systems and subsystems within the OTA. The thermal control system uses thermostat controllers and 792 heaters bonded to the OTA structure to prevent distortion in the primary mirror by maintaining a stable temperature. The EP/TCE also collects thermal data for transmission to the ground, enabling engineers to monitor the changing thermal environment. Information such as this could be helpful if anomalies are detected with the SSM or OTA or to the science instruments.

Discussed earlier in the context of their application to the Pointing and Control System, the three Perkin-Elmer Fine Guidance Sensor (FGS) units are attached at 90° intervals around the circumference of the Focal Plane Structure, located between the structural frame and the

Main Ring. Each FGS assembly contains a guidance sensor and a wavefront sensor, the latter being elements of the Optical Control Subsystem (OCS) which is used to align the optical system of the Telescope. When two sensors lock on to a target the third can measure the angular position of the star for astrometric observations as described in the preceding section of this chapter.

The detectors in the three Fine Guidance Sensors consist of a pair of Koester prism interferometers coupled to photomultiplier tubes. Because each detector operates in only one axis, two are needed. The interferometers compare the wave phase at one edge of the Telescope entrance aperture with the phase at the opposite edge and when the two are equal the target star is centred, any phase difference detected indicating a pointing error requiring correction.

The FGS enclosure is a large structure containing the array of mirrors, lenses and servo-actuators to locate the image, prisms to fine-track the image, beam splitters and four photomultiplier tubes. A photomultiplier consists of a glass tube containing a high vacuum, a collection of dynodes and an anode. As light

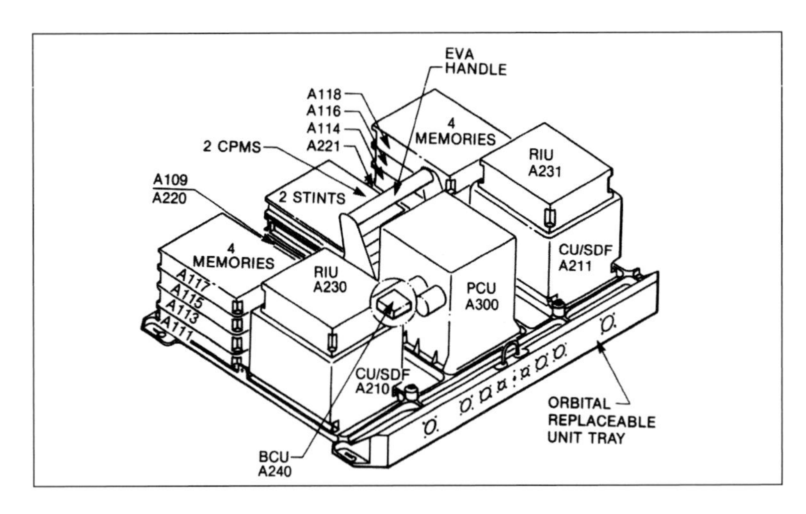

ABOVE Mounted on the door of Bay 10 in the Equipment Section of the SSM, the Scientific Instruments Control and Data Handling unit consists of a tray of components built on to an Orbital Replacement Unit with grab handles and quick-lock levers for ease of removal and replacement by a service crew on orbit. *(Lockheed)*

BELOW The command flow into the Scientific Instruments Control and Data Handling unit (SI C&DH) accepts commands from the Command Data Interface (CDI) or the Data Interface Unit (DIU), at the bottom right in this diagram, where they are reformatted and sent either to the Remote Modules (RM) or to the NASA Standard Spacecraft Computer Model-I or to the DF-224 computer in the SSM. *(Lockheed)*

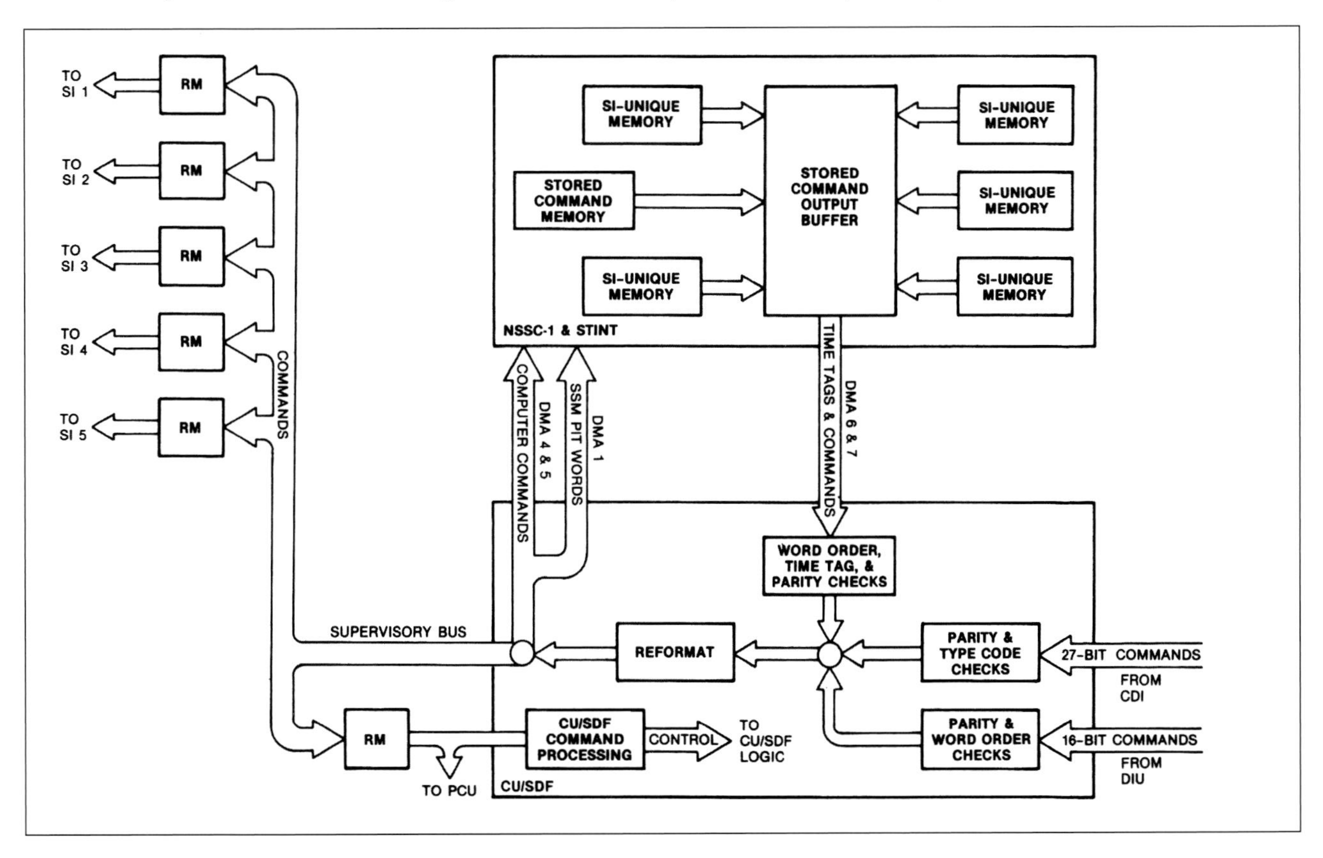

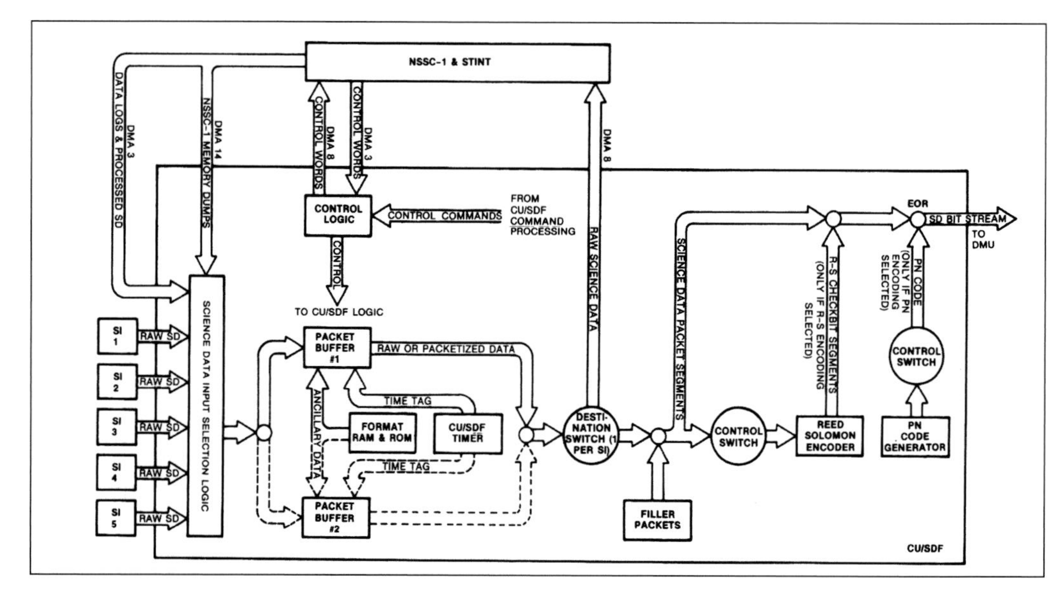

RIGHT The science data flow in the Scientific Instruments Control and Data Handling unit moves from the separate science instruments to the packet buffer and thence to the destination switch for routing as shown. *(Lockheed)*

photons hit the photocathode material placed as a thin deposit on the entry window, electrons are produced due to the photoelectric effect. The electrons are directed to the focusing electrode toward the electron multiplier, where they are multiplied by secondary emission.

The multiplier has a number of dynodes (essentially electrodes) which hold progressively more positive voltage in a cascading increase in the number of electrons produced, each dynode liberating four electrons for each one received so that the great number reaching the anode triggers a pulse of an easily detectable electrical current that signals the arrival of the photon at the photocathode which had occurred about a nanosecond earlier. In its application to the HST, each sensor has a large 60 arc-min^2 field of view in which to search for and to track stars plus a 5.0 arc-sec^2 field used by the detector prisms.

Each FGS uses a 90° field of view outside the main science field of interest which has the greatest astigmatism and curvature distortion, an area deemed large enough to increase the probability of finding an appropriate guide star at the lowest population zones near the poles of the Milky Way galaxy. A pick-off mirror takes the incoming image and projects it on to the sensor's field of view which, having identified it as the target star, locks on to the signal to stabilise the image. The RGS has a pair of star selector servo-motors, one moving in a north–south axis and the other east–west, steering the smaller (5 arc-sec^2) field of view of the detectors to the required position within the field. Indigenous encoders send back information on the precise coordinates of the detector field centres at any point.

In the optical path from the initial acquisition to the detector there are additional elements that turn the beam to fit within the FGS enclosure. This also allows correction to any astigmatism that may be sensed as well as any field curvature. The optical elements are attached to a thermally-stabilised graphite epoxy composite optical bench. Each FGS enclosure is sometimes referred to as the radial bay module.

Each sensor package is 5.4ft x 3.3ft x 1.6ft

RIGHT The Flight Support Structure holds the Telescope in the Shuttle payload bay for orbital placement and for capture and servicing at intervals. It allows for rotation and umbilical connections to provide power and data feed while the HST is attached. Movement off and back on to the FSS is made using the Remote Manipulator System attached to the port (left side) of the Orbiter as viewed looking from tail to nose. *(Lockheed)*

(1.5m x 1.1m x 0.5m) in size and weighs 485lb (220kg). Precision of the FGS is 0.002 arc-sec[2] with a capability of measuring ten stars in ten minutes at a magnitude range of 4–18.5 and a wavelength range of 4670–7000 angstroms. Each FGS carries a filter wheel for measuring stars with different brightness levels and to classify the stars being observed. This wheel has a clear filter for guide-star acquisition and faint star astrometry for levels below $13m_V$. It has a neutral density filter used for near stars and two colour filters for determining a star's colour index. This also allows discrimination between close stars of different colour or differentiating background light from star nebulosity.

An Optical Control Electronics (OCE) unit controls the Optical Control Sensors (OCS), which consist of white light interferometers that continually monitor the optical quality of the OTA and send this data to the ground for engineering analysis. There is one OCS for each Fine Guidance Sensor but all are controlled by the OCE.

The OCS consists of three wavefront sensors and the actuators that give ground controllers the means to orientate the Telescope. Attached inside the FGS assembly they act as interferometers but are designed so as to measure minute imperfections in the stellar wavefront as it transits through the OTA. These provide an indication of misalignment in the secondary mirror as well as the optical quality of the primary mirror. This data is telemetered to the ground where a computation is made for any corrections to the optical system. These are then uplinked and converted into actuator commands for reorientating the OTA.

Machined from beryllium, the bench to which the wavefront sensors are attached is thermally controlled to prevent any disturbances due to flexing or expansion and contraction of the support structure. These sensors are used only occasionally, unlike the FGS sensors, which are in continuous use. Their most frequent application came within the first month of the HST's orbital operations, a period during which the Telescope was correctly aligned and became thermally stabilised on its orbital day/night cycle.

ABOVE The Flight Support Structure consists of a circular mooring saddle inside a horseshoe-shaped yoke within which it can pivot and rotate. This view in the clean room shows the reflective coating on the thermal insulation. *(NASA)*

RIGHT The Aperture Door in the open position revealing impacts and ageing. Note the prolific array of handholds and rails. *(NASA)*

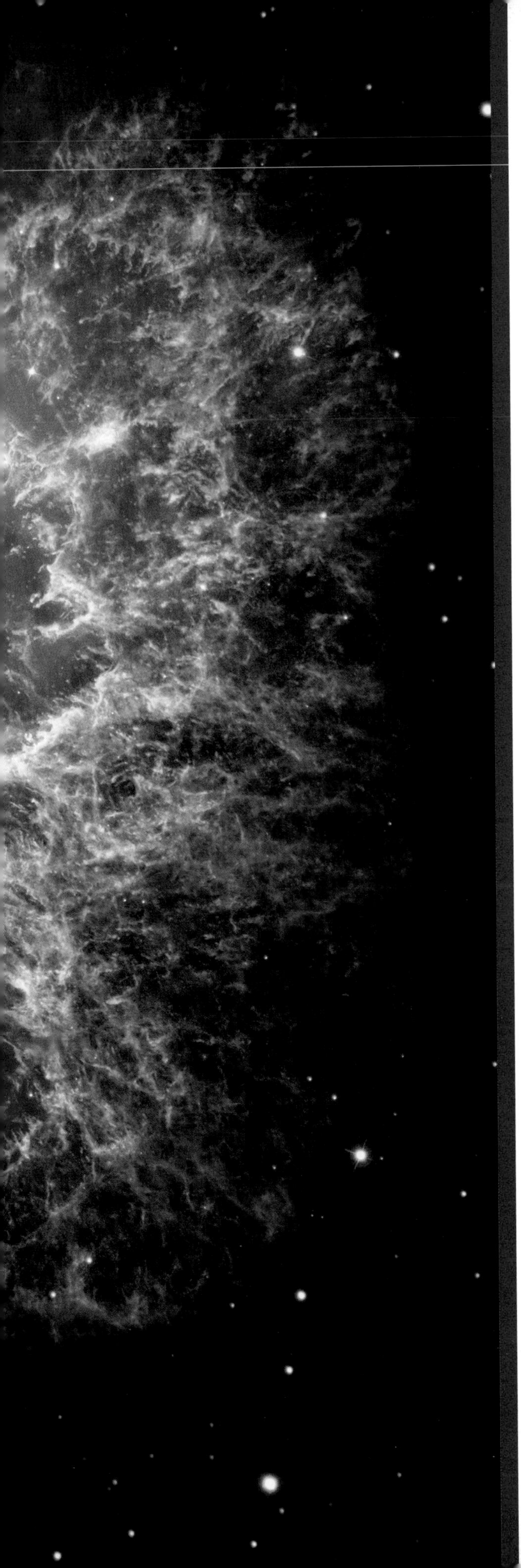

Chapter Four

Science instruments

Selected in 1977, the science instruments installed in the Hubble Space Telescope as originally equipped consisted of the Faint Object Camera (FOC), the Faint Object Spectrograph (FOS), the Goddard High Resolution Spectrograph (GHRS), the High Speed Photometer (HSP) and the Wide Field/Planetary Camera (WF/PC). The first four of these instruments are located parallel to the optical axis, with incoming light falling directly into their respective apertures. All share the same dimensions so that they can be placed in the focal plane interchangeably.

OPPOSITE The remnants of a supernova that occurred in 1056, the Crab Nebula is 6,000 light years away, which means that this event, seen by Hubble, happened around 7,000 years ago, when early farmers were experimenting with crops. *(NASA)*

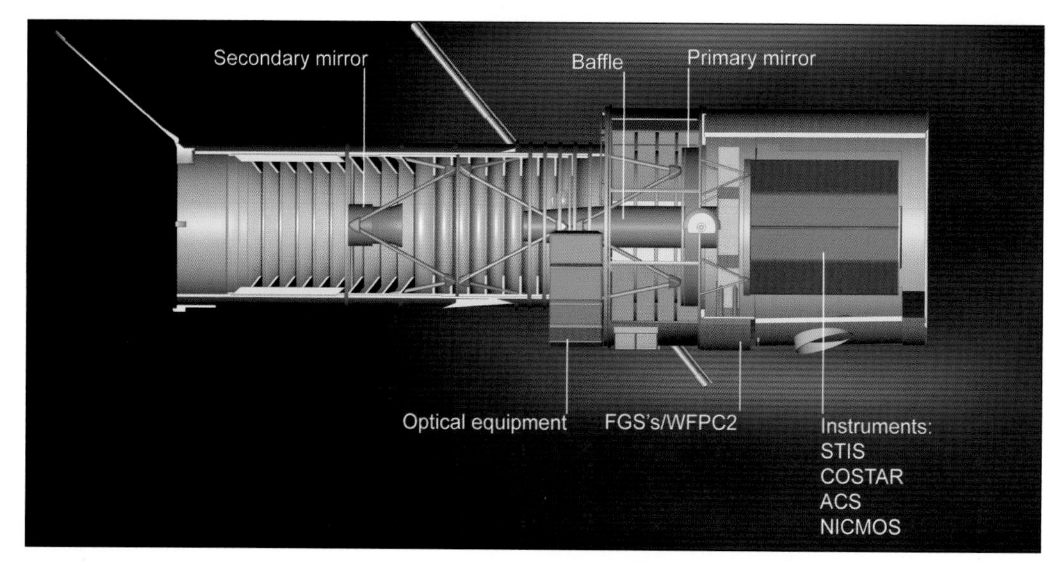

RIGHT The optical path of the HST provides a light route to the science instruments, some indicated here from later servicing and update missions. *(NASA)*

The WF/PC and the three Fine Guidance Sensors (see the OTA subsection) are placed forward of the Focal Plane Structure and mounted at right angles to the optical axis. These four radial instruments employ pickoff mirrors placed directly in the optical path to deflect light into their respective apertures. The alignment of the apertures for each instrument means that no one instrument receives all the light entering the focal plane but each can make the necessary calculations to adjust the position of the Telescope so that incoming light falls directly on to the instrument's optical detection system; this mode is known as target-acquisition.

Although each instrument was fixed in its assigned location the investigating teams could select between filters, gratings or prisms within individual instruments to break the light into spectra as required for investigations. The optional settings and mechanical adjustments or movements within each instrument were handled by the Scientific Instrument Control and Data Handling (SI C&DH) unit and its computer (see later in this section). Some instruments had their own unique computers for operating separate mechanical and motor-driven devices for monitoring and collecting data.

Use of a specific instrument by its team would commence by first choosing the appropriate guide stars and locking on to those with the Fine Guidance Sensors. The next step would be to use a large aperture in the selected instrument and acquire the target before selecting and commanding the appropriate filters and associated devices required to modify the incoming light. Only then

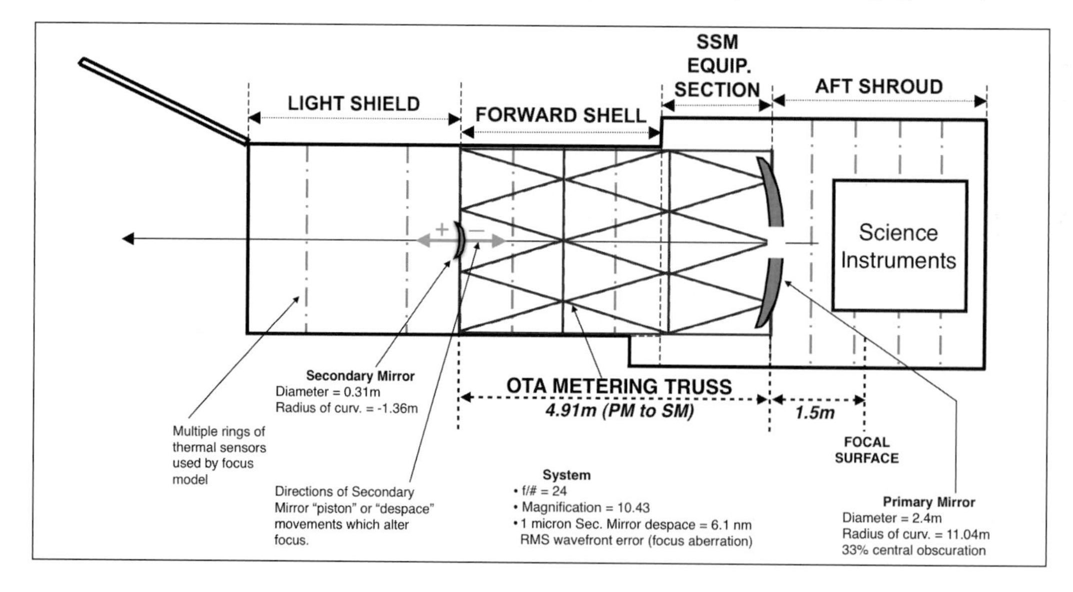

RIGHT The specification of the HST with details of the primary and secondary mirrors and associated equipment. *(Lockheed)*

could the instrument's detector be exposed to the beam of light for the required duration of the experiment. Once acquired the data is then collected up and bunched with other data for transmission to Earth, where the principal investigator and his or her team analyse it and for the possible selection of further observations requested in accordance with the tasking procedure (see Chapter 6).

Faint Object Camera (FOC)

For scientists the Faint Object Camera provides an unprecedented, still unparalleled, access to deep-space objects down to $28m_V$, sources of light so distant and so far back in time that ground-based telescopes were unable to capture light from that distance and age. For the general public the FOC has produced some of the most stunning imagery ever seen, producing a visible image of events and phenomena only surmised by astronomers and physicists. Along with the Wide Field/Planetary Camera, the FOC is the jewel in the crown of observational astronomy and takes the physics of the telescope to new heights. Because of that it has produced enduring and lasting impressions of awe and wonder at the sheer scale and spectacle of the observable universe. But for scientists it has been an exemplar of the value in lifting large astronomical objects beyond the atmosphere.

The FOC was undertaken by the European Space Agency and built by Dornier System in (the then West) Germany and by the then British Aerospace in the UK. It has an elongated dimension of 3ft x 3ft x 7ft (0.9m x 0.9m x 2.2m) and weighs 700lb (318kg). It consists of four primary subsystems: the load-carrying structure which carries the optical elements and the photo-detector head; the opto-mechanical assembly which carried the primary optical bench supporting the main optical and mechanical equipment; the electronic bay with data-processing and data-handling equipment; and the photo-detector assembly composed of

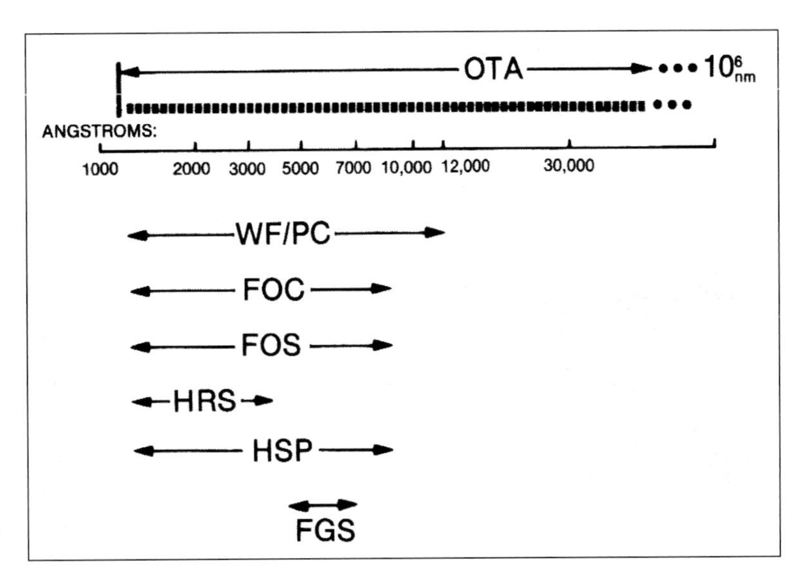

ABOVE The initial suite of science instruments span the visible portion of the electromagnetic spectrum and into the infrared, shown here with their spectral range. *(Lockheed)*

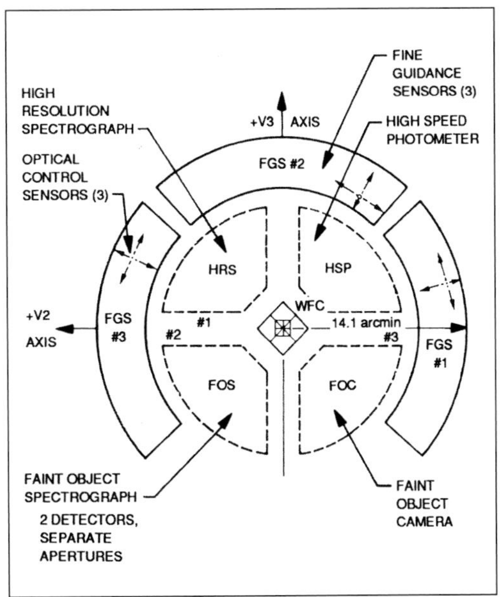

LEFT Sensor fields of view for the four axial instruments and the Fine Guidance Sensors, each of which has a 60 arc-min² FOV in a 90° sector. *(Lockheed)*

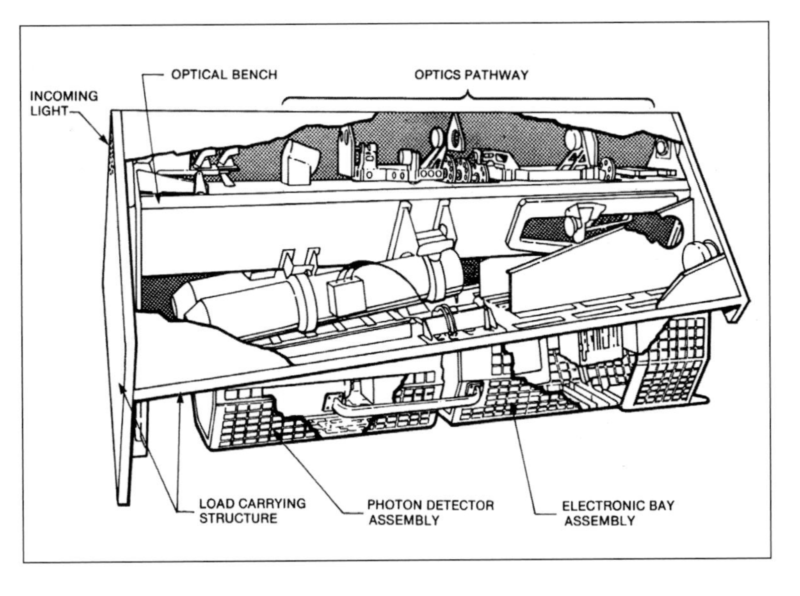

RIGHT Major systems in the Faint Object Camera, which was a major contribution from the European Space Agency. *(Lockheed)*

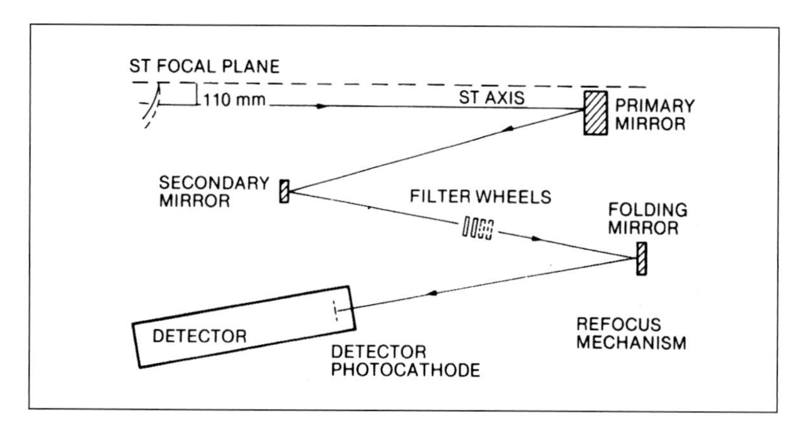

ABOVE The layout of the Faint Object Camera optical relay system, which consists of two independent systems with different apertures and focal ratios. *(Lockheed)*

the detection system, its associated processing electronics and the power supply.

The FOC receives light focused on its aperture, essentially a stream of photons that are channelled down into one of two optical paths. Each photon entering the FOC detection device is identified and placed in a data store where it can be intensified by a longer exposure. This data is then transmitted to Earth where it can be enhanced by computer graphics into an image of the celestial object, be it a star, a galaxy or a nebula.

The optical system consists of two separate and independent systems, each using unique apertures and focal ratios of f/96 and f/48, which are the same as equivalent f-stops on a typical hand-held camera. Each system works in the same way. Incoming light passes through the selected aperture and into a small Cassegrain telescope which works in the same way as the HST itself. The light reflected from a primary concave mirror is passed to a secondary convex mirror on a different plane and then to a folding mirror that directs the photons to a detector tube for enhancement and processing. Both optical paths have filter wheels that are used to isolate critical wavelength beams for particular studies.

The folding mirrors increase the focal length but without increasing the FOC's physical length and they move to adjust the focus and to correct for any astigmatism in the image. Readers will recall that fine-scale adjustments such as this can be performed at the instrument level to compensate for the natural astigmatism inherent with the hyperbolic primary mirror of the telescope itself. Each optical system also has a shutter that remains closed, blocking light until the FOC is needed for observations. When the shutter closes a mirror attached to it can reflect a beam of light from a calibration source along the optic path to measure the light-detecting capability of the detectors from this known source. The response of the instrument to visible light and to geometric light distortions can be measured this way and both systems have a zoom capability to double the field of view at the same spectral resolution.

The f/96 optics system provides the best optical angular resolution of any instrument on the HST and increases the Telescope's f/24 resolution fourfold. A special device inserted into the path of the instrument can raise this to f/288. The quadrupling of resolution is only achieved by reducing the field of view to 22 arc-sec but which does, however, enable astronomers to differentiate two objects only 0.01 arc-sec apart. This feature allows fine-scale discriminatory analysis of densely packed stars, notably those in globular clusters where stars are so tightly packed that ground-based telescopes are unable to visually separate them.

To block out unwanted light, the f/96 system has two coronographic fingers in the aperture and an apodiser mask that can be moved into the f/96 light path. The two fingers are 0.4 and 0.8 arc-sec wide and can be used to block light from a bright image to resolve a less bright image alongside, each consisting of two 2mm protruding opaque metal strips. This allows detailed study of the brightness of a minor optical pair and could be used to determine whether it is a planetary object or an independent but intrinsically luminous object such as a faint star. Apodising is a change in the transmission properties of the aperture that affects the intensity distribution and the diffraction pattern generated.

The high-resolution apodiser is embedded in another small Cassegrain telescope located near the edge of the f/96 field of view but including a coronographic obscuring finger. The mask blocks stray light from the HST secondary mirror and its support structures, and it is this that expands the resolution to f/288, but at the cost of an even smaller field of view. Nevertheless, with this mask applied the system increases focal range to f/288 and allows the FOC to detect objects 17 magnitudes fainter than a bright companion separated by only 1 arc-sec.

The f/96 optical path also has four filter wheels that contain 48 filters, prisms and other optical devices. These can be made to direct the beam through a magnesium-fluoride prism to enhance the far ultraviolet portion of the spectrum, a useful tool for seeing stars either very early or very late in their mainstream life cycle, where they are hotter and less visually luminous than usual – stars which are either collapsing down under gravity and reaching very high temperatures, or stars at the end of their life and expanding into giants where most of the energy is in the ultraviolet. Stars in these phases of their life cycle may be visually obscured to ground-based observatories due to gas or dust and some very distant sources bright in the far-UV may be galaxies.

The f/48 optics system carries the obvious advantage of a wider field of view of 44 arc-sec but at the cost of reduced resolution, less than that with the f/96 optics. This system is reserved for spectrographic work as well as for target searches and also as a secondary optics relay. The main pointing angle of the HST will have to be changed to redirect the light path into the f/48 aperture but the optic system is identical in its overall concept and design to the f/96 system and it carries the same calibration device.

The f/48 system has a 20 arc-sec spectroscopic slit in its aperture, and a mirror rotated into the optical path of the instrument diverts light through the slit and on to a diffraction grating that breaks the light into a spectrum of its component wavelengths. This dispersed light is reflected into the detector. The slit has only a limited spectral range, dependent on the order chosen. This ranges from 3,600 to 5,400Å for the first order to 1,200 to 1,800Å for the fourth order. Because the spectral resolution of the diffracted light is 2,000Å the detectors register separate spectral lines only one angstrom apart in the 2,000Å range.

The quality of this resolution lies between that of the Faint Object Spectrograph and the High Resolution Spectrograph, and the special function of this instrument is to measure light spreading across a spatial distance of many lights years, from sources such as entire galaxies. The f/48 optics also include two special filter wheels which can be used with the main optics or for specialised viewing purposes. These filters can block certain wavelengths or be used to select specific wavelength ranges.

The Photo Detector System (PDS) includes two identical detectors in the FOC that are sensitive to radiation between 1,150 and 6,500Å. Photons from a target source enter an image intensifier and are converted into electrons, which are accelerated to high voltage in three steps. The detector reconverts the electron output to photons, providing 100,000 photons for every original electron. A high-sensitivity TV camera tube intensifies the image, which sends an amplified video signal to the FOC processing unit. This is stored in the science data store. Typically, a photon burst

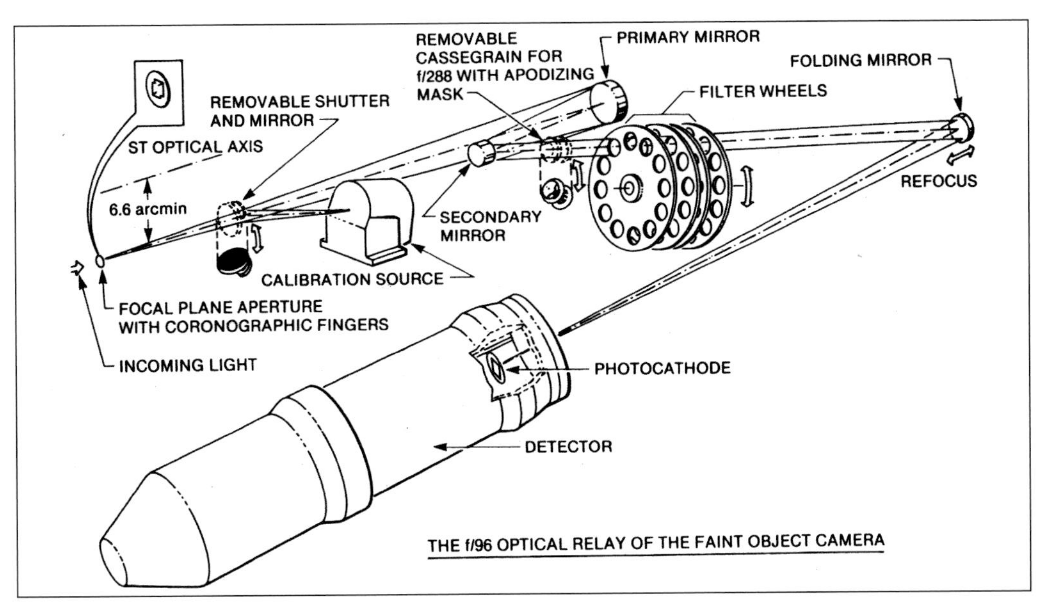

LEFT The best angular resolution is obtained by the f/96 optical system and increases the Telescope's focal ratio four fold. A special device inserted in the light path can increase that ratio as high as f/288. *(Lockheed)*

is a spot 75–100 microns in diameter, and each time an incoming photon bursts on to the diodes at the same location the recorded light is intensified at that position.

A useable image can be produced in a period from ten minutes to ten hours depending on the brightness of the object being observed. The data store will count up to 65,535 photon bursts at each location depending on the length of the observation. Only the most distant objects at $28m_v$ will take the full imaging duration and the size of the completed image is proportional to the light size originally selected. The standard light format is a square measuring 512 x 512 pixels. Each pixel is 0.025nm x 0.025nm in size. Other formats are available for larger, smaller or different shapes of lights and the pixel size can be doubled on selection, although this will affect the total number of pixels the FOC can count because of data storage and transmission limits.

A great advantage with this system is that the light in the science data store can be viewed at any time without destroying or contaminating the data being collected at the same time. This is a useful tool for the intermediate studying of the developing light, ensuring a continuous monitoring process. The light is stored on magnetic tape by the tape recorders in the SSM before it is sent to the Space Telescope Operations Control Center via the Tracking and Data Relay Satellites.

The FOC electronics system provides physical support, data processing and transmission as well as thermal control to maintain a thermally stable FOC system protecting the optics and the electronics. Command and communication units connect the FOC to the Telescope and to the NASA standard spacecraft computer (NSSC-I) and the Science Instrument Control & Data Handling unit's computer which, with the FOC's on-board computer, operates as a back up.

There are four operating modes for the FOC: targeting, imaging, occultation and spectrographic. The target mode is a first step to either the coronographic or the spectrographic instruments with a special exposure of the target within the FOC processed through the SI C&DH to locate the target within the field of view. The HST is then orientated to place the light on the coronograph finger or on to the spectrograph slit at the edge of the FOC field of view. The imaging mode provides the option of selecting any of the various formats and filters to change image size or wavelength using either of the optical systems.

The occultation mode is the eclipsing of a celestial body by another and this uses the coronographic fingers and the high-resolution apodiser to block out a dominant source of light so that the observer can see a fainter object close by. A typical example could be the observation of a galaxy by occulting a quasar to determine whether it resides in a galaxy temporarily obscured by the brightness of the quasar. The

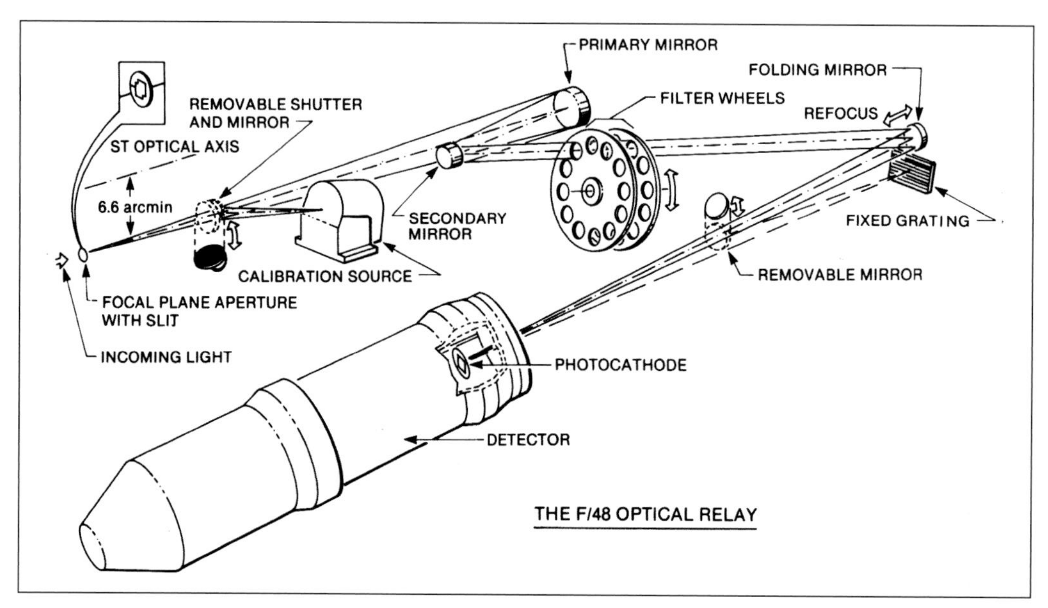

RIGHT With a wide field of view (44 arc-sec), the f/48 system is relegated to spectrographic observations as the resolution is less than the f/96 system in the Faint Object Camera. *(Lockheed)*

spectrographic mode uses the long slit in the f/48 aperture to diffract incoming light into its composite wavelengths with associated filters and a cross-dispersing prism, enabling astronomers to separate the light into narrow wavelengths for particular study of a specific source.

The Faint Object Camera was a useful observing instrument on its own, but used in conjunction with other scientific instruments on the Telescope it could provide a unique opportunity to study interstellar gas clouds, to measure the distance of faraway galaxies, to examine the structure of globular clusters and to pick apart the detail of normal and irregular galaxies. It could also be used to conduct high-resolution observations of various phenomena in the solar system.

A specific contribution to the study of star formation was provided by its ability to discriminate fine structural detail in the ejected cloud of gas and dust that surrounds a proto-star in the early days of pre-forming into an independent main-sequence star. Condensating objects within the swirling cloud may be indications of planets forming, and the availability of both coronographic and spectrographic observations opened new possibilities for observing the birth pangs of stars and attendant planets, still forming in the swirling gas clouds.

With high angular resolution and to see objects unobserved from the ground, the FOC provided a new tool for precisely measuring the distance of remote galaxies as well as measuring the distance between stars. By examining the parallaxes and the motion of star clusters, together with the speed at which material is dispersed, astronomers searched for ways to more accurately pin down these distances. One target for achieving this was the Large Magellanic Cloud, one of the small group of galaxies of which the Milky Way is the dominant member. By observing Cepheid variables within the LMC, astronomers were able to calculate the absolute magnitude to a far higher degree of accuracy, and because these variables are yardsticks on measuring distance, observing them in more distant systems opened new levels of refinement.

Globular clusters are of great interest to astronomers because they frequently contradict logical explanations for their observed characteristics. These clusters, and sources of energy within ordinary galaxies which seemingly produce more energy than the total output of the surrounding galaxy, bring questions which astronomers seek to answer by using the FOC along with other instruments. Before the launch of the HST, astronomers were particularly interested in the centre of the elliptical galaxy NGC 4486 (Messier 87), which appeared to display the characteristics of a massive object created by stellar collisions and collapsed matter.

It is now known that the centre hosts a super-massive black hole, and this object has so attracted the attention of astronomers that other instruments on the HST have been used in an extensive survey to determine the behaviour of this black hole and to identify and map associated disturbances, including a jet of matter spewing out across a distance of 5,000 light years. The nature of this jet has been the subject of attention from scientists using the Wide Field/Planetary Camera 2 instrument installed on the HST during the first servicing mission (see SM1 in Chapter 7).

While astronomers in general are reluctant to employ high-value assets such as the Hubble Space Telescope on anything less than the most exotic objects in the universe, the HST has proved valuable in observing less distant objects within the solar system and the FOC has had a prime role in that activity. Several NASA planetary missions have received support from pre-imaging using the FOC, and for some events, such as the impact of the fragmented comet Shoemaker-Levy on to the outer

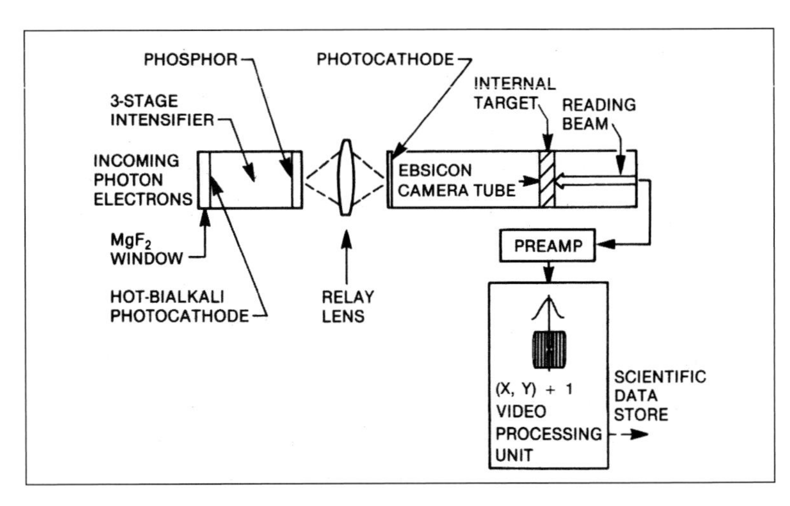

ABOVE One of two identical detectors in the Faint Object Camera, each providing an intensified image produced by an Ebsicon camera tube that is available for intermediate viewing to facilitate snapshot observation without compromising the data. *(Lockheed)*

atmosphere of Jupiter in 1994, it has had a unique role to play.

The Faint Object Camera was one of the most important instruments for portraying to the general public the power and the capability of the Hubble Space Telescope. With a capability of recomposing the light of an image so that it was 100,000 times brighter than the photons detected at the instrument, it provided a unique tool for astronomers around the world. In 1993 the first servicing mission brought equipment to the HST which changed the optical mode of the camera, and this is discussed in Chapter 5. In 2002 the FOC was replaced by the Advanced Camera for Surveys (ACS) during the second component of the third servicing mission (see SM3, Chapter 8). Returned to Earth in the Shuttle, it is now on display in the Dornier Museum in Friedrichshafen, Germany.

Faint Object Spectrograph (FOS)

The Faint Object Spectrograph is a companion instrument to the Goddard High Resolution Spectrograph (GHRS) and is designed to measure the faint light of stars from very different locations or exceptionally dim stars from closer distance. The GHRS does just the opposite and measures light from very bright sources. The FOS and the GHRS overlap to some degree in the mid-range of their capabilities but each is specifically equipped to capture light at their extremes. The FOS is classed as a medium-resolution instrument and has a broad spectral range for objects down to 22–26m_V and a spectral resolution of 250, which means that in the 1,500Å range it can differentiate spectral lines as close as six angstroms apart. For brighter objects the resolution increases to 1,300, separating spectral lines as close as 1.2 angstroms. In either case, the FOS has a spectra range of 1,100Å in the ultraviolet to 8,000Å in the near infrared.

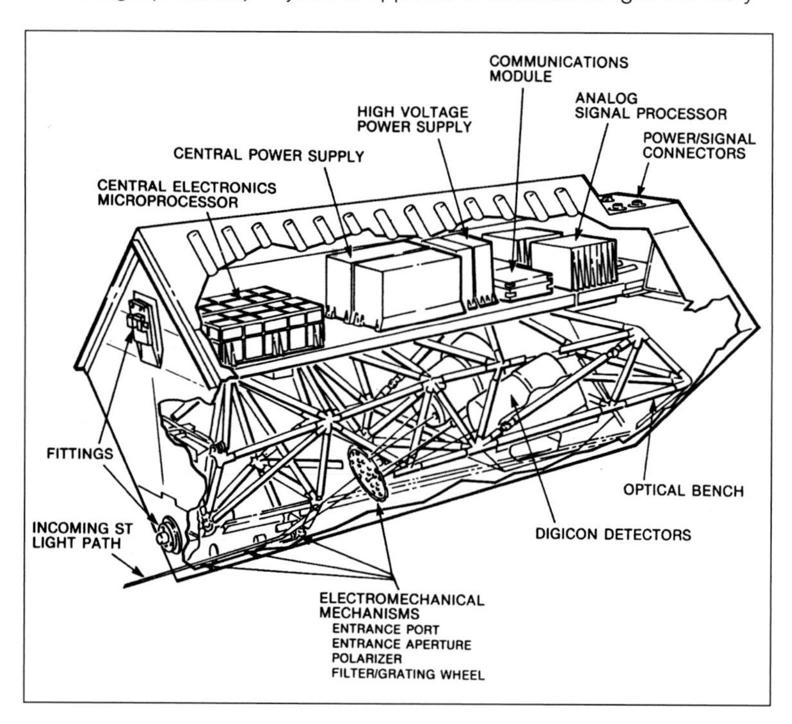

BELOW A companion of the Goddard High Resolution Spectrograph, the Faint Object Spectrograph takes measurements of the most distant and faintest of objects across a wide spectral range. *(Lockheed)*

This spectral range can produce very broad compositional portraits of targeted objects because it includes the emissions of most of the chemical components of stars. But in addition to measuring the chemistry of stars, the FOS will also determine their velocity. The FOS also has a capability of measuring both the wavelength and intensity (spectrophotometric) and the wavelength and polarisation (spectropolarimetric) of light sources, and with the former this can be useful in discovering the interaction of hot, x-ray emitting binary stars and their cooler companions bombarded by x-rays from the primary. The spectropolarimetric capability tells about internal processes in interstellar dust clouds that polarises light passing through them.

The FOS is a box-shaped structure 3ft x 3ft x 7ft (0.9m x 0.9m x 2.2m) in size and weighing 680lb (309kg). It was built by the then Martin Marietta Corporation. The instrument has two optical paths, one into the red-sensitive Digicon detector for sources emitting longer wavelengths and the other to a detector for objects emitting shorter (blue) wavelengths. Digicon detectors (light detectors) use magnetic and electric fields operating in a vacuum to count photons from weak sources and do this by forcing electrons released from a photocathode on to a group of silicon diodes. Displaying its design age, the use of Digicon detectors in general went quickly out of favour due to the electric fields required and the complex vacuum equipment, and have been replaced by CMOS active-pixel detectors.

The optical system of the FOS contains

special apertures, mirrors, a filter/grating wheel and the separate detectors. There is a light entrance port leading directly into the aperture assembly but the port remains closed until the FOS is ready to begin operating. By pointing the HST at a specific orientation the light from the target enters one of two optical paths and into one of 12 apertures. The largest of these is 4.3 arc-sec in diameter and is used for acquisition of the target. The smaller apertures are used for various observations, with two occulting apertures capturing faint light surrounding a bright target. This could, for instance, be the faint light from a distant galaxy saturated by a particularly bright quasar.

Four apertures observe the target and the sky background with different fields of view. One aperture serves for wide targets, perhaps galaxies, and one is blank to act as a light shield. Three apertures specialise in spectropolarimetry and send light through a polarising analyser. Because light waves oscillate in a particular plane if they pass through a magnetic field the analyser is rotated so that only light polarised in that plane passes through. The planes are selected in increments of 22°.

The light travels through the selected aperture on to a prismatic mirror which reflects the beam 22° away from the optical axis of the HST and through an order-blocking filter attached to the filter/grating wheel. The filter allows light to pass in the desired spectral range. For instance, if that range is 4,500–6,800Å the filter blocks all light waves shorter than 4,500Å or longer than 6,800Å. This light reflects off a collimated mirror that sends the beam back to the grating wheel. This wheel is a carousel that contains a range of gratings. Six gratings cover the range 1,150–9,000Å with a moderate resolution of 1,300. Two gratings provide low spectral resolution (250) for the ranges 1,150–2,500Å and 3,500–7,000Å. A prism is available for observing long, visible wavelengths with a limited spectral resolution of 100. Target aiming is achieved with a clear mirror that passes undispersed light.

The light continues on beyond the grating wheel to the Digicon, which is sensitive to the selected wavelength spectrum. For the blue-sensitive detector that range is 1,100–5,000Å and for the red-sensitive detector it is 1,700–8,000Å, but only one Digicon detector can be used at any one time. Each detector has a photocathode that is designed to release electrons only from a specific central range of photons. A magnetic field boosts and focuses the incoming electrons on to a string of silicon diodes. Each diode records and amplifies a pulse as each electron hits the diode. The diodes can record each pulse and pass it along immediately or it can accumulate the data according to the coded instructions from the observer. The FOS, when held steady, collects at least 99% of all the pulses from a target over four hours on 99% of the operating diodes.

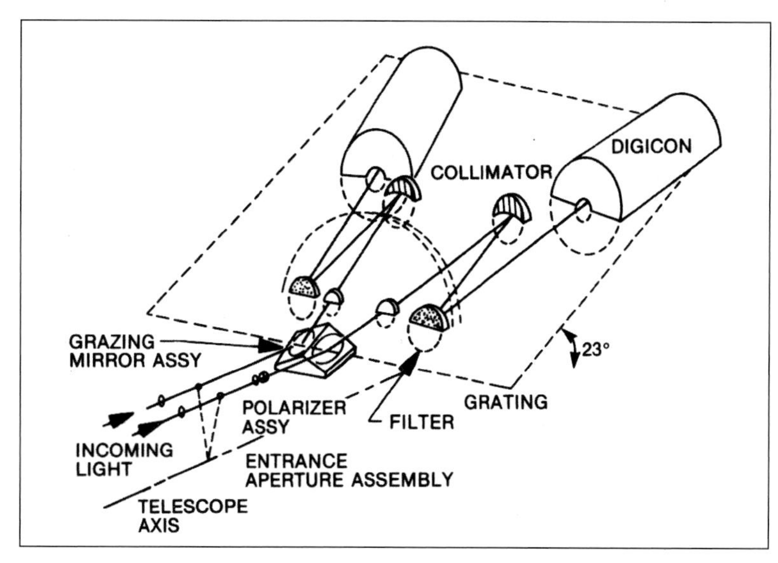

ABOVE Diagrammatic depiction of the optical path for the Faint Object Spectrograph with observation times of up to four hours on a single target. *(Lockheed)*

The electronics, power and communications equipment supporting the FOS is an essential component of the system's operating capabilities and includes a signal processor, central electronics assembly and a remote communications system. These components as well as the power supply are all located on the optics bench support structure. The signal from the detector diodes passes through to an analogue processor where it is shaped into spectra before being sent to the central electronics assembly, where it is processed and packaged for transmission.

The central assembly contains two microprocessors and its own support electronics, one dedicated to each detector. The small computers rotate the filter/grating wheel to the required setting based on commands uplinked from the STOCC. These operate the magnetic focus system to direct the electrons toward the diode array and control the pulse-counting mechanism. The central electronics assembly then sends the data through the communications unit to the SI C&DH in the Support Systems Module. The FOS is also responsible for collecting its own engineering data on temperatures, voltages and the processing of data from the microprocessors. All this information is either stored on magnetic tape or transmitted direct to the ground in real-time.

There are four operational modes for the FOS: spectroscopic; time-resolved; time-tagged/rapid-readout; and spectropolarimetric. The spectroscopic mode is the one usually applied with observations by selecting a grating for the wavelength range and the spectral resolution preferred. The incoming light will be directed to the appropriate Digicon detector for a predetermined observation duration that can be from several minutes to four hours. The data are read to the HST or ground communications links at regular and pre-planned intervals. The time-resolved use of the FOS is primarily for objects with radiation that pulses at reliable and predictable intervals, anywhere from 50 microseconds to 100 seconds. The detector specified takes samples during each period the radiation is on for a preselected number of periods, the data being added together later.

The time-tagged and rapid-readout operations are designed for studying objects with irregular and highly unpredictable pulsations. The time-tagged operation allows the FOS to count the spacecraft clock until the first photon arrives at the FOS detector and then it freezes the count. The rapid-readout cycle accumulates the incoming data but produces frequent readouts, up to every 20 milliseconds, so it can capture the start and stop points of the pulsations. Spectropolarimetric observations measure the polarisation of light to test for magnetic effects. The light passes through the polarising analyser and a combination of plates and prisms that separate the light into different polarisations, each of which can be measured and compared. The FOS has a total magnitude range of 19–26.

The observational value of the Faint Object Spectrograph lies in many fields of astronomical study but especially in the study of galaxy formation, how supernovae can be applied to distance formulae and the composition and origin of interstellar dust. Using the FOS to study quasars and exploding galaxies can provide valuable information on the way galaxies evolve and how quasars may have been an integral part of galaxy formation from the very early stages in the evolution of the universe. Using the FOS to examine the chemical and physical relationship between ejected gas clouds discharged from energetic sources and the early stages of galactic formation helps astronomers understand the mechanisms that power the observable universe today.

Using the FOS's spectrophotometer, astronomers sought to examine supernovae to measure their luminosity during the event itself and afterwards at periodic intervals. By measuring precisely the apparent luminosity and determining from that its absolute luminosity, astronomers were able to refine the precise distance between Earth and the phenomenon under observation. This technique was also useful in estimating the value of the Hubble constant – the rate at which the universe appears to have been expanding since its origin. With the ability to measure distant planetary nebulae down to a magnitude of 22, it is possible to compile accurate maps of temperature, chemical composition and the total mass of these enigmatic sources of light. Studies such as this contribute greatly to an understanding of stellar evolution.

By measuring ultraviolet energy passing through dust clouds and looking for the chemical composition of the dust, the FOS can relate the presence and abundance of dust in relation to its magnetisation. Interstellar material is impossible to study from the surface of the Earth as it is obscured by the atmosphere, and the use of the FOS ultraviolet detectors adds greatly to the scientific understanding of this trace material, free from gravitational attraction to any star in the vicinity yet tied inexorably to an understanding of how matter has moved through the galaxy and between stars as a product of earlier activity in stars long dead. Before the launch of the HST, astronomers theorised that this dust is related to giant magnetic fields, which is why studying the polarised light through such sources is a major step forward in understanding its characteristics.

Goddard High Resolution Spectrograph (GHRS)

The second ultraviolet-detecting instrument carried aboard the HST, the GHRS is another tool in the astronomer's armoury fighting to understand the composition, temperature and density of the stellar objects and vast gas clouds that populate the universe. The instrument was developed by NASA's Goddard Space Flight Center (GSFC) and was built by the then Ball Aerospace Systems company. It continues the study of ultraviolet radiation made by earlier astronomical spacecraft including the International Ultraviolet Explorer (IUE), which had been launched in January 1978 and would continue operating until September 1996, when it was deliberately shut down due to a lack of support funding.

The IUE was a joint endeavour between NASA, the UK Science Research Council and the European Space Agency and it was one of the most successful astronomical space missions ever mounted. Some say it was the best of them all and it certainly paved the way for the ultraviolet surveys which would be undertaken by the GHRS on the Hubble Space Telescope. The IUE had been proposed by the British astronomer Robert Wilson in 1964 and was studied as a mission when the European Space Research Organisation (ESRO), the precursor to ESA, was considering a Large Astronomical Satellite, as the project was called. When the LAS was cancelled on the grounds that it was too big and costly, Wilson took the concept to NASA in 1973 and with the support of American astronomers it was redeveloped as an idea and approved.

Building on success with the IUE, the GHRS is more sensitive to faint objects, is more accurate and can resolve images down to $17m_V$, although it is much less sensitive to dim sources than the Faint Object Spectrograph. The two instruments overlap to some extent and function as a pair, for instance when measuring stars and gas clouds in the ultraviolet. They differentiate valuable functions due to their different levels of sensitivity and in their differences in spectral resolution.

The GHRS can resolve down small differences between spectral lines up to a resolution of 100,000, but only for objects $13m_V$ or brighter. The Faint Object Spectrograph is more limited in its resolution but it can detect objects down to $26m_V$. The GHRS is aligned parallel to the HST optical axis and contains an optical system for producing spectra, support electronics and thermal control systems and a structure housing the enclosed systems in a package 3ft x 3ft x 7ft (0.9m x 0.9m x 2.2m) and weighing 700lb (318kg). The SI C&DH handles all GHRS computer functions.

The optical system has two apertures of different size, a rotating carousel with gratings for separating wavelengths, mirror to place the

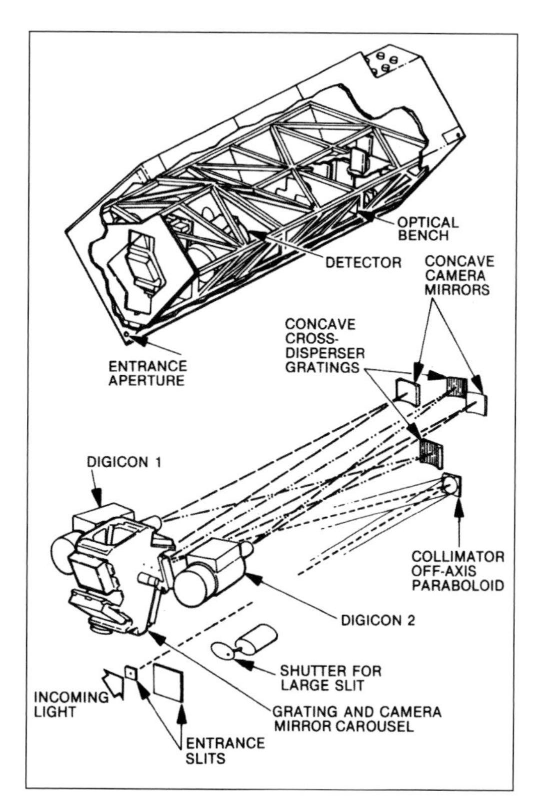

LEFT The overall structure and the optical system for the Goddard High Resolution Spectrograph, which to some extent duplicates the functions of the Faint Object Spectrograph but which has much finer resolution while being less sensitive to faint objects. *(Lockheed)*

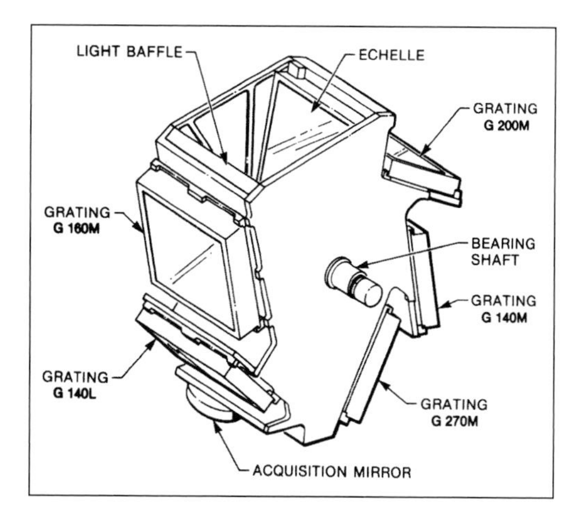

LEFT The GHRS carousel contains seven diffraction gratings and can move into several positions to obtain the appropriate grating or mirror. *(Lockheed)*

light into a specific detector and two Digicon light detectors, one for 1,050–1,700Å and the other for 1,150–3,200Å. The larger of the two apertures measures 2.0 arc-sec across its field of view and the smaller science aperture measures 0.25 arc-sec. Because the apertures are not on the HST optical axis, the light coming through each aperture is astigmatic, and to adjust the divergent points of light each aperture has two slits set at right angles so that the incoming lights merge again.

The larger of the two science apertures is used to locate the target, observe galaxies and perform spectrographic observations where precise spectral resolution is not required. The smaller aperture captures the full light of single objects, such as a star, and is used to obtain the instrument's maximum resolution of 100,000. This means that the GHRS will display features separated by 0.02 of an angstrom. When the smaller aperture is operating, a shutter automatically blocks light from entering the large aperture. Two slits in the aperture provide wavelength calibrations and, for measuring the incoming starlight with sufficient accuracy, lamps shine ultraviolet light through the two slits. The resulting wavelengths can be measured precisely and are referred to as the calibration standards. When compared with incoming starlight they produce very accurate ultraviolet wavelength readings.

The incoming light reflects off a collimating mirror that directs the beam on to the carousel, which consists of a rotating wheel with seven diffraction gratings and three acquisition mirrors that are used to select specific wavelengths. The carousel is rotated by an encoder and electric motor through coarse and fine positioning steps so as to direct a desired wavelength to within 0.03 arc-sec. The carousel can be made to move in any direction to position the appropriate grating or mirror assembly.

There are three mirrors in the carousel. Each has four settings that are used to lock the GHRS on to the desired object. Two settings reflect faint objects with no diffraction and the other two settings allow bright objects up to $-1m_V$ to be targeted. Theses latter settings moderate the brightness level so that the sensitive detector diodes are not overloaded. Each of the seven carousel gratings spreads out the incoming light into separate wavelengths for analysis. These gratings can be selected to observe the ultraviolet spectrum with low (2000), medium (20,000), or high (100,000) spectral resolution. The highest setting is obtained with two echelle gratings where the GHRS can discriminate between spectral features as close as 0.02Å.

Echelle gratings derive their name from the French word for 'step'. They are uniquely beneficial where overlapping orders of resolution are sought, which they achieve by taking advantage of the natural tendency in a grating to scatter light in longer wavelengths so that they overlap. Like standard diffraction gratings the echelle grating comprises a number of separate slits with widths close to the wavelength of the diffracted light; but the overlap, known as 'blazing', is deliberately used to permit multiple overlaps, which produces a spectrum of stripes with wavelength ranges running cross the imaging plane in an oblique angle.

As the resolution increases the portion of the overall spectrum in one frame decreases. For example, the medium-range grating G140M displays a spectral range just 29Å wide while the low-resolution grating G140L has a spectral band 290Å wide. There is a trade-off between spectral resolution and brightness, which is measured by the number of electrons striking the detection diodes per second for a particular

BELOW **Grating spectral ranges and spectral resolutions for the Goddard High Resolution Spectrograph.** *(Lockheed)*

MODE	WAVE-LENGTH RANGE (Å)	SPECTRAL ORDER(S)	SPECTRAL RESOLUTION (X 10^4)	WAVE-LENGTH AMOUNT COVERED IN ONE FRAME (# OF Ås)
MEDIUM RESOLUTION				
G140M	1100-1700	1	1.8-3.0	27-29
G160M	1150-2100	1	1.5-3.0	33-36
G200M	1600-2300	1	1.8-2.7	38-41
G270M	2200-3200	1	2.1-3.3	45-49
LOW RESOLUTION				
G140L	1100-1700	1	0.15-0.22	290-291
HIGH RESOLUTION				
ECHELLE A	1100-1700	53-33	8.0-10	5.5-9.5
ECHELLE B	1700-3200	33-18	8.0-10	8.5-17.9

wavelength range. The more electrons counted per second, the more intense the source for that wavelength band. As the spectral resolution increases, the intensity decreases and the electron count drops. Therefore, at higher spectral resolution an object must be studied longer to collect enough energy to record.

The spectral resolution of the sample spectral range depicted ranges from 2,000 (top) to 20,000 (middle) to 100,000 (bottom). The number of spectral features measured in the range 1,475–1,480Å increases dramatically as the wavelength range shortens. The grating or acquisition mirror selected on the carousel sends light to concave mirrors that focus the dispersed grating light or the collected targeting light on to the appropriate detector.

Two special concave gratings, known as cross-dispersers, pass along the highly resolved reflection from the echelle gratings. Diffracted wavelengths cast many light orders and these can overlap in places on the same horizontal spectrum; if they do, that would make the data unreadable. Because the echelle gratings only reflect lower orders it is important to keep the orders separate. To do that the cross-dispersing gratings place each light order on parallel horizontal bands with all the first orders on one band, all second orders on another band, and so on. Only then can scientists examine any light order desired.

The Digicon detectors are photocathode tubes that count photons in the incoming light. Each detector has a maximum wavelength range to which it is sensitive: 1,400Å for Digicon 1 and 2,000Å for Digicon 2. Each tube contains a photosensitive window, a magnetic focus assembly that accelerates and focuses the incoming electrons, 500 diodes arranged in a line to record the spectral data, and preamplifiers to transmit the data electronically. Twelve extra diodes aid in focusing and also monitor background and unwanted light. A magnetic deflection system can shift the incoming bombardment of photons horizontally or vertically across the diodes to concentrate on particular bands of the spectrum or resolve one band more clearly.

Photons strike the window of the Digicon and direct electrons inward which are accelerated in the magnetic field and land on the diode array. Each diode counts the electron hitting it as a single event and this signal then goes to the amplifier and from there to an accumulator that records all the diode pulses. This information is replayed either immediately or after data build-up over a preset exposure time. The magnetic deflection system can realign shifted data to compensate for the motion of the Telescope. This data then goes to the accumulators in the SI C&DH memory from where it passes to the formatter and to the ground for analysis. If real-time transmission is not possible the data are stored in the HST's science tape recorders.

The software controlling observation is not in the GHRS equipment itself but instead resides in the NSSC-I computer in the Science Instrument Control and Data Handling

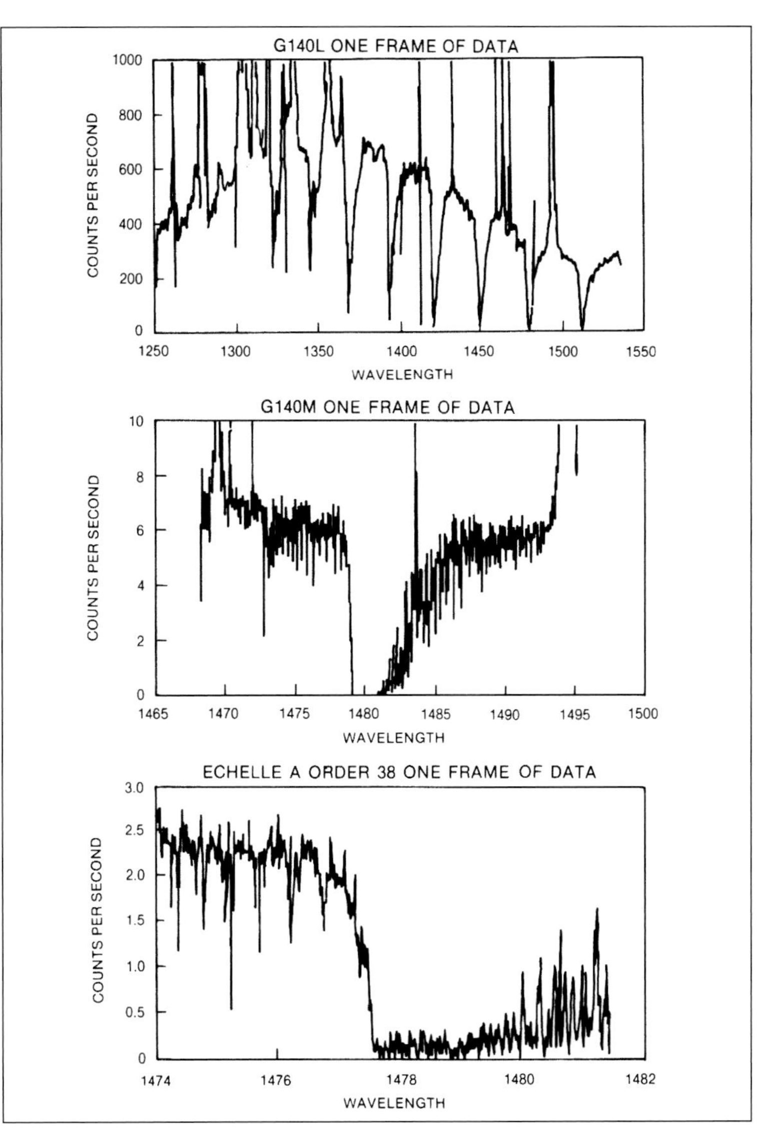

ABOVE Data from three separate grating settings as observed in low (top), medium (middle) and high resolution. Compare with the table opposite. *(Lockheed)*

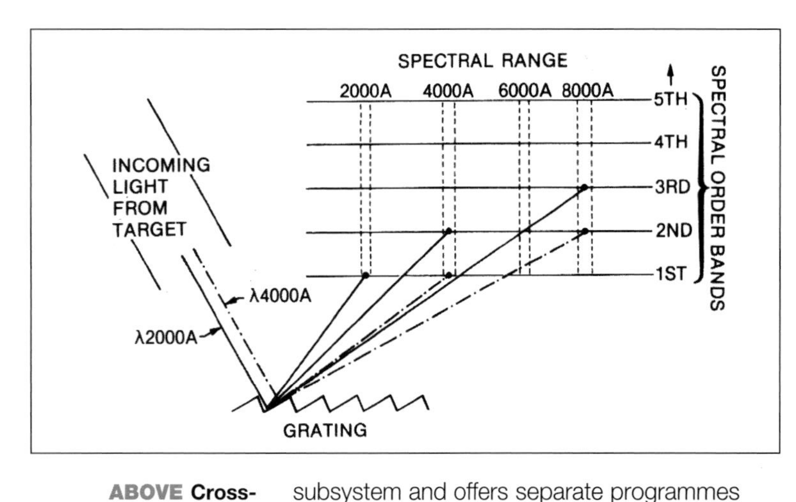

ABOVE Cross-dispersion gratings place each light order in parallel horizontal bands so that observers can select whatever light order they desire. *(Lockheed)*

BELOW Photocathode tubes count incoming photons with a magnet that focuses the incoming electrons, with 500 diodes arranged in a line to obtain spectral data. *(Lockheed)*

subsystem and offers separate programmes for the operational phase: observation set-up, which selects the detector and aperture, rotates the carousel to the relevant setting and sets exposure time for the observation; targeting, where the HST searches for and pinpoints a target by sending pointing manoeuvre requests to the SSM Pointing and Control System; mapping, where the full field of view is mapped with the data sent to the STOCC if commanded; and data collection, which controls the detector magnetic fields, collects diode data, stops observations that exceed under- or over-exposure limits, compensates for motion of the Telescope and checks the quality of the incoming data.

The GHRS has two basic operating modes: target acquisition and science data acquisition. Target acquisition is achieved using the science aperture lock to fix on to the target and make the necessary adjustments to the Telescope to place the target light into the small aperture. The basic procedure centres the target in the larger of the two apertures and measures its location relative to a brighter object. The target is then centred in the smaller aperture by small manoeuvres of the HST using the brighter object as a guide. The different acquisition mirrors are then selected for specific targets.

There are also two variations of the target acquisition mode. In the first an image is initially taken with the Wide Field/Planetary Camera before using the large GHRS aperture to locate the target from a difficult background of similarly bright objects which could confuse the target searches. The other uses the field of view of the large GHRS to map the sky up to 10 arc-sec in field, which takes less time to process than the WF/PC but covers less area. If the observer knows the precise location of the target image it can be positioned blindly, but in this case precise centring is less reliable and much less accurate. This is acceptable if the task is, for instance, a repeat observation using known calibrations or when the observer is studying large projects such as galaxies where any segment of the target will suffice.

The second operating mode is science data acquisition, for which there are two options. The first of these is the accumulation of data method, which is the normal one used with the GHRS for gathering spectral information. The GHRS software suite accommodates this option, which is useful for extended or protracted observations lasting from several seconds to many hours and particularly beneficial because the software can respond favourably to interruptions. The rapid readout option is the second available with this mode. It provides data for very short observation bursts between 50 milliseconds and 13 seconds. An individual block of data goes to the ground with each burst before the next observation capture begins. Again, here too data can be stored on tape directly as the information bypasses the SI C&DH computer.

For astronomers, the GHRS is particularly useful for atmospheric composition and dispersion analysis of stars, studying the content and material in the interstellar regions, observing star formation and binaries and observing quasars and objects outside the Milky

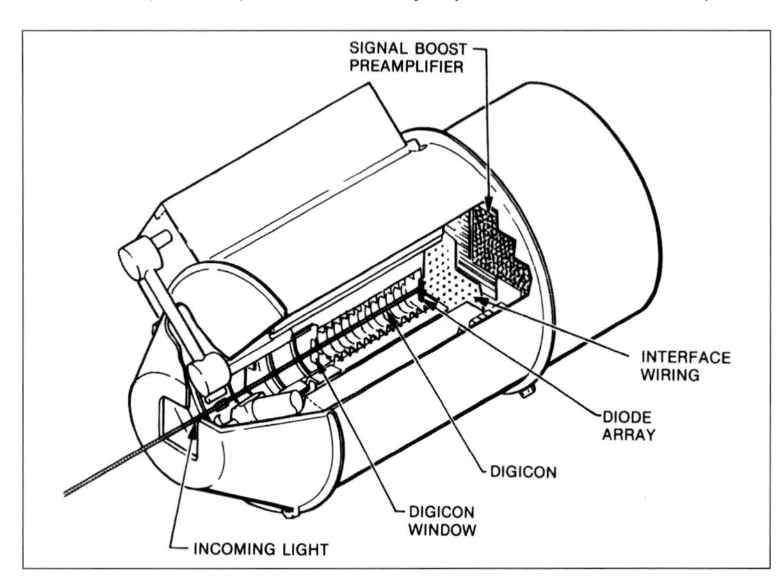

Way galaxy. The material envelope surrounding stars can tell astronomers a great deal about the compositional nature of the object and, especially when coupled with data from other instruments, it can help clarify the chemical composition of the outer atmosphere that has unique wavelengths which can be detected by the GHRS.

One particular example was the cool red giant star Zeta Aurigae some 790 light years from Earth, where matter is being pulled from its atmosphere by an invisible companion which was thought to be a younger, hot star. It is now known that this is indeed a binary system with one eclipsing the other on each revolution. Another example is the planet Jupiter in our own solar system where the moon Io displays volcanic eruptions of sulphurous material 'squeezed' out as the great mass of the planet pulls and tugs at the moon's structure. Material discharged in this way is trapped as a ring around Jupiter by the planet's strong magnetic field, and as well as observing the ejected material the GHRS can also observe the hot, invisible ring of gas containing sulphur, sodium and oxygen in the ultraviolet spectra.

For studies of the interstellar environment, astronomers use the GHRS to examine material in the spiral arms of the galaxy where the highest resolution is preferable for extended periods to seek out high-pressure regions that would be indicated by the intensity of spectral lines. Also, starlight shining through the spiralling galactic arms would reveal the chemical make-up of the components by revealing the missing spectra caused by absorption within the gas, the width of the absorption lines revealing the temperature of the elements.

And so it is with the study of star formation and binaries, where dust and other obscuring matter seen in images through ground-based telescopes prevents a full and clear understanding of the object under study. Even the most dynamic and violent events in the galaxy are not immune from attention with the GHRS. Quasars emit prodigious quantities of matter from their centre and the GHRS can observe the matter to measure its velocity while other instruments on the HST can look at other phenomena associated with these dramatic events.

High Speed Photometer (HSP)

Built by the University of Wisconsin, this instrument, also on the axial alignment of the Telescope, measures the intensity and colour of light and detects variations in intensity over periods of observation as short as ten microseconds on a spectral scan from ultraviolet to infrared. This allows the HST to measure the brightness of stars, to test ideas about black holes by detecting surrounding discs of gas, and to search for visible pulsars, which at the time the Hubble Space Telescope was launched were mainly observed as sources of radio signals.

With no moving parts, the HSP is the simplest of all five primary science instruments on the HST. It is about the same size as the three other axial instruments aligned with the optical axis, with exterior dimensions of 3ft x 3ft x 7ft (0.9m x 0.9m x 2.2m) and a weight of only 600lb (273kg). The structure is in the form of a box beam running the length of the instrument with power, electronics, thermal control and communications equipment mounted to bulkheads. The core of the instrument as a science tool is the optical detector subassembly, which is located at the forward end of the HSP.

The optical detector system has four entrance holes in the forward bulkhead, directly in front of four filter aperture assemblies. Four light detectors, one photomultiplier detector and

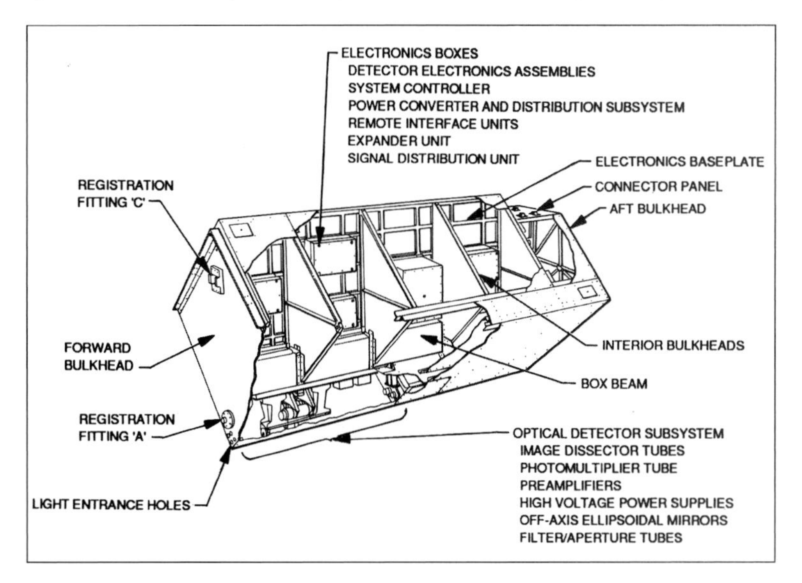

BELOW The High Speed Photometer is unique among all four axial science instruments in having no moving parts; with a relatively simple design the entire Telescope manoeuvres to select a filter rather than having a filter wheel do the alignment. *(Lockheed)*

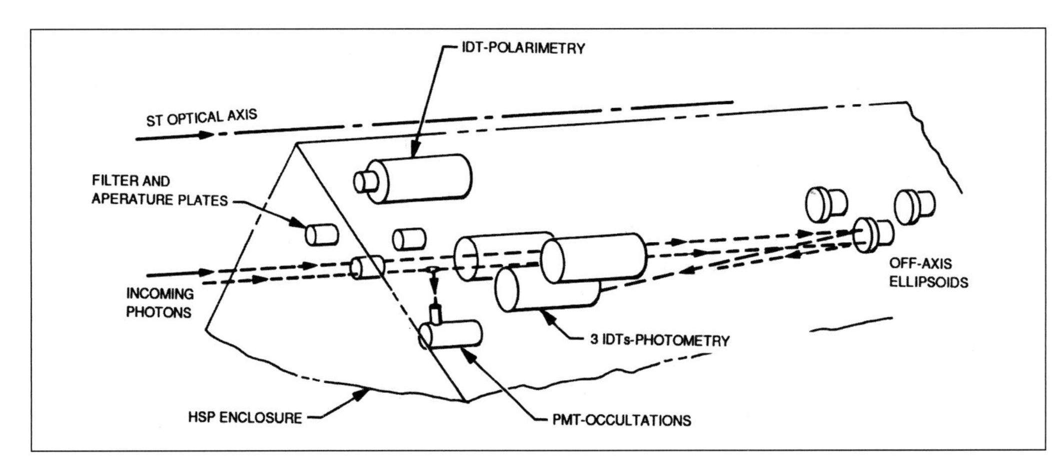

RIGHT The configuration of the HSP with filter and aperture plates and off-axis ellipsoids. *(Lockheed)*

three off-axis ellipsoid relay mirrors complete the optics. Incoming light from the target object passes through one of the entrance holes and falls on to a particular filter/aperture assembly. Each filter plate contains 13 coloured filters that isolate certain spectral ranges in the incoming light. The selected assembly directs a portion of the beam through one of three apertures, 0.4, 1.0 or 10.0 arc-sec in diameter. The smallest aperture removes much of the background light, the medium aperture is the most accurate, and the large aperture is used for locating a target object.

The light passes through a filter/aperture assembly and reflects off the ellipsoid mirrors, which sharpens the light. The light then enters an image dissector tube (IDT), of which two are sensitive to light in the 1,600–6,500Å (ultraviolet to visible) range and two are sensitive to ultraviolet wavelengths from 1,300–3,000Å. The electrons emitted by the IDT photocathode are focused by a magnetic field into the entrance of the 12-section photomultiplier section where the electron signal is amplified. Electrons from any area of the photomultiplier tube can be amplified, which allows the sky background to be measured.

A photomultiplier tube sensitive to the red end of the spectrum measures light in the near-infrared close to 7,500Å, and after passing through a filter the light is diverted by a beam splitter into the photomultiplier tube, with ultraviolet light going to a light dissector tube so that red and blue light from the same object can be studied separately and at the same time.

Polarimetric measurements can also be made for measuring the intensity of polarised light via one of four ultraviolet filters covered with a 3M Polacoat material and thence to an image dissector. This allows astronomers to study magnetic fields and light reflected through stellar dust, aspects which are not visible through photometry, and the accuracy of the HSP was designed for greater accuracy than could be achieved in the 1980s through ground-based instruments due to the turbulence in the atmosphere.

The general operation of the HSP begins with a decision about an object and how many bands of wavelength will be necessary for measurements and for how long, after which the Telescope is positioned for the observation to begin. The Payload Operations Control Center puts the software commands into the SI C&DH computer and this is usually assigned a period when the Telescope is already aligned with the same general area of the sky for some other experiment. The software reads the

RIGHT This exploded diagram shows the filter/aperture tube in the High Speed Photometer. *(Lockheed)*

commands and with the Pointing and Control System finds and locks on to the guide stars and then on to the target.

First, light passes through the 10-arc-sec aperture for centring the target, with the precise angle required to allow light to fall on the first filter calculated by the NSSC-I computer model, from where coordinates are passed to the Pointing and Control System. There may be several small adjustments to position the light so that it passes through different filter/aperture combinations as desired. Most filters have two pairs of 0.4ndf 1.0-arc-sec apertures for simultaneous measurements taking observations of the target star and the background sky at the same time without moving the Telescope. After passing on to the image dissector the amplified signal is passed to the SI C&DH via the HSP electronic data system and is then transmitted to the ground, or stored on board for download at a later time.

The operational function of the HSP provides the observer with several optional ways of using the instrument in addition to the basic target acquisition mode, with targeting using the largest aperture. Calculations necessary for repositioning light from the target to the appropriate image dissector depend upon the task chosen by either using previous observations to pinpoint the target or interaction with special targeting software so as to locate the target within the crowded and densely populated field of background stars.

Thereafter there are several options for the astronomer, selecting either a single-colour mode using only one aperture/filter, a star-sky mode that captures brightness and background over a period of time using several apertures, a double-IDT mode in which one is used for brightness and the second for sky background study, or the photopolarimetry mode which uses the standard polarising filter.

Wide Field/Planetary Camera (WF/PC)

While scientists and many amateur astronomers have eulogised over the images returned by the Faint Object Camera, the WF/PC promised a remarkable window on to the universe with images expected to eclipse anything seen through any other instrument – in space or on the ground. Unfortunately, problems with the primary mirror prevented the full value of this instrument from being realised at first. The information below refers to the instrument as installed prior to launch. For details about its successor, WF/PC II, installed during the first servicing mission, see Chapter 7.

The installation of the WF/PC and the Faint Object Camera (see above) provided two instruments with four different functions, and these lay at the core of the Telescope's visual imaging capabilities. Although the WF/PC was designed to take broader images of faint objects, with its narrow field of view the FOC has much sharper resolution and can concentrate on detail in galaxy clusters, allowing the WF/PC to capture the complete cluster set. Because of this, the instrument was expected to be the most frequently used instrument aboard the HST and was expected to support a wide range of astronomical observations and requirements.

The WF/PC has two camera modes: Wide-Field mode and Planetary Camera. The WF mode exposes a relatively wide field and uses a focal ratio of f/12.9 with a large field of view of 2.7 arc-min for magnitudes down to $20m_V$ but with this it has a limited angular resolution. The PC mode has a much better angular resolution, separating objects just 0.01 arc-sec apart, but it has a smaller field of view of 66 arc-sec and a focal ratio of f/30.

The instrument itself was contained within a structure 3.3ft x 5ft x 1.7ft (1m x 1.5m x

BELOW The optical path for the WF/PC passes to a pick-off mirror where it is sent through a filter carousel to a pyramid mirror for splitting the beam to a set of CCDs. *(Lockheed)*

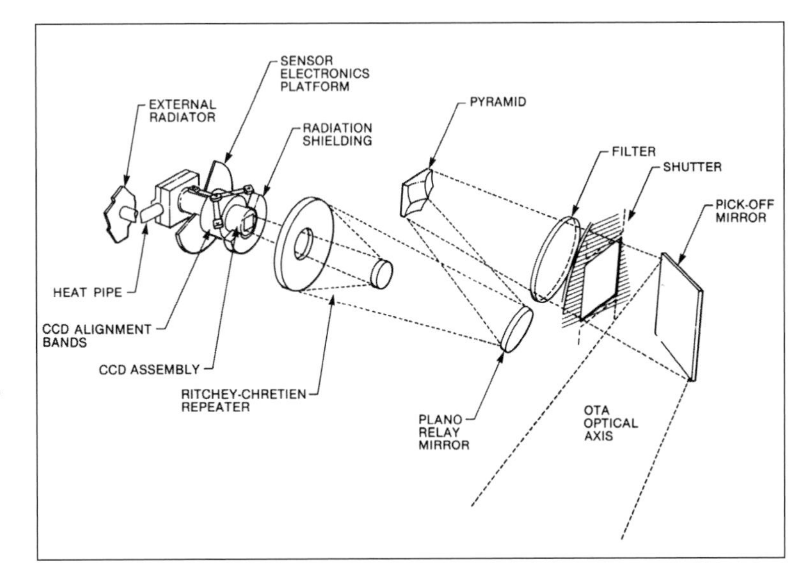

BELOW **The Wide Field/Planetary Camera combines three operating functions in one instrument: images, photometric measurements and polarimetric measurements. This instrument and its successors installed on subsequent servicing missions have produced some of the most stunning and impressive images of the universe.** *(Lockheed)*

0.5m) in size attached to which was an exterior radiator 2.6ft x 7ft (0.8m x 2.2m) shaped so as to conform to the cylindrical structure of the HST. The complete assembly had a weight of 595lb (270kg). It was designed by J.A. Westphal and a team at the California Institute of Technology and was built by NASA's Jet Propulsion Laboratory. Along with the Fine Guidance Sensors, the WF/PC is the only science instrument mounted at right-angles to the optical axis, in front of the focal plane structure. A special pick-off mirror mounted in the centre of the focal path reflects the middle of the light beam into the camera assembly. At 1,150–11,000Å, the WF/PC has the widest spectral range of any of the six instruments on the Telescope, with a resolution of only 0.1 arc-sec.

The design of the camera is based around an optical system that consists of eight charge-couple devices (CCDs) in two camera configurations and with a cooling system, and a processing system to operate the instrument and handle the data to the SI C&DH. The central optical structure directs light from the pick-off mirror to an aperture with a shutter and into a carousel that has filters, gratings and polarisers. There is a pyramid mirror to split the light and special folding and relay optics to locate the light on to the CCDs.

The pick-off mirror is fixed diagonally in the Telescope's light path and deflects the beam at 90° to the entrance aperture of the instrument, the shutter located behind the aperture controls the length of the exposure which can range from about 0.1 seconds to more than 27 hours, although a more typical exposure time was around 45 minutes, which represents one

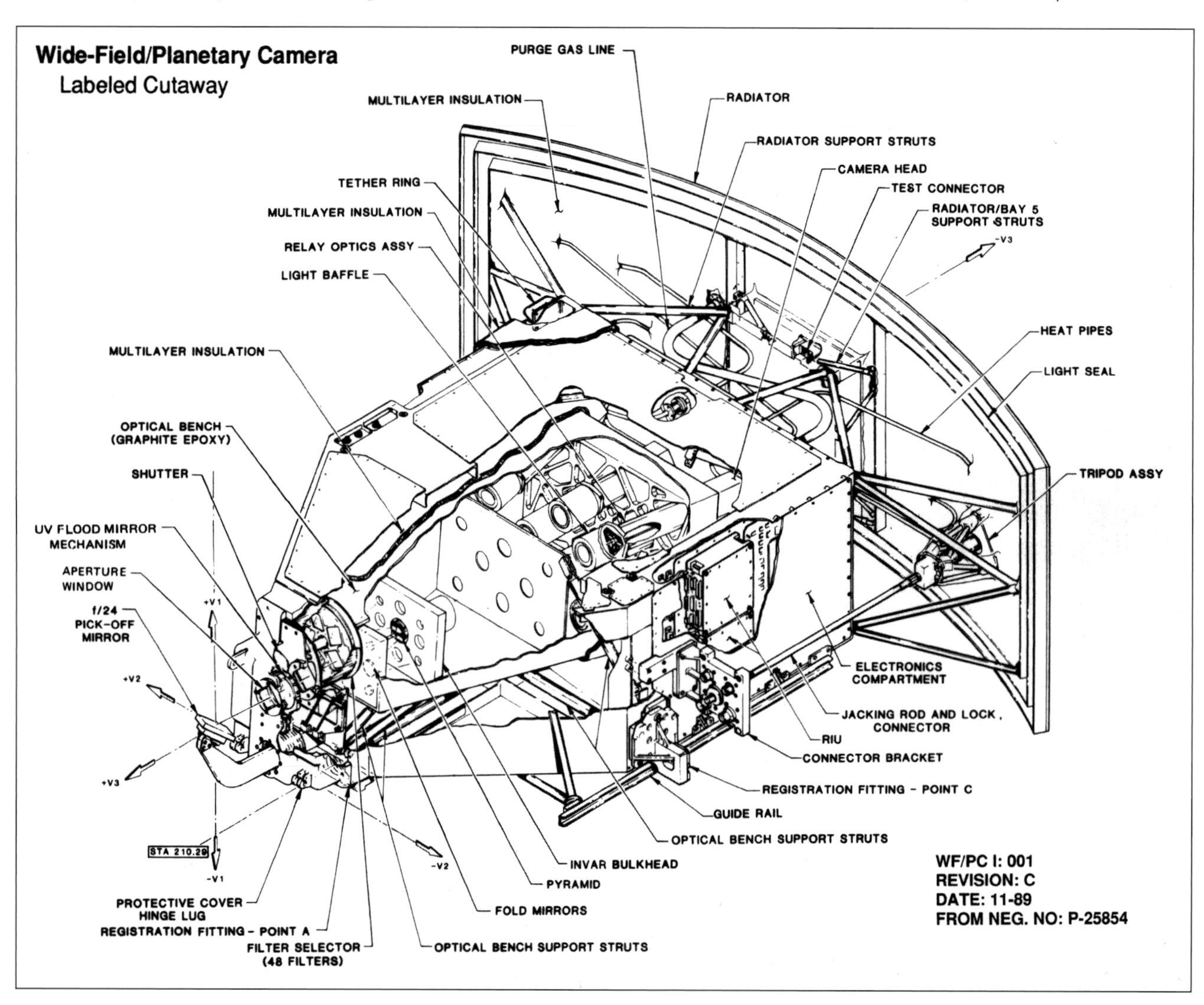

half-orbit of the Earth. This light passes directly to the filter carousel with 48 filters, four clear lenses for undispersed targeting light, three diffraction gratings, and three polarisers. The instrument can place either one or several filters into the light path for dual functions, which might be when it is perhaps taking a picture and a light intensity measurement with the same exposure. This filter assembly affords great versatility to the WF/PC.

From the filters the light falls on to the moving pyramid mirror and this can be set to one of two 45° angles. The pyramid splits the light into four parts on the path going to a particular set of CCDs. This light is reflected from the mirror back up to fold mirrors and from there down past the pyramid mirror to re-imaging mirrors that focus the beam on to the selected CCD. Each CCD consists of a silicon chip with an array of pixels, which are position detectors, with 800 on each side. As the photons bombard the array each pixel records an electrical charge proportional to the number of photons striking it. This reproduces the light intensity pattern when it is reconstructed later. The charge produced by each pixel is passed to the electronics system.

The CCDs used in the WF/PC cover a wide spectral range and are free from background noise from heat and electronic interference because they are cooled in a system of pipes and conduits that maintains a constant -95°C temperature. Excess heat is ejected across the radiator, which is exposed to the vacuum of space. The spectral range of each CCD is wide due to the phosphor coating that converts ultraviolet photons to visible photons, with 64,000 signals from each CCD composing a single image. To enhance and increase CCD sensitivity to short wavelengths, each CCD is routinely bathed in ultraviolet light using a special unit within the assembly.

Camera operations and data transfer is handled by an integral processing system that also transfers data to the SI C&DH unit. This microprocessor sets shutter exposure time, selects the required filter combinations, rotates the pyramid mirror for a specific optical path and reads the signals from the CCDs. The assembly operates as two cameras, either a wide-field or a planetary camera, the latter for close-up work. The Telescope's guidance system is used initially to find the required target but there are several operating modes for observers to use, including photometry, spectroscopy and photopolarimetry.

While the basic operating mode is imaging, the CCDs provide a wide range of applications that support the primary function for which the instrument was designed, spectroscopy using a grating to capture ultraviolet spectra. The camera can also be used to capture ultraviolet spectra using a grating and filters specifically provided for wavelength photometry as well for filter polarimetry in the spectral range of 2,500–8,000Å.

ABOVE Albeit viewed through the uncorrected optics of the Telescope as launched, this comparison between images of NGC 4261 from a ground-based observatory and the WF/PC on the HST in 1992 shows the degree of resolution available from its orbital vantage point devoid of atmospheric aberration. *(STScI)*

Fine Guidance Sensors (FGS)

While the five instruments described above are the prime science payload aboard the Hubble Space Telescope, one of the two Fine Guidance Sensors on board can serve as a sixth instrument for astrometry, a measurement of the position of selected stars in relation to others. The age-long history of the measurement of stars is described elsewhere in this book, but the provision of a third FGS allows for its application to an important role performed by the Hubble Space Telescope, that of determining stellar masses and distance to a high degree of refinement and tolerance. The design and application of the sensors

RIGHT Organising allocation of time and Telescope activity between observers is the job of the Space Telescope Science Institute in cooperation with the Association of Universities for Research in Astronomy (AURA) in processes that take the initial request through to an assigned slot on the HST. The Science Institute was set up on the Homewood Campus of the Johns Hopkins University in Baltimore. *(STScI)*

BELOW Distribution of data from the science instruments passes from the spacecraft to a Tracking and Data Relay Satellite, from there to the ground facility at White Sands, New Mexico, and from there – probably via domestic communication satellites – to the Space Telescope Science Institute. *(STScI)*

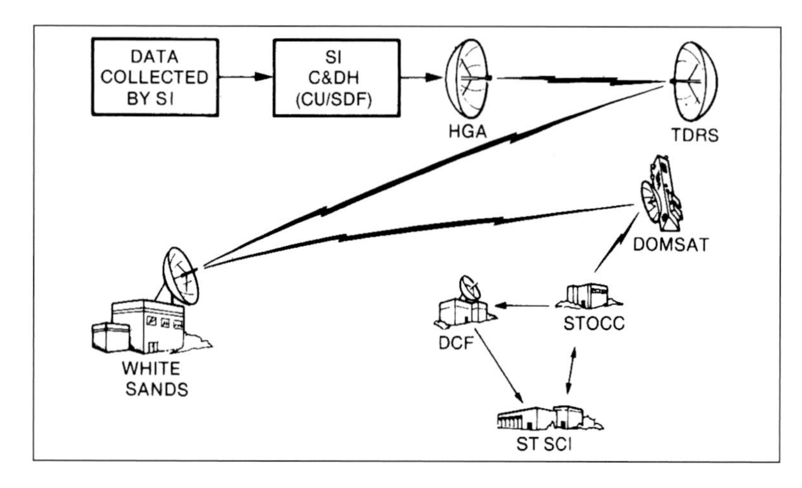

to the functional operation of the Telescope are described in the section on structure and layout but its use for astrometry has a range of operating modes.

The three standard modes are: position, transfer-function, and moving-target. The position mode allows the astrometric FGS to calculate the angular position of a star relative to the guide stars and to do that up to ten stars will be measured over a span of approximately 20 minutes. This keeps the pointing stability of the guide star within an accuracy of 0.04 arc-sec.

The transfer-function mode provides a means of measuring the diameter of the stellar object through either a direct analysis of a single-point object or by scanning a diffuse target. These could be planets within our own solar system, double stars visually closer together than 0.1 arc-sec, or targets surrounded by nebulous gas. Measurements of such binary stars that are visually separated by this angular dispersion can provide useful information about the precise measure of gravitational attraction between bodies and so improve understanding about the evolution of stars and planetary systems.

The moving-target mode allows measurement of a rapidly moving target relative to other targets when it becomes impossible to lock on to the target in relative motion to the primary. This might be the measurement of the relative motion of a moon orbiting a planet. In this way, the highly precise differences in relative motion can provide a degree of detail about the orbital characteristics and the relative masses of the two bodies.

Each of the three Fine Guidance Sensors has a filter wheel for astrometric measurements of stars with different brightness. This wheel has a clear filter for guide-star acquisition and faint star (>$13m_V$) astrometry. A separate neutral density filter provides for observation of nearby bright stars and two coloured filters are used for estimating the colour of the target, a product of its chemical index. It can also be used to increase the contrast between close stars of different indices or for reducing the background light from stellar nebulosity.

The Fine Guidance Sensors are an example of how equipment and instrumentation built into the Telescope to support its suite of science instruments can, in many instances, become a supplement to those and, in their own right, perform valuable and supporting measurements to the primary requirements of the investigators.

LEFT The TDRS series of data relay satellites evolved during the 1970s as a replacement for several ground stations around the world. Built by TRW, the initial series fed directly into the Hubble Space Telescope programme as a suitable programme for almost continuous contact wherever the spacecraft was around the globe. *(NASA)*

Scientific Instrument Control and Data Handling Unit (SI C&DH)

Control of the science instruments is handled by the SI C&DH which watches all the equipment in those instruments to keep them synchronised and working, and operates with the data management unit to process, format and communicate all science and engineering data created by the instruments. The equipment was built by the then Fairchild Camera and Instrument Corporation and by IBM.

The SI C&DH is a collection of electronic elements attached to an Orbital Replacement Unit (ORU) and mounted to the door of Bay 10 in the Support Systems Module equipment section. Elements of the unit are distributed among the separate science instruments and connected to the central modules. The components of the SI C&DH include the NASA standard spacecraft computer I (NSSC-I), two Standard Interface (STINT) circuit board units for the computer, two control unit/science data formatter (CU/SDF) units, two central processor unit modules (CPMs), a power control unit (PCU), two remote interface units (RIU), together with various memory, data and command communication lines (buses) connected by bus coupler units.

The NASA standard spacecraft computer model I had a central processing unit, which was essentially built around an Intel 386

RIGHT The TDRS system supported satellites placed in geosynchronous orbit where they appear to remain stationary in the sky as viewed from Earth, based on satellites over the Atlantic and Pacific oceans. The gold-coloured 'umbrellas' are folded antennae and the large, dark panels are solar arrays that will unfold when on station. *(NASA)*

ABOVE Lifted into low-Earth orbit by the Shuttle, the TDRS satellites were propelled to geostationary altitude of 22,300 miles (35,900km) by a solid propellant Inertial Upper Stage (IUS), seen here attached to the satellite. *(NASA)*

microchip, with eight memory modules each holding 8,192 18-bit words. The embedded executive software program runs the computer and moves the data, commands and operations program, which is the application for individual science instruments, in and out of the PCU. These applications programs monitor the specific instruments and control and analyse the collected data. The memory stores operational commands for activation when the HST is operating autonomously and out of contact with the ground. Each memory unit also has five areas reserved for commands and requirements unique to each specific science instrument.

Central to the SI C&DH is the Control Unit/ Science Data Formatter (CU/SDF), which is connected to the computer by the STINT, in effect a bridge. This formats and sends the correct commands and handles all data passing between the ground, the data management unit, the NSSC-I computer, the Support Systems Module and the science instruments. It has a single microprocessor for each of the control and formatting functions.

The control unit receives ground command, data requests, science and engineering data and systems signals and information. Two examples of these signals are time-tags that send controlling impulses to synchronise the entire spacecraft, and processor interface tables, which provide the dedicated communication codes. The unit itself transmits commands and requests after formatting them so that the unit at the receiving end can read the signal. This is necessary because commands sent from the ground have a different electronic signal format to those sent from the Support Systems Module and use a 27-bit signal versus the SSM commands' 16-bit format. The unit also reformats the engineering and science data to send it back.

The Power Control Unit distributes and switches power between the various components of the SI C&DH unit, modulating the specific power level required by each unit. These can vary widely. The computer memory boards variously need +5V, -5V and +12V while the control unit requires +28V of electrical energy. The unit controls and maintains the voltage levels within specification.

The remote module units transmit the commands, the clock signals and signals from other systems together with engineering data between the science instruments and the SI C&DH. The modules themselves do not send the science data but there are six distributed around the Telescope. Five are attached to the science instruments and one is dedicated to the control and power units in the SI C&DH. Each module contains a unique remote interface unit which in turn contain one or two expander units.

Signals and data moved between the SI C&DH and the science instruments are conveyed on data bus lines, with each bus multiplexed – that is, one line sends system messages, commands and engineering data requirements to the module units for activation, and a reply line transmits the requested information and science data back to the SI C&DH. The bus is attached to each remote module by a bus coupler unit, which isolates the module should the remote interface unit fail. The SI C&DH coupler unit is located on the orbital replacement unit tray.

In operational mode, the SI C&DH handles the science instruments' systems monitoring, command processing and science data processing functions.

Systems monitoring provides vital engineering data on whether instrument systems are functioning and operating as required. At varying times, and at regular intervals of between 500 milliseconds and 40sec, the SI C&DH scans each monitoring device and passes that information to the NSSC-I computer or, as appropriate, to the SSM computer. The computers either process or store the information, and through constant analysis of any spurious anomalies a potential failure could trigger a safing hold (see 'SSM systems and subsystems').

The command processing function checks and reformats commands which go either to the remote modules or the NSSC-I for storage. Commands enter the CU/SDF through the command data interface for ground commands, or the data interface for SSM originating commands. These are time-tagged and stored in the computer memory, or they are sent to the remote modules. Each separate command is interpreted as a real-time event, even if the command has been time-tagged, and many such stored commands are activated by certain situations or activities; an example being when, for instance, the Telescope is positioned for an observation with the Faint Object Spectrograph. The SI C&DH activates the systems necessary to perform the required function for that activity, which includes manoeuvring the Telescope for the correct alignment.

Science data processing is a complex integration task where one or all of the science instruments can be sending data at the same time. The control unit transfers this incoming stream through computer memory locations known as packet buffers. Each buffer is filled in sequence, switching between them as they fill, and each packet is then sent to the NSSC-I for further processing, or directly to the SSM for storage or transmission to the ground stations. Data returns to the control unit after it has been processed by the computer. When transmitting, the control unit must send a continuous stream of data, either full packet buffers or empty ones called filler packets. These maintain a synchronised link with the SSM, and special check codes (Reed-Solomon and pseudo-random noise) can be added to the data as an option.

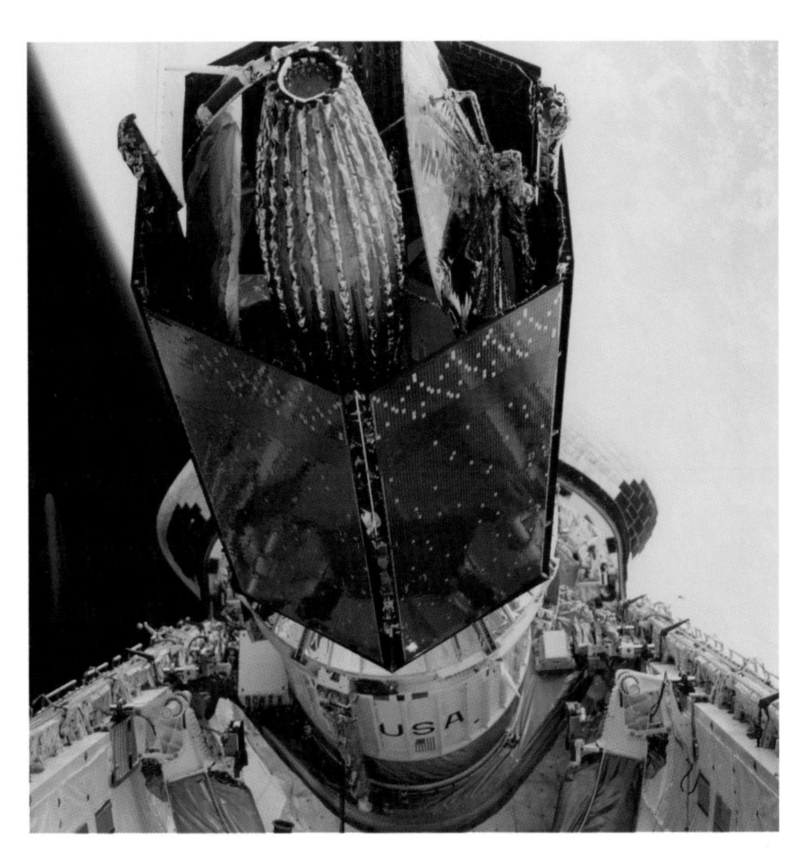

ABOVE Viewed looking aft from the flight deck of the Shuttle Orbiter, a TDRS satellite and its IUS is about to be released, from where it will be placed in geostationary orbit. *(NASA)*

BELOW Located at NASA's Goddard Space Flight Center, the Space Telescope Operations Control Center (STOCC) coordinates activity associated with planning and executing an agreed set of observations. *(NASA-GSFC)*

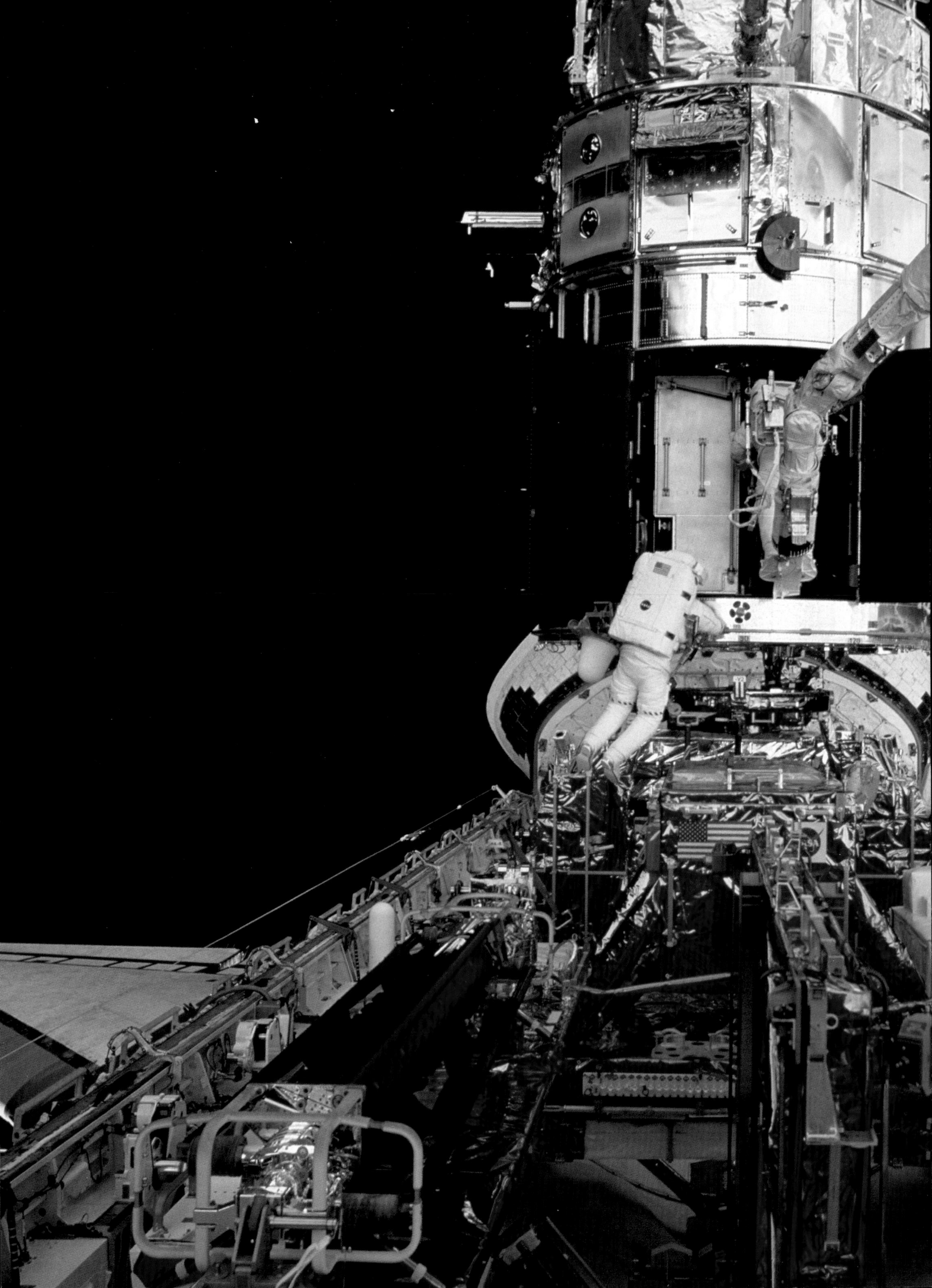

Chapter Five

Support and maintenance

The Hubble Space Telescope was unique when designed in that it was specifically configured for orbital maintenance and refurbishment through routine servicing by several visits from the reusable Space Shuttle. The Shuttle was in development from January 1972, when that programme was officially approved by the White House, and sanctioned by Congress later that year. It would not fly until April 1981 but the Shuttle and the HST literally grew together during the decade of the 1970s.

OPPOSITE Astronauts on the STS-109 mission give Hubble an upgrade while the Shuttle Orbiter acts as a mobile service shop visiting the Telescope in orbit. From the outset, the HST was designed for routine servicing and upgrade. *(NASA-JSC)*

ABOVE Extensive training is the key to successful EVAs and the Neutral Buoyancy Facility (NBF) at NASA's Johnson Space Center provides a close approximation to weightlessness. This mock-up of the Equipment Section of the SSM allows the spacewalkers to evaluate how best to carry out maintenance, repair and replacement activities. *(NASA-JSC)*

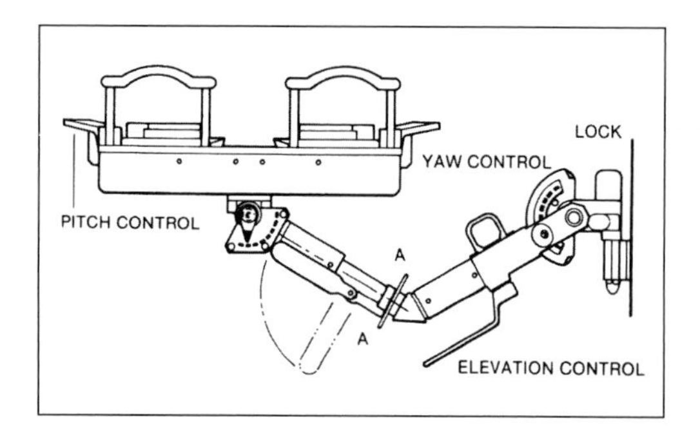

RIGHT Key to having hands free for work while the body remains tethered and secured are restraints and tethers, but the portable foot restraint is a useful way of taking this to the site without having numerous duplicate aids to clutter the outside of the Telescope. *(Lockheed)*

The opportunity afforded by routine access was a fundamental aspect of HST design and mission operations planning, and the opportunity for serviceability had a great influence on the acceptability of the HST by the professional astronomical fraternity. Many astronomers were not convinced by promises of performance capabilities and some questioned the amount of money the HST would cost to build, launch and operate and referred to what that could buy in telescopes and instruments for ground-based observatories.

Before the commitment to build a reusable transportation system it was believed the Telescope would be launched by an expendable Titan rocket, and if some form of on-orbit access had not been developed its operational life would have been much shorter. As it was, the original design life of 15 years was to be interspersed with visits to the Telescope at approximately five-year intervals, which would allow a progressive improvement taking advantage of technology developments and the opportunity to replace some science instruments with completely new instruments developed along the way.

Had an expendable rocket been used to launch the HST it would still have been possible to service it periodically using the Shuttle, but the desire to have this reusable system replace all launch vehicles bar the very

RIGHT One astronaut is held firm by a portable foot restraint on a work pedestal attached to the end effector of the Shuttle Remote Manipulator System (RMS) while the other floats free to work his way around the worksite. Working in teams the crew can maximise their efficiency while assigning specific tasks to individual crewmembers. *(NASA-JSC)*

smallest (Scout) rocket, made the choice of a Shuttle launch inevitable. Moreover, in the early 1970s NASA made a very conscious effort to blend together the very different worlds of human and unmanned space exploration. It suited the overall objective to shift all payloads, including satellites and spacecraft for science, off expendable launch vehicles and on to the Shuttle, but there was certainly some resistance to this within the science community.

Plans to launch the HST with the Shuttle and then to have it routinely serviced required some hardware to support the HST inside the cargo bay for launch and deployment. There would need to be a suite of special equipment to ease the task of spacewalking astronauts in servicing it. Some consideration had been given to using the Shuttle to bring the HST back to Earth for a complete refurbishment before returning it to orbit, but the cost of running two missions for a single task argued in favour of in-orbit servicing.

The whole argument for the Shuttle had been built – erroneously, as it turned out – on the premise that a reusable launch system (the Shuttle) would be much cheaper over time than continued use of expendable launch vehicles over a similar period, even if some individual missions appeared more expensive. That premise, in itself, turned out to be true, but after the real costs of running the Shuttle became apparent it was clear that only for flight rates of more than 45–50 flights per year would the Shuttle average out costing less than an equivalent programme using expendable rockets. Nevertheless, intangible costs and savings from being able to repair and restore spacecraft in orbit had no equivalent comparator for expendable rockets and it was here that the Shuttle came into its own.

The Hubble Space Telescope could never have been repaired and its life both extended and made more data-productive had the Shuttle not been an integral part of mission operations, and implicit within that was the ability for astronauts to access the HST in orbit and conduct servicing and repair operations. But in the early to mid-1970s when these decisions about the serviceability of the HST were made, the challenges for the NASA astronaut corps, while welcome, were largely unpractised. Apart from the very limited experiences with EVA during the Gemini mission, during which the difficulties of working outside a spacecraft in the weightlessness of space had been encountered and overcome, there was only one programme in which NASA gained real confidence in being able to apply standard EVA (extravehicular activity) techniques to the Hubble Space Telescope.

ABOVE The exterior of the HST was equipped with a permanent set of handholds and rails, enabling an astronaut to work his or her way across all accessible surfaces and work at locations where Orbital Replacement Units (ORUs) could be removed. *(NASA-JSC)*

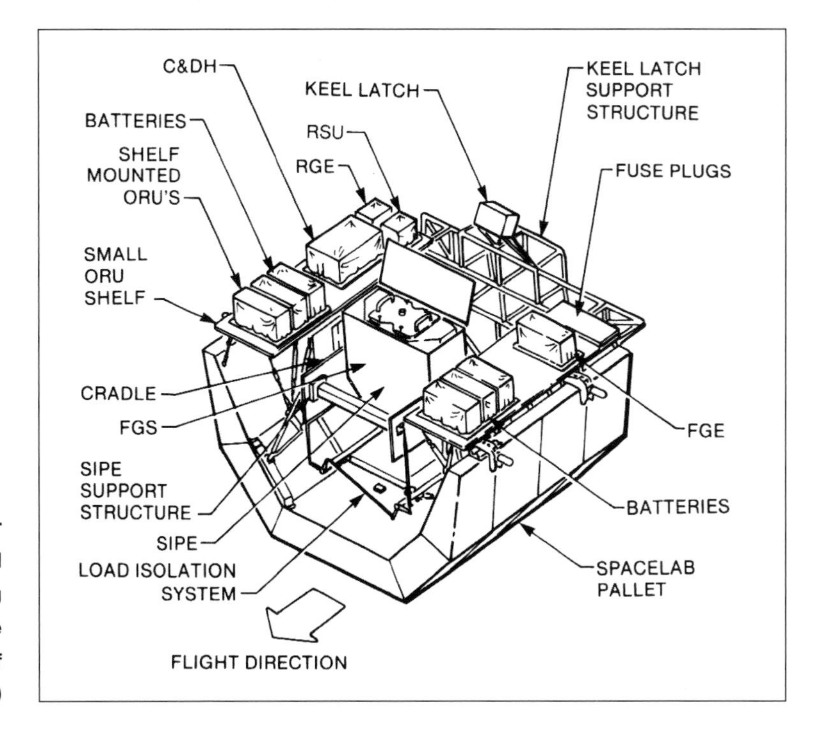

RIGHT The Orbital Replacement Unit Carrier (ORUC) adopted a Spacelab pallet, designed and built by the European Space Agency for supporting a range of external payloads and science equipment, ideally suited to carrying a range of equipment for servicing the HST. *(Lockheed)*

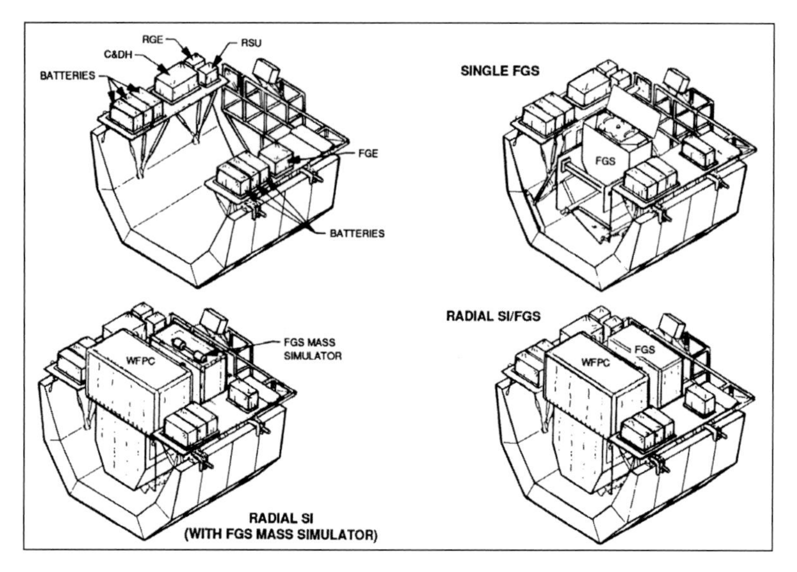

ABOVE Before the launch of the HST, NASA developed a set of standard ORU packages to fly as an integrated ORUC unit, with optional configurations of equipment, science instruments and tool packages. *(Lockheed)*

The five Gemini missions in which EVA was conducted were carried out in 1965 and 1966, and apart from one dual EVA on Apollo 9 in 1969 there had been only three other brief periods of working outside in weightlessness. During the Gemini missions, astronauts learned how difficult it is to work in weightlessness without tethers and restraints securing the torso and feet. Only with a major development effort mounted by astronaut Edwin 'Buzz' Aldrin toward the end of 1966 was the equipment and the practice for a successful EVA demonstrated successfully in a neutral-buoyancy water tank. Aldrin applied these techniques when he flew the last Gemini mission in November 1966.

The next EVA occurred during Apollo 9 when Russell 'Rusty' Schweickart stood on the 'porch' of the Lunar Module while it was still docked to the Apollo Command Module, from which David 'Dave' Scott emerged halfway out the hatch used to access the spacecraft on the ground. This Command Module hatch was used for the only other weightless EVAs, conducted in 1971 and 1972, when canisters of film were retrieved from the Service Module on three Apollo flights during the return flight to Earth.

These limited exercises in weightless EVA conducted during the Gemini and Apollo missions provided some reassurance that servicing the Hubble Space Telescope was a plausible proposition, until the Skylab programme that supported three missions launched in 1973 demonstrated full work capability under stressful conditions. The Skylab missions are a subject in their own right, but suffice to say that after running into trouble it took several EVAs to get the space station back to a satisfactory operating capability. It demonstrated that astronauts could conduct real work outside in space and would be fully capable of servicing the Hubble Space Telescope. How paradoxical, then, that the HST would also be 'rescued' by astronauts conducting EVA!

The experience with Skylab raised confidence that astronauts could fulfil the expectations placed upon them by HST design engineers and missions operations planners. A comprehensive programme developing tailored crew aids resulted in a wide range of tools, equipment bolts, connectors and other hardware which were developed for the HST, for Shuttle-related HST activity and for a host of other spacecraft being designed at the same time. For this was a time when NASA confidently expected not only to replace all expendable launch vehicles with the Shuttle but also to introduce a series of measures which would revolutionise space operations.

Engineers were able to install a total of 225ft (68.6m) of handrails encircling the HST, painted yellow for visibility. In addition there are guide rails for the crew to hold on to as well as the trunnions and scuff plates fore and aft. These

BELOW Across the five Servicing Mission (SM) visits to the HST between 1993 and 2009, several major systems and science instruments were upgraded or changed out, evidence that had it not been for the serviceability of this spacecraft and the availability of the Shuttle, the Telescope would have been severely limited in its applications. *(Lockheed)*

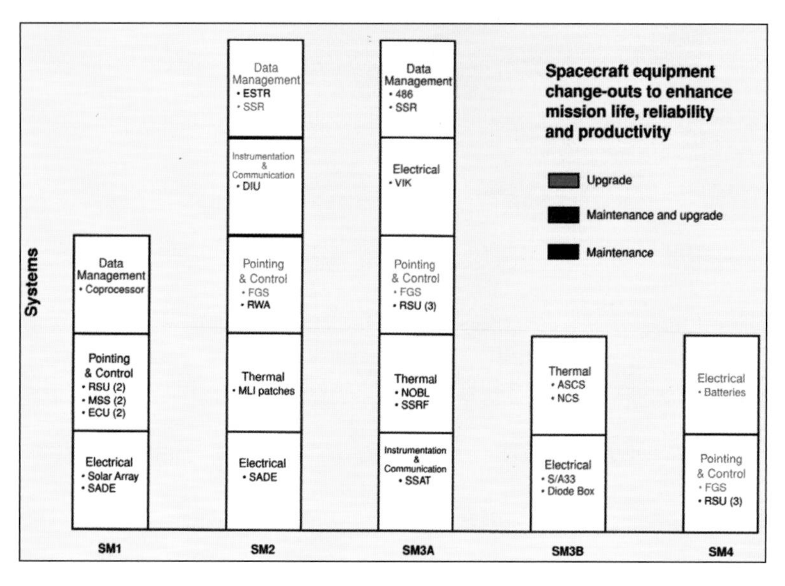

aids were designed for multiple servicing tasks and for access to the sections of the Telescope specifically designed for access or for removal. A systematic analysis of probable failure rates was fed into a model of how best to arrange the access rails and handholds. At the core of it all was the Orbital Replacement Unit (ORU), a set of components that might need lifting to the HST for use by astronauts on servicing missions.

The concept of the ORU applied to the HST grew out of the desire for a common set of hardware and support equipment that could be used for a variety of spacecraft, and the concept of the 'bus' and the 'payload' emerged. The bus was the basic platform providing electrical power, communications, attitude control and thermal control; the payload was the unique suite of instruments or equipment required for particular missions. Thus a manufacturer would provide a common bus, to which government and other users could attach the payload. When the HST was designed in the early 1970s it was confidently expected that a common bus could be used for a variety of different mission payloads, eliminating the cost of designing a separate bus for each spacecraft type.

From this template idea of commonality emerged the Multi-Mission Modular Spacecraft (MMMS), marketed by the then Fairchild aerospace company. The MMMS was selected for the Solar Maximum Mission (SMM) and launched by a conventional rocket in February 1980, a year before the first Shuttle flight. With an expected life of three years, pointing control problems just eight months later seriously curtailed observations, and a plan was raised to use the Shuttle to accomplish an on-orbit repair.

Delays to the Shuttle programme postponed this plan, but in April 1984 the Shuttle Orbiter *Challenger* successfully conducted a repair mission whereby the satellite was grappled down on to a service structure inside the Orbiter payload bay and in two periods of EVA was repaired and released for further work, studying the Sun. The flexibility of this overall scheme encouraged discussion of the retrieval of the SMM by Shuttle in about 1990, involving jettisoning the solar panels and securing it in the Shuttle payload bay for return to Earth. Refurbished by Fairchild, the MMMS would receive a different payload and be sent back into space two years later by another Shuttle. This was not pursued, not least because of major delays caused by the loss of *Challenger* in January 1986.

The SMM re-entered the atmosphere and burned up in December 1989, but the failed attitude control and pointing system that had been returned by the Shuttle in 1984 was itself refurbished and used for the UARS (Upper Atmosphere Research Satellite) launched by the Shuttle *Discovery* in September 1991. While the successful refurbishment and upgrade capability was a key aspect of several science missions developed during the 1970s and early 1980s, sustained operations in Earth orbit were no less important, and that was where the Hubble Space Telescope became a pivotal reason for developing the Orbital Maneuvering Vehicle (OMV).

The OMV was a reusable propulsion stage conceived as a means of moving satellites and orbital payloads from one orbit to another. The Shuttle was capable of reaching a relatively low Earth orbit and some satellites would need

ABOVE Solar cell arrays retrieved from the HST display micrometeoroid impacts, which marginally reduce power production and threaten reliability. *(ESA)*

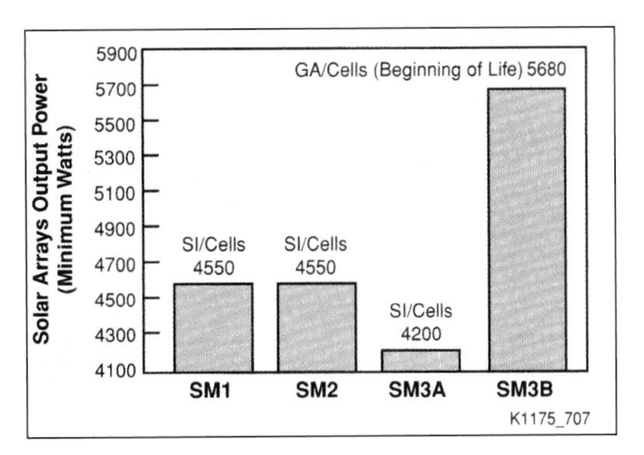

LEFT During the life of the Telescope, the efficiency of solar cells increased markedly and four servicing missions were able to increase the power production capacity significantly. *(Lockheed)*

ABOVE Astronauts working at the SSM access the instruments in the Aft Shroud, one crewmember working on the RMS arm and tool pedestal, the other inside. *(NASA-JSC)*

to be delivered to higher orbits, or returned to the lower altitude of the Shuttle for repair or return to Earth. Moreover, the OMV would have sufficient potential to boost the decaying orbits of some satellites and it was specifically aligned with the Hubble Space Telescope as being effective for that purpose.

The OMV had a diameter of 14.6ft (4.47m), about as large as could be accommodated in the Shuttle Orbiter payload bay, with a length of 6ft (1.8m) and an empty weight of 10,175lb (4,614kg), or 20,875lb (9,468kg) fully loaded with propellant for propulsion and attitude control. It would be carried to and from orbit in the Shuttle but operate semi-autonomously for mission operations via communications with the ground via the Tracking and Data Relay Satellite System (TDRSS), crucial to the operation of all orbiting satellites including the HST. The OMV was cancelled in June 1990 for cost-cutting reasons.

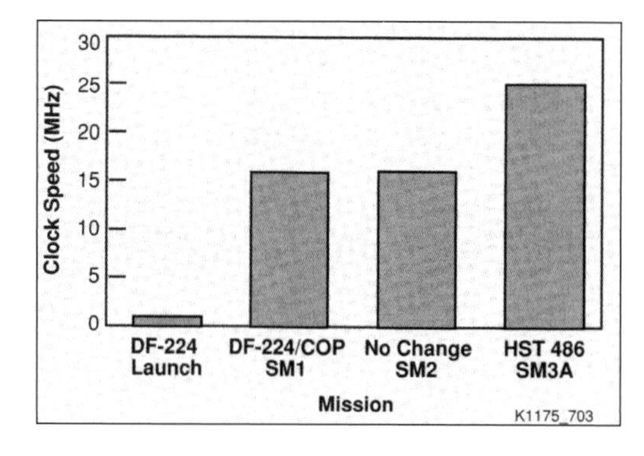

RIGHT Upgrades to the central processing unit increased appreciably during the various changes that raised the clock speed to allow more efficient and quicker management of engineering and science data. *(Lockheed)*

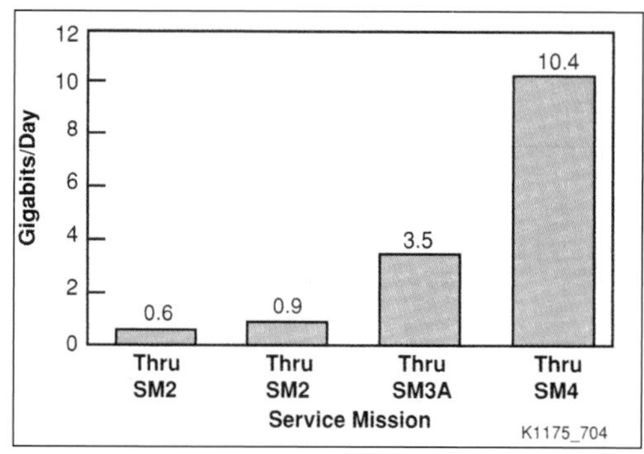

RIGHT Improvements to the data storage capability enhanced the pace of scientific operations and expended the overall volume of data within set time frames. *(Lockheed)*

Of more than passing interest here, but as evidence of the extent of the plans for space hardware when the Hubble Space Telescope was in development during the 1970s and 1980s, NASA's Marshall Space Flight Center was planning development of a much more powerful Orbit Transfer Vehicle (OTV) as another component of the proposed Space Transportation System of which the Shuttle formed the first part. The OTV would be used for moving payloads weighing up to 15,000lb (6,800kg) from the low Earth orbit of the Shuttle to geostationary orbit 22,300 miles (35,890km) above the equator. For this it would be equipped with one or two Pratt & Whitney RL-10 cryogenic rocket motors.

In 1988 this was renamed the Space Transfer Vehicle (STV) and given a more robust mandate to also carry landers from Earth orbit to Moon orbit, from where they would separate and descend to the surface, leaving the STV to propel itself back to Earth. It was to use aerobraking (decelerating in the outer layers of the atmosphere) in ellipses, ending up in a stable and circular low Earth orbit from where it would be used again or returned to Earth in the Shuttle. Aerobraking would cut the amount of decelerating propellant required by 35%. All work on the STV stopped in 1993.

While neither the OMV nor the OTV/STV were ever built and launched the concept of the Orbital Replacement Units underpinned the hardware developed for launching and supporting the Telescope throughout its planned 15-year mission, a period which has been almost doubled already! NASA engineers selected 70 ORUs consisting of 26 different components, some of which were duplicated, and varying in size from items as small as fuse plugs to the Faint Object Camera the size of a telephone box. The ORUs were kept in storage on the ground but were available from a bank for specific requirements. Each was designed so as to be easily transported to the HST and used as necessary.

To get ORUs up to the HST they would be attached to the Orbital Replacement Unit Carrier (ORUC) for installation in the Shuttle Orbiter payload bay. This was a Spacelab pallet filled with shelves and containers and comprising the cradle itself, support structures, latches to hold the equipment within the carrier, and closed-door compartments, tethers and crew aids for the EVA. Spacelab was the European Space Agency science pallet designed for instruments and equipment carried in the Orbiter payload bay for exposure to space.

Manufactured by the then British Aerospace at its Stevenage works, the U-shaped Spacelab pallet is 13ft (4m) wide and 9.8ft (3m) long, supported inside the payload bay by two locations either side and a single keel location. Each pallet weighed 2,685lb (1,218kg) and could be used singly or in trains of two or three attached together to support a major external payload or several smaller instruments. The first Spacelab mission was flown in November 1983 involving one pressurised module and one pallet, although an engineering model of the pallet alone had been launched on the second Shuttle mission in November 1981. Serving the HST programme, only a single pallet would be used for carrying the ORUC, but pallets were used for several duties on other missions.

Of unique design for the HST missions, the Flight Support Structure (FSS) provides a platform for the Telescope and rotates and tilts it as necessary for deployment. The superstructure has three main components: the exterior horseshoe-shaped cradle with a supporting latch beam, a pivot arm, and the rotating FSS platform itself, to which the HST is mounted. The cradle is two-thirds of a horseshoe shape with supporting beams for strength. It is attached to the Orbiter payload bay perpendicular to the long axis of the Shuttle and secured in place with latches.

Mounted to the end of the FSS cradle, the pivot arm supports and manoeuvres the rotation platform to move the HST to intermediate positions up to 140° toward the payload bay. The rotation platform located at the end of the cradle holds the HST and incorporates a small TV camera in its base to guide the HST into the correct position on the platform. This platform itself contains two umbilical units that carry power from the Orbiter to the HST and to support equipment during servicing operations.

LEFT ESA astronaut Nicollier works on a storage enclosure with a power tool specially designed to prevent torque turning the astronaut around! *(NASA-JSC)*

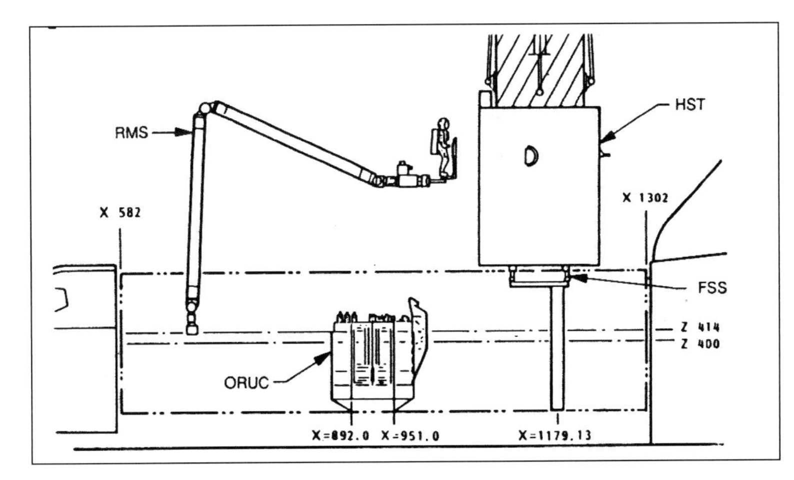

BELOW Access to the HST is facilitated by the FSS that holds the Telescope in a fixed position relative to the EVA astronauts and the equipment carriers for a wide range of ORUs. *(Lockheed)*

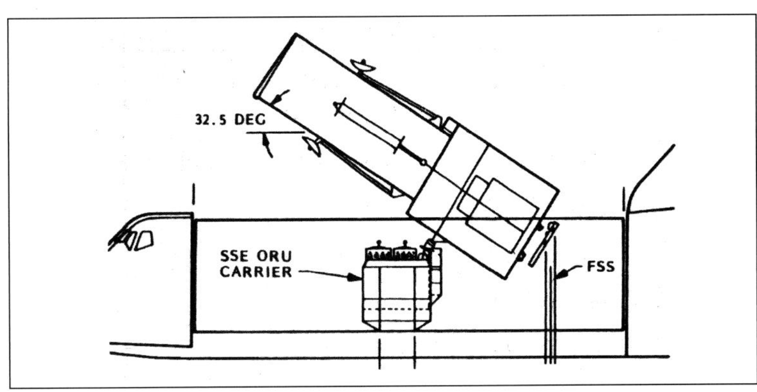

RIGHT The Shuttle would also be used to raise the orbit of the Telescope during routine servicing visits, with the FSS pitching the angle of the HST at an inclination calculated to transmit minimum stress loads to the structure of either the Shuttle Orbiter or the Telescope. *(Lockheed)*

Chapter Six

Launch and deployment

Delayed four years by the grounding of the Shuttle following the loss of *Challenger* in January 1986, plus a number of technical issues, the deployment to orbit of the Hubble Space Telescope was assigned to Shuttle Orbiter 103 (*Discovery*) commanded by astronaut Loren B. Shriver with pilot Charles F. 'Charlie' Bolden (later to become the NASA Administrator).

OPPOSITE The two solar arrays have been deployed with the tension rods in place. The apparent curvature is an optical effect of the fisheye lens. *(NASA)*

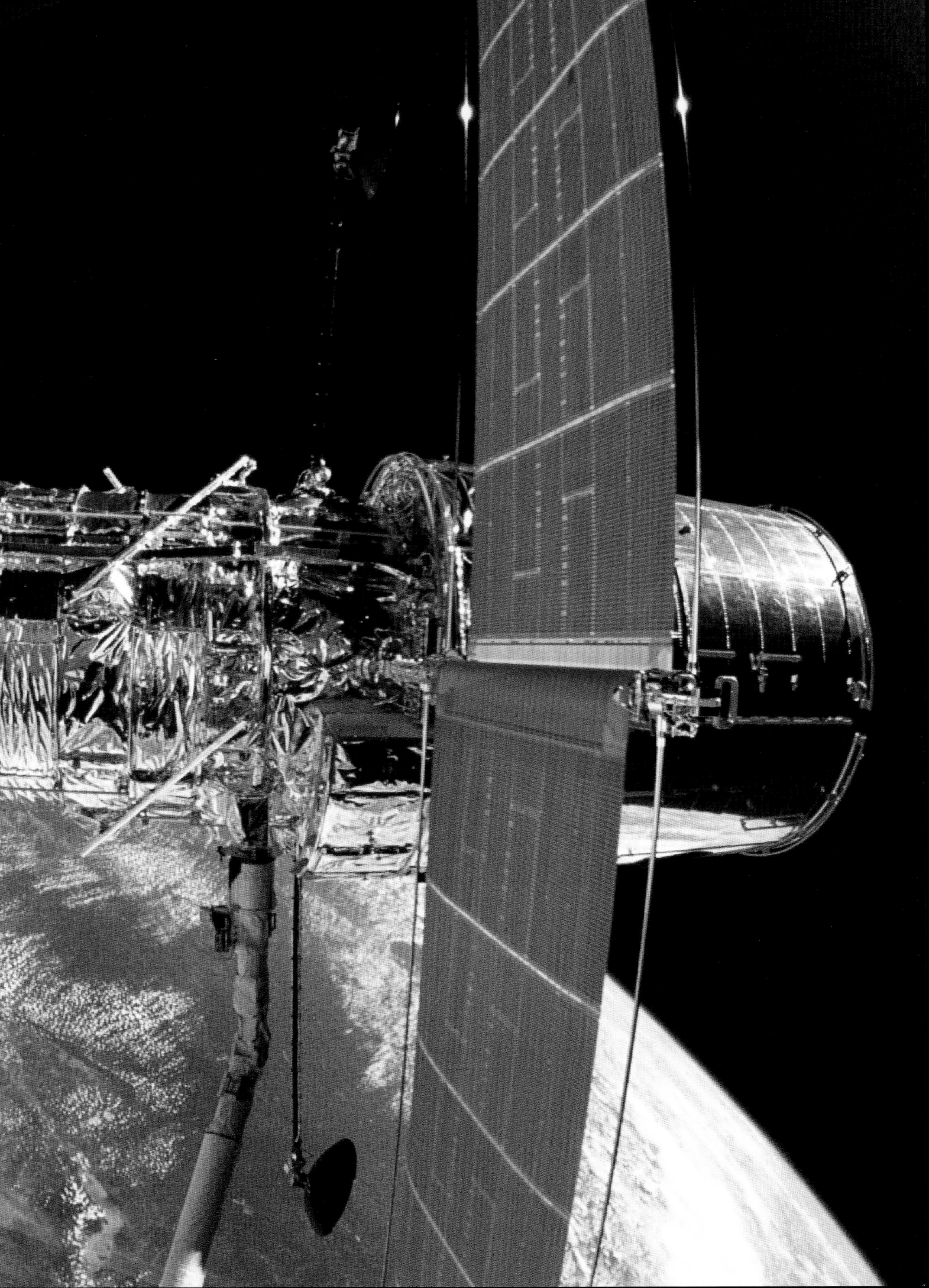

RIGHT Launch of the Hubble Space Telescope was delayed first by problems within the programme and then by the loss of the Shuttle Orbiter *Challenger*, an event which had a significant effect on this and many other science programmes as delays and postponements built up. Some satellites and spacecraft could be off-loaded to expendable rockets, but not large observatories such as the HST. *(NASA)*

The three Mission Specialists on board were Steven A. Hawley, Kathryn D. Sullivan and Bruce McCandless II. Scheduled for launch on 12 April 1990, it was planned that the mission would last 5d 1hr 15min. At a flight readiness review on 31 March the launch date was advanced by two days to 10 April (see 'Pre-flight preparations' below).

After reaching orbit the separate facilities responsible for the various elements of the HST would be directly involved in checkout and deployment. Pre-deployment procedures would be managed by the STOCC at NASA Goddard Space Flight Center in Maryland but only after the standard Shuttle post-orbit insertion procedure of opening the payload bay doors. Several hours are required to allow air to vent from the internal elements of the HST into the vacuum of space, to remove the threat of arcing when main power is applied from the Orbiter.

Only when the STOCC is satisfied that this has been achieved can the command be given to apply power from the Shuttle to the Telescope. That comes when the STOCC clears the Huntsville Operations Support Center (HOSC) at NASA's Marshall Space Flight Center in Alabama to authorise the crew to switch electrical power to the HST. Shuttle operations meanwhile are under the command of Mission Control, Houston, at the Johnson Space Center in Texas. Goddard mission operations would command data from the HST that tells the STOCC the health and status of the Telescope after its dynamic flight up through the atmosphere when acoustic levels and vibration are high. At the same time the technical support team in the HOSC will analyse the engineering ('housekeeping') data to determine the health of the bus systems.

The next stage for the Orbital Verification team at the STOCC is to thermally stabilise the Telescope while the crew prepare for their first night in space. Thermal safing activates various heaters in the HST and the on-board command computer is switched on and its performance verified. During the time the crew are asleep, the STOCC control centre monitors, interrogates and manages the systems on the Telescope to fully characterise its performance and response to several diagnostic commands and procedures to ensure that it is in a healthy condition for the following day. Both teams will have confirmed that the appropriate safing-mode systems have been activated, which puts the HST in a safe contingency posture should it unexpectedly lose contact with the ground or the Shuttle Orbiter.

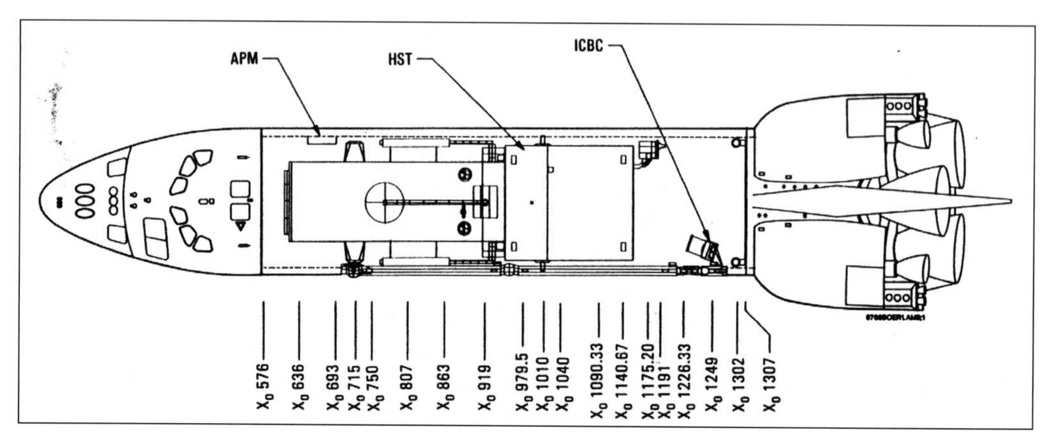

BELOW RIGHT The internal arrangement of the Shuttle Orbiter payload bay shows the enormous volume taken up by the Telescope. The Ascent Particle Monitor is an experiment retained within the payload bay designed to measure particulate contamination at various stages of the flight, samples obtained as separate doors are opened and closed being analysed after the return of *Discovery*. The numbers along the X-axis are inches of length from the datum point at the front of the External Tank, indicating a payload bay length just short of 61ft (18.56m). *(Rockwell International)*

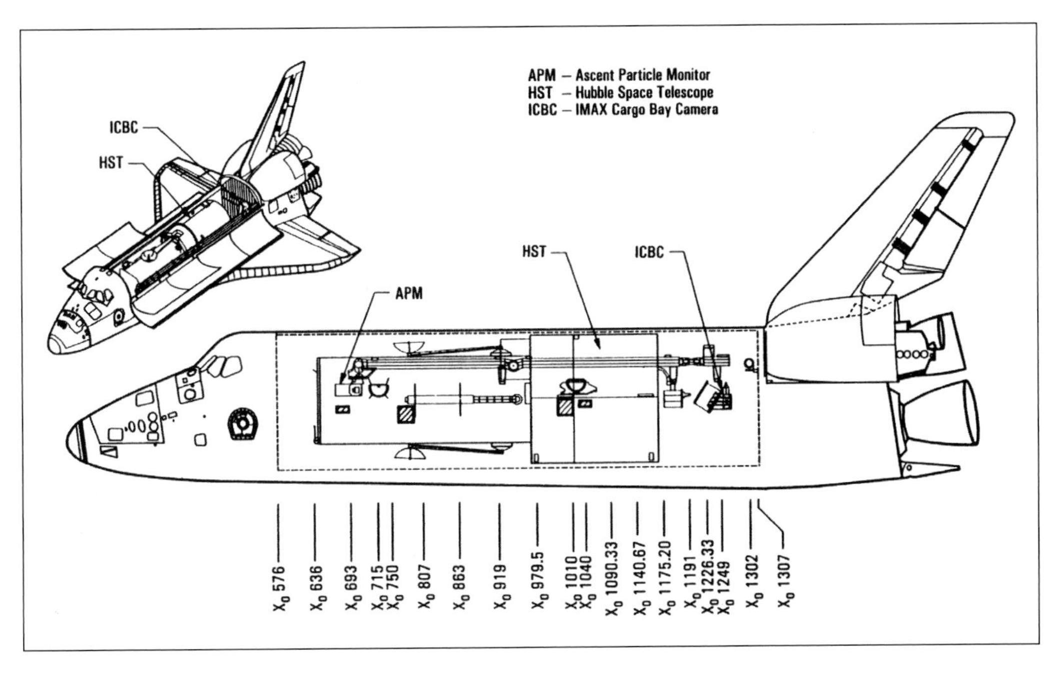

LEFT The maximum dimensions of the HST were set by the size of the payload bay, which was fixed by the time the Shuttle received the official go-ahead in January 1972. While the diameter of the primary mirror was largely determined by the technical limitations of the day, the size of the supporting structures was set by the 15ft (4.62m) diameter of the payload bay. Note the angle of the IMAX camera. *(Rockwell International)*

Deployment was scheduled for the second day with the Orbiter remaining close by for a further two days in case HST activation uncovered a problem and the Orbiter had to re-rendezvous for contingency EVAs, in which case McCandless would be EV1 and Sullivan EV2. In any event those two astronauts would be on stand-by in the Extravehicular Mobility Unit (EMU) suits during the deployment phase ready to conduct an EVA and address the problem manually.

Nominal deployment would begin with HST systems checked out using power from the Shuttle, followed by manually controlling the Remote Manipulator System (RMS), or robotic arm, to move it from the stowed position to the grapple fixture on the forward shell to which the end-effector would be attached. The crew would manually switch electrical power from the Orbiter to the HST's batteries and then switch off the Telescope's heaters to save power and reduce the drain until the solar arrays could be deployed. The Orbiter power umbilical would be remotely disconnected and the HST positioned for deployment of the solar arrays' wings by raising it into space at the end of the RMS.

This begins a crucial three-hour period during which the STOCC will remotely command a sequence of events to deploy the solar arrays and the high-gain antennas. First, the HST is positioned so that when deployed the arrays face the Sun, and that can cause a wait of up to 30 minutes. Then the forward and aft latches securing the arrays' arms are released using the SSM mechanism control unit. That takes approximately five minutes, after which the masts are raised with the primary deployment mechanisms, another eight minutes.

Next step is to unfurl the +V2 solar array blanket using the secondary deployment mechanism in an action that takes around five minutes. During this period the electrical power subsystem switches on the current charge controller directing electrical energy to the batteries. When this is accomplished the OTA and battery heaters are switched back on. With this completed, the –V2 solar array blanket is unfurled, taking a further five minutes. Only now can the STOCC start up the Pointing and Control System's magnetic sensing system, and when the crew on the Orbiter give the word the STOCC will command simultaneous erection of the two high-gain antenna booms in a deployment sequence taking ten minutes.

With data flowing through the two TDRSS antennas, the STOCC will monitor, along with the HOSC, up to 6,200 telemetry points to check that all are within specification and that both the science-related team at Goddard and the engineering team at Marshall are satisfied that the two elements of the Telescope are functioning correctly. Only then can full deployment begin. While the antennas are being extended the aperture door latch will be released, but the door will not be opened until the coarse Sun sensors are operating satisfactorily to protect the aperture from excessive sunlight. When they are the STOCC will start slew tests to check that the solar arrays are moving and positioned properly. The Orbiter crew will be constantly watching to confirm this visually.

When the solar arrays are satisfactorily deployed and providing power to the six batteries the remainder of the Telescope's equipment will be turned on and these items will be checked. Only then will the RMS arm adjust the HST in its proper release orientation whereupon the attitude control system can be activated. After a 15-minute dwell period to confirm that this is operating correctly the end effector on the RMS arm will spool back, releasing the grapple fixture to allow the Telescope to separate as the Orbiter gradually moves away.

Next up is the command from the STOCC to turn on the DF-224 computer's 'keep-alive' monitors to begin pointing procedures. Then the SI C&DH computer will be powered up and after 24 hours the aperture door will be opened while the Shuttle remains close by in case it is needed. But this is only the first phase of several months during which the Telescope will be brought up to full operational status and released for experimenters to begin their work.

The mission verification is in four phases and began with testing before the Telescope left the Orbiter. The second milestone was when the Telescope was stabilised in orbit. The Orbital Verification (OV) phase began with the opening of the aperture door, a period during which the STOCC would commence a rigorous sequence of housekeeping tests to ensure the Telescope was fit to begin the fourth and final phase, Scientific Verification (SV). The OV phase will start with the secondary mirror being adjusted until the images are sharp and precise. This step may take some time, as refinement of the focus will have to compensate for the somewhat imprecise contraction of the focal plane metering truss during outgassing of water vapour.

One of the more arduous start-up procedures involves the Pointing and Control System, components of which will have to be tested to calculate and transmit data to move the HST into an optimum orientation for communications and telemetry. To do this the star trackers will begin mapping the general location of stars while the magnetic sensing system will start to accumulate data on the orientation of the Telescope with respect to the Earth's magnetic field. The rate gyro assemblies will assess the pointing attitude and the rate of angular motion as it orbits the Earth while the coarse Sun sensors keep track of the Sun's position.

When this data has been acquired by the ground computers, an activity which can take up to 12 hours, it will generate updated position coordinates for the gyroscopes and the Telescope can adjust its pointing attitude as a result. This enables the high-gain antennas to take this information and be similarly positioned for aligning with the Tracking and Data Relay Satellite System and to transmit the larger volumes of data necessary for determining a wider range of parameters. This will feed back again to the STOCC, enabling it to take the HST out of Software Sun Point Control mode and permit the rate gyro assemblies to control the orientation of the Telescope.

At this stage the NSSC-I computer will be switched on to enable the ground to verify the performance of the instruments, and over the following day the rate gyro alignments are sharpened to improve the telemetry and generally prepare for the operation of the Fine Guidance Sensors.

By the third day at the latest the aperture

BELOW The crew for STS-31, the mission to place the HST in orbit, comprised (left to right) Charles F. Bolden, Steven Hawley, Loren Shriver, Bruce McCandless and Kathryn Sullivan. Charlie Bolden would be appointed NASA Administrator, in charge of the civilian space agency from 2010. *(NASA)*

door will have been opened and will transfer activities to checking the capabilities of the instruments, by which time the HST will be ready to test the Fine Guidance Sensors, which will be turned on at the start of the fourth day. It is at this point that the Shuttle will end its standby role and manoeuvre away so that it will be far from the HST when the de-orbit burn occurs and ejects lots of chemical products into space through its rocket motor – not something to do close to a finely tuned optical system.

For the following three days the FGS course tracking and mapping calibrations will take place, after which the STOCC practise adjusting the secondary mirror actuators in the Optical Telescope Assembly. This activity will continue into the 11th day together with tests to ensure that the guidance sensors can align with respect to the Telescope's axis and each other. An important part of this exercise would be to determine how well the HST absorbs the continual day/night cycle, where temperatures range across a wide spectrum and could compromise the Pointing and Control System. It will also determine how well the optics cope with the quality of light coming off the primary mirror. This will be used for the STOCC to determine different focal adjustments by moving the mirror actuators and then re-testing them. This methodical process would continue until the Pointing and Control System meets performance specification.

This pre-planned sequence was a nominal process with built-in flexibility for changes and extensions to perfect the operating performance of the Telescope. During the Orbital Verification process the STOCC at Goddard hosted an engineering team from Marshall, with their team leader being the Director of Orbital Verification (DOV). Each step of this carefully scripted plan would be approved by the DOV before final go-ahead for the next stage. A second team at the Marshall Space Flight Center monitors activity at Goddard and stands ready to provide additional technical and engineering back-up. Final commands to the HST are, however, conducted by Goddard personnel.

The four objectives for the Orbital Verification phase include: fine-tuning the pointing accuracy; focusing the Telescope; activation of the science instruments; and evaluating

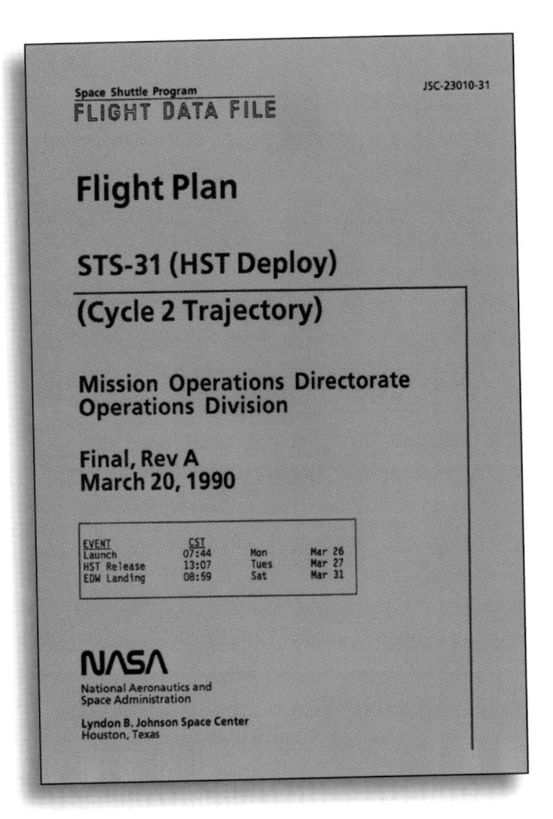
Space Shuttle Program
FLIGHT DATA FILE

JSC-23010-31

Flight Plan

STS-31 (HST Deploy)

(Cycle 2 Trajectory)

Mission Operations Directorate
Operations Division

Final, Rev A
March 20, 1990

EVENT	CST		
Launch	07:44	Mon	Mar 26
HST Release	13:07	Tues	Mar 27
EDW Landing	08:59	Sat	Mar 31

NASA
National Aeronautics and
Space Administration

Lyndon B. Johnson Space Center
Houston, Texas

LEFT The cover of the author's copy of the STS-31 Flight Plan. *(David Baker)*

the integrated performance of the Telescope support systems, the science instruments and the ground control elements. Only when this was accomplished was command handed totally to the STOCC at Goddard for the Science Verification phase, with Marshall standing by in a support role should problems be encountered that required specialised engineering help.

The Orbital Verification of the HST would last several months and begins with aperture calibrations to determine the very precise location of that in each instrument. A set of refinements begins which will set the alignment of each aperture to a few thousandths of an arc-second and set it within that instrument's portion of the Telescope's focal plane field of view. During this early period of verification the instruments will be used to monitor the effects of the magnetic perturbations of the South Atlantic Anomaly (SAA) and from this data the sequences for turning on the high voltage to see if the instruments could continue to obtain data during this passage.

During the first week of the OV phase the WF/PC commands an activity to purge any contamination that may have formed on the CCD, power being applied to the Thermal

Electric Coolers to reduce their temperature to the proper operating region for science observations. Only then would attention turn to the FOC for its first external observations on a specified star for aperture alignment. Monitoring of the SAA continues during all these activities to obtain a proper understanding of how the Telescope and its instruments react to different situations.

The second week of the OV phase sees continuation of instrument activation with the FOS performing its first external target observations of a star, to check its alignment and to demonstrate that the Telescope can perform an accurate and continuous scan. Sustained observations with the WF/PC are the focus of the third week in the SV phase and this enables the ground controllers to determine the level of sharpness in images and to test the ability of the camera to recognise two closely spaced images. Also, the GHRS will take its first external observations to align its apertures.

Finally this week, a thermal stability test is conducted to characterise the response of the Telescope and to establish the capability of the Fine Guidance Sensors to perform astrometry.

The fourth week was reserved for DFGS to FGS alignment, checking to verify that the sensors could be taken to a new and higher level of accuracy than was achieved during the Orbital Verification phase. This set the calibration of the science instruments at a level that would allow a more stringent calibration of the instruments. This phase was considered the midpoint at which the overall boresighting activities associated with determining the Telescope's guidance system alignment, the optical truss alignment and the alignment of the individual instruments converged. The FOS spectrum would be taken during the fine aperture alignment calibration and the spiral search target acquisition capability of the GHRS would be verified.

The Science Verification phase begins a

BELOW The pre-launch flight plan schedule covering the events associated with deployment of the HST on FD2 (Flight Day 2). The black and white bars denote the periods of light and dark as the Shuttle orbits the Earth. *(David Baker)*

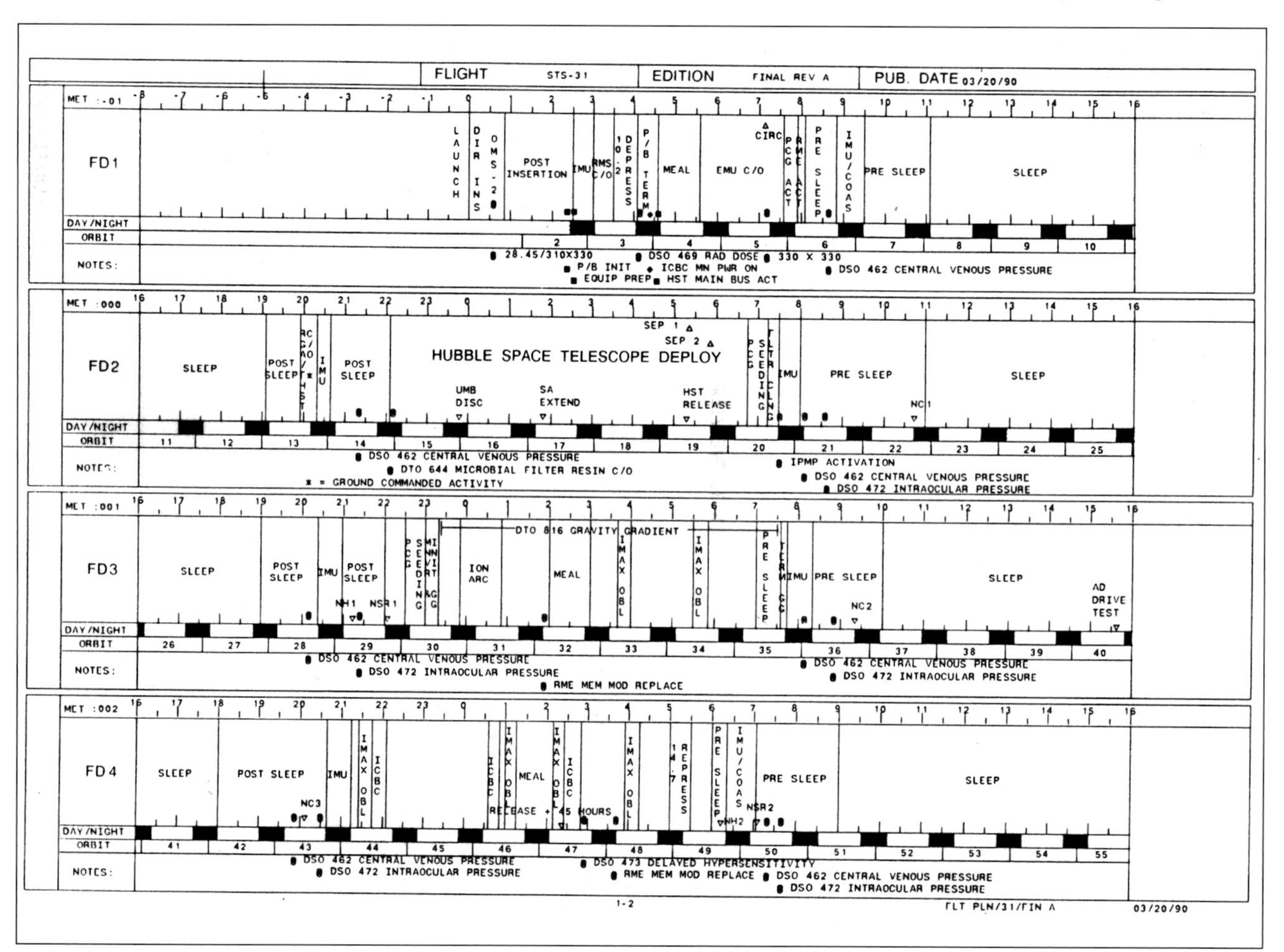

process of using the aligned instruments to test their performance and capabilities, and this is undertaken through the Space Telescope Science Institute (STScI), with astronomers previously involved with the design of the Telescope given the opportunity to conduct preliminary research. Calibration standards common to astronomy helped measure the ultraviolet and visible light, a test during which the Telescope is pointed at celestial objects with known calibrations so that these readouts can be compared with those of the Telescope.

The entire Science Verification process is necessarily long and meticulous, with several months elapsing before the HST was considered to have been characterised; that is, understood as a spacecraft operating in a settled environment with a measured range of responses and reactions which may be marginally different to those which had been predicted prior to the flight, or indeed designed into the Telescope from the outset.

Only after this lengthy period of interrogation could the HST be handed over to the STScI located on the Homewood campus of the Johns Hopkins University, Baltimore, Maryland. Mission control remained at the Space Telescope Operations Control Center (STOCC), situated on campus at the Goddard Space Flight Center, Greenbelt, Maryland, from where operational commands would be uplinked to the Telescope. Operational commands related to the equipment provided by the European Space Agency went through the Space Telescope European Coordinating Facility (ST-ECF).

Pre-flight preparations

The HST had been at the Lockheed Sunnyvale, California, facility since 1986, when it had originally been scheduled for launch, and it was shipped to the Kennedy Space Center in a Lockheed C-5A transport aircraft of the US Air Force for flight aboard the Shuttle Orbiter *Discovery*, arriving in Florida on 4 October 1989. Prelaunch testing took place in the Vertical Processing Facility, where it was powered up on 28 October via a command sent by satellite from Lockheed's HST facility in Sunnyvale. There followed 40 days of

ABOVE The launch of *Discovery* on the morning of 24 April 1990 heralded the beginning of more than two-and-a-half decades of observations attracting an additional five Shuttle missions to improve and upgrade the Telescope. *(NASA)*

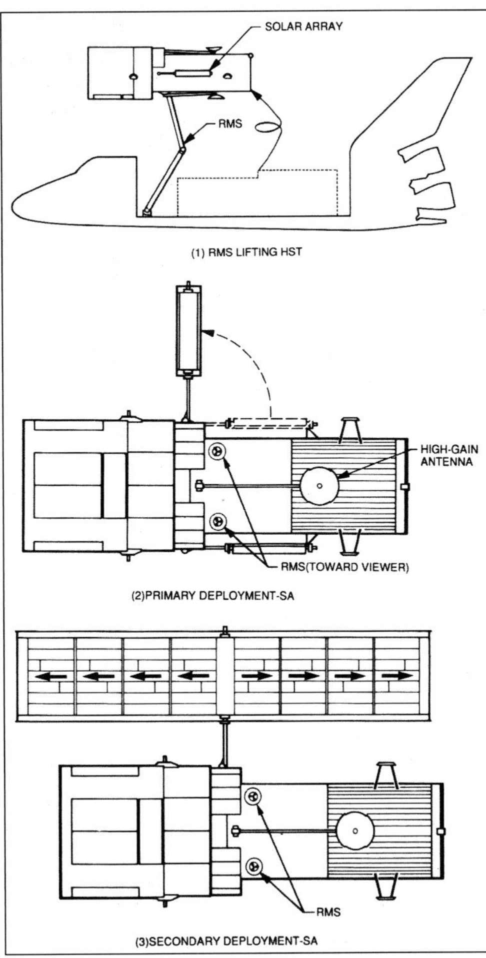

LEFT Deployment of the Telescope first required the raising of the spacecraft above the payload bay, followed by unfurling of the solar arrays. *(Lockheed)*

checkout involving the operating systems and the science instruments. For 11 of those days a variety of simulated mission commands tested the satellite links between Lockheed and the STOCC.

Discovery had returned from its previous mission – its ninth – on 27 November 1989 and been in the Orbiter Processing Facility (OPF) for approximately three months during which it underwent several modifications, including the fitting of new carbon brakes. It was moved to the Vehicle Assembly Building for stacking with its external tank and the two solid rocket boosters. The stack was moved to Launch Complex 39B on 16 March, where it received the HST in its payload bay, heading toward a launch on 12 April at 9:21am local time.

BELOW The Telescope is lifted above the payload bay prior to unfurling of the arrays and deployment to orbit. This view was shot from one of the two aft-facing windows in the aft flight deck area of the pressurised cabin. *(NASA)*

Payloads are installed in Shuttle Orbiters either in the OPF, when the Orbiter is horizontal, or at the pad when it is in a vertical stacked position. The HST was installed on the pad but there was a delay of two days when midges invaded the Payload Changeout Room (PCR), the vertical enclosure raised to the height of the Orbiter and drawn up close to form a seal between itself and the payload bay. It had been a hard winter and citrus groves had a lot of rotting fruit that brought a spate of midges. Insects inside the PCR could contaminate the sensitive optical devices on the HST, so six traps had to be set to eliminate the midges before removing the protective covering around the HST.

Apart from that, the only concern at that point was a lock nut on the nose-wheel landing gear which might have had to be replaced after a similar lock nut on *Atlantis* showed thread wear – a small matter with big implications. Had the decision not been made to continue with processing (the lock nut was determined to still have a few more flights in its life), the stack would have had to be returned to the Vehicle Assembly Building, delaying the mission.

On 31 March NASA held its customary flight readiness review and agreed to speed up the launch by advancing the flight day to 10 April at 9:24am, saving precious electrical charge in the six batteries aboard the HST. It was an important decision since the batteries were in the critical path to getting the Telescope operational. Nominally charged to 90 amp-hours each, they start to lose charge on a continuous basis, and this determined the amount of time that the Telescope could sit inside the Orbiter waiting for launch. There must be a charge of 45 amp-hours in each battery at the time of lift-off to allow sufficient energy to remain should the deployment be delayed to Day 3 of the mission.

If the launch was delayed by three days the next launch opportunity would come on 21 April, allowing time for technicians to get inside and recharge the batteries. This procedure required the payload bay doors to be kept slightly open for cooling purposes while the batteries were charged. It only takes 58 hours to charge the batteries but the procedure to open the bay, gain access to the Telescope, carry out the charging and reverse the process, sealing the bay once more, absorbed the rest of the time.

Preparations for the launch went ahead for a flight on 10 April but the countdown had to be stopped at T-4min when a pulse control valve on a Sundstrand Auxiliary Power Unit (APU) apparently failed to close as predicted, allowing in more hydrazine fuel which produced an over-speed condition triggering an automatic shutdown to the count. Each Orbiter has three APUs, activated at T-5min to provide hydraulic power, and all must be operating for the launch to go ahead. The replacement required a fix to be made with the Shuttle in a vertical position and this had never been done before, so a completely new launch date had to be agreed that allowed NASA to remove all six batteries and take them to the battery laboratory at KSC where they could be recharged at room temperature, raising the charge level to 90 amp-hours from an estimated 63 amp-hours had that been done at the pad.

With a new launch date of 25 April the APU was replaced and hot-fired on 18 April, which allowed the launch to be advanced by a day when a predicted 76 amp-hours would be in each battery at the time of lift-off. The unexpected delay to *Discovery* had come at a busy time for KSC, with two Orbiters being processed at the same time. On adjacent pad 39A, *Columbia* was being prepared for STS-35. Planned for a morning lift-off at 8:31:00am local time, STS-31 proceeded toward launch but the automatic sequencer stopped the count at T-31sec when newly written software sensed that an oxygen fill and drain valve on the External Tank was open. It was not, and it took just a couple of minutes for technicians to spot the error in the system and restart the count.

STS-31 *Discovery*

24 April 1990
Mission duration: 5d 1hr 16min 6sec
EVA: None
Commander: Loren J. Shriver (Flt 2)
Pilot: Charles F. Bolden (Flt 2)
MS1: Steven A. Hawley (Flt 3)
MS2: Kathryn D. Sullivan (Flt 2)
MS3: Bruce McCandless II (Flt 2)

Lift-off for the 35th Shuttle flight came at 8:33:51 local time (12:33:51 UTC) on 24 April 1990 on what was a unique trajectory for the Shuttle programme. The Hubble Space Telescope would operate from a higher Earth orbit than that usually accessed by the Shuttle and this required a special ascent and insertion phase, aiming for a 380-mile (612km) circular orbit inclined 28.5° to the equator.

ABOVE With the Remote Manipulator System robotic arm holding the Telescope, the solar array booms are rotated 90°. *(NASA)*

The two main SRBs shut down at 2min 6sec and the three main engines continued to fire until an elapsed time of 8min 38sec. During ascent, and because of the unique trajectory, there was an additional emergency landing option at Hoedspruit in South Africa should the Orbiter have encountered a problem with two of its main engines.

At main engine shutdown *Discovery* was in an orbit of 380 x 55 miles (611 x 89km) with

BELOW A close-up view of the bistem solar array booms. It would prove difficult to get both arrays to unfurl, and with limited battery life there was an urgency to begin producing electrical power from sunlight. *(NASA)*

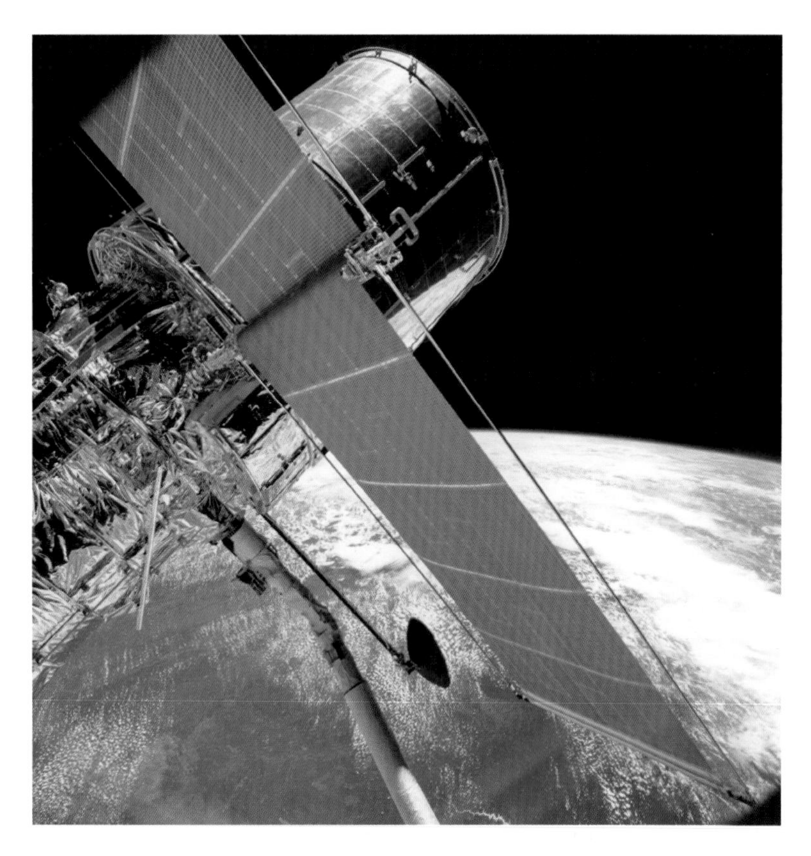

ABOVE **A clear shot of the deployed booms together with one of the High-Gain Antennas.** *(NASA)*

BELOW **Just over 32 hours into the mission, the Hubble Space Telescope is released to free flight and will not be captured again for more than 43 months.** *(NASA)*

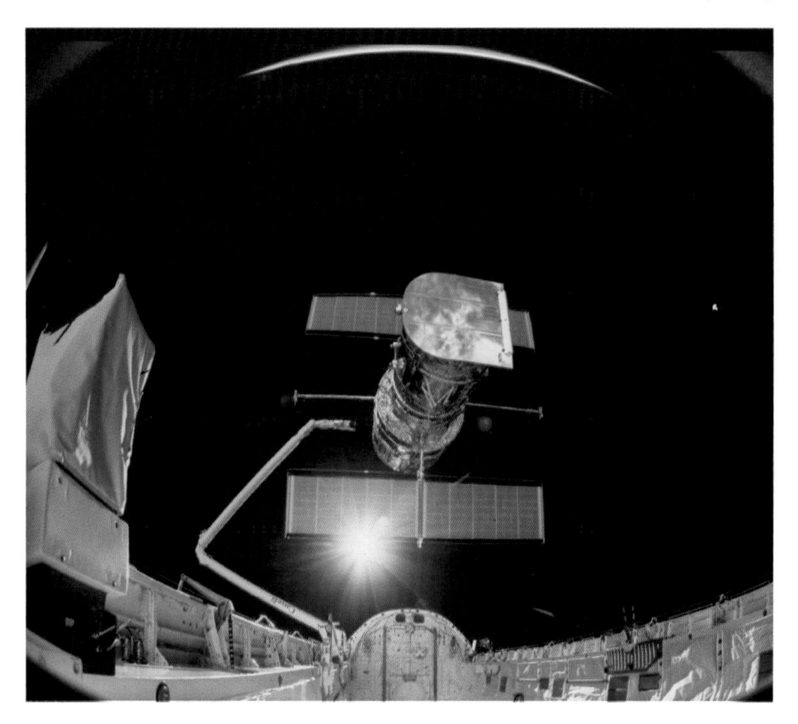

an inclination of 28.45°. From its low point the Orbiter drifted up toward apogee (the high point) and at 42min 36sec elapsed time the two Orbital Manoeuvring Systems (OMS) were fired for 5min 4.8sec. This was the longest burn thus far in the Shuttle programme and, changing velocity by 496.7ft/sec (151m/sec) it achieved an almost circular path of 380.5 x 357.6 miles (612.2 x 575km).

To place the Orbiter in precisely the right orbit for deployment of the HST, the reaction control system thrusters were fired for a continuous burn of 2min 17sec at an elapsed time of 7hr 9min 14sec. This velocity change of 33.5ft/sec (10.2m/sec) produced an obit of 382 x 380 miles (612 x 611 km), slightly higher than planned but positive in that it benefitted the orbital life of the deployed Telescope.

Steve Hawley had unstowed the RMS arm at 3hr 20min, and during a nightside pass using a TV camera found the HST to be in excellent condition. The STOCC approved power transfer from the Orbiter into the Telescope so that numerous heaters in the Optical Assembly could be activated for thermal protection from the cold of night. With initial checks good the crew went to sleep at about 11 hours into the flight, higher above the Earth than any previous Shuttle crew, higher in fact than anyone had been since Apollo 17 returned from the Moon in December 1972.

With much larger windows and a virtually panoramic view the mission provided an opportunity for high-altitude Earth photography, and *Discovery* carried cameras and special film for that purpose. It also provided an opportunity for studies into the radiation effects on a human skull, which had been carried into space on the STS-28 and STS-36 missions – Shuttle flights were assigned a mission number according to their payload and were frequently out of numerical sequence. Several other peripheral experiments were carried as well in addition to a 70mm IMAX camera to film the HST deployment.

Activities to deploy the HST got under way on the second day in orbit, with the RMS being unstowed again by Steve Hawley at 22hr 11min to attach the end effector to the grapple fixture on the HST just as dawn was breaking over the Gulf of Mexico. In the period since reaching orbit the STOCC had sent

18 commands to the Telescope. Flying over Florida one orbit later in full daylight, Hawley gradually lifted the HST from the cradle to which it had been secured in the payload bay. It was 24hr 11min since lift-off and Hawley slowly lifted the Telescope above the Orbiter and turned it base-end to the Sun to minimise thermal soak, the front of the Telescope pointing toward the tail of the Orbiter.

In case they were needed for a contingency EVA, McCandless and Sullivan had been pre-breathing oxygen and the pressure in the Orbiter had been reduced to 10lb/in^2 (68.95kPa) so that they could move into the airlock module to suit up. While the Shuttle had a sea-level atmosphere of oxygen and nitrogen normally pressurised at 14.7lb/in^2 (101.35kPa) the EVA astronauts would breathe pure oxygen in their EMU suits; to prevent them getting the 'bends' this pre-breathing was a necessary step to purge nitrogen from the blood.

The next stage in the deployment sequence was to rotate the solar array arms 90° from their launch positions, and the STOCC at Goddard sent the appropriate signals. However, telemetry indicated that although they appeared to have extended as planned they were not locked in position. McCandless and Sullivan were ordered into the airlock module to don their EMU suits, following which they would reduce the pressure in the module to 5lb/in^2 (34.5kPa) ready for a spacewalk to conduct a manual lockdown. As a last attempt to get the mast locked controllers transmitted a command to energise both motors, and this time it locked in place. As expected, the two high-gain antennas deployed prior to the next stage, unfurling the solar arrays from their stowed positions.

There was a time-critical element to the deployment of the arrays. The solar cells had to be providing power to the HST within eight hours of going on batteries prior to severing the umbilical link to *Discovery* and any delay would drain the batteries below the point of recovery. Because of problems with the motor drives there was concern – particularly from the ESA Space Telescope European Coordinating Facility – that when the bistem solar array booms unfurled the blankets of solar cells they would not stop as they were designed. As a precaution, Hawley and Shriver were advised to send a backup 'stop' command even if the automatic system appeared to be working.

Command to start the port-side array unfurling went out at an elapsed time of 27hr 51min with *Discovery* over South America. Slowly, the blankets began to unfurl, and in 5min 20sec had reached the end stop in the fully open position; but when the backup command was sent it cut off all data from the arrays. Goddard advised to send another signal that had originally been intended for transmission only after both arrays were deployed. It solved the problem, and with time running out the command was sent to deploy the starboard-side array at 28hr 14min. After about 1in (2cm) it stopped, and a conference was held between all engineering teams.

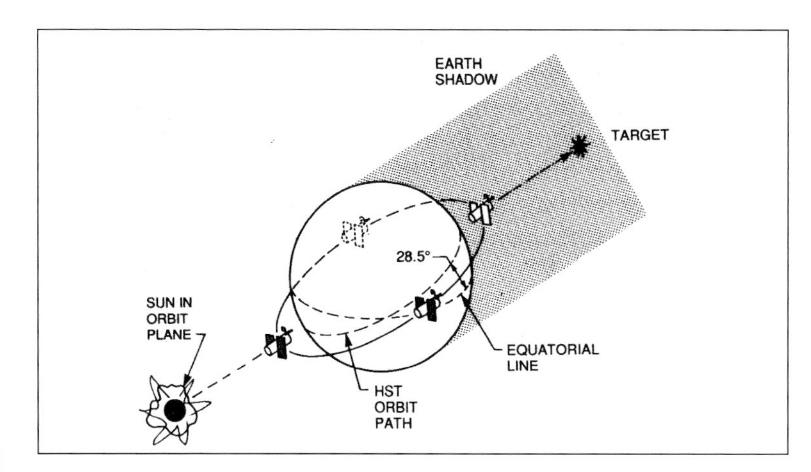

ABOVE The normal orbit for the HST places the plane of the orbit at 28.5°, which is the tilt of the Earth's rotational axis and therefore normalises the angle with respect to the Sun and maintains it in the solar plane. Faint-object viewing is best done when the Telescope is in the Earth's shadow. *(Lockheed)*

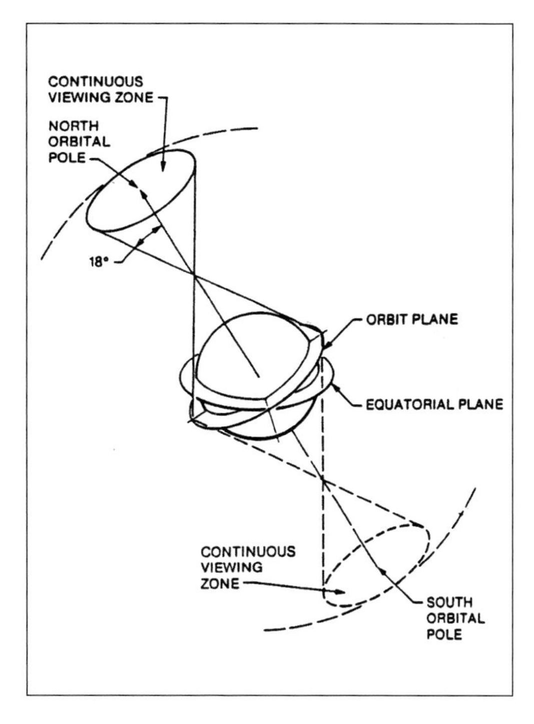

LEFT The Telescope completes one orbit in approximately 97 minutes with a maximum of 61 minutes in sunlight and a maximum of 69 minutes in darkness, depending on the cycle of the Earth's orbit around the Sun. For extended periods of observation lasting several hours, a zone of continuous viewing will exist up to 18° either side of the poles of the orbit plane. *(Lockheed)*

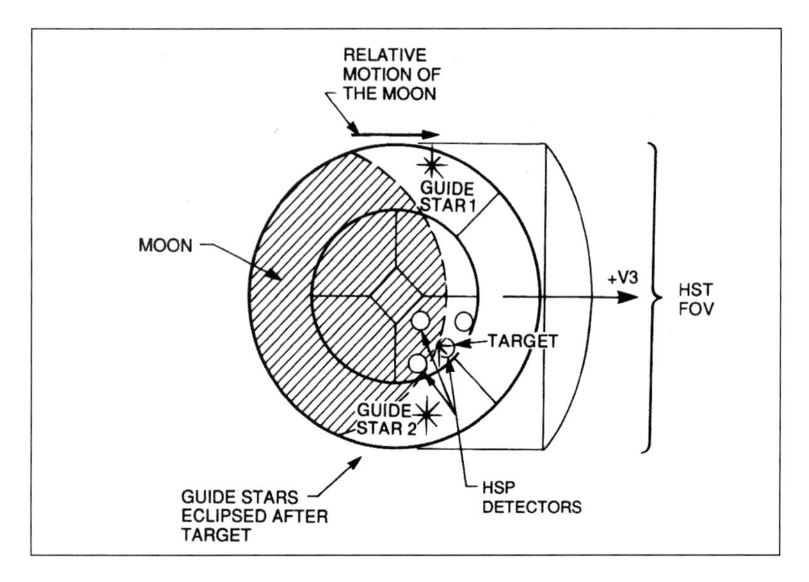

ABOVE Lunar occultation viewing is permissible when controllers unlock the Fine Guidance Sensors so that they do not, as programmed, occlude their view when the bright reflected light from the Moon comes within 10° of their centreline. In this way, and by using the phase between New Moon and Quarter Moon so that the occulting edge precedes the illuminated part, the lunar sphere can enhance an observational target. *(Lockheed)*

When the second attempt to deploy was made, it too was unsuccessful, stopping after 30 seconds. Facing high drama and with options appearing limited, consultation resulted in a command to switch off the tension limiter,

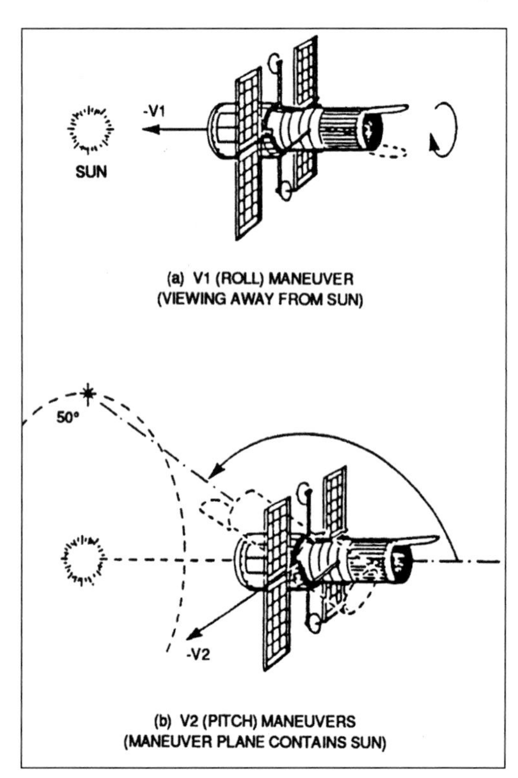

RIGHT Various manoeuvring profiles are available for selected directions. Viewing targets away from the Sun the reaction wheels roll the spacecraft as seen in (a), while for viewing up to the 50° exclusion zone the spacecraft will pitch up and roll around the radius of the prescribed circle. Manoeuvres are made at a rate of 0.22°/sec, taking 14 minutes to move through an arc of 90°. *(Lockheed)*

a force switch which would cut motor drive to fully unfurl the array if it overran a tension greater than 10lb (44.5N). With McCandless and Sullivan in their EMU suits, and anticipating a need to go to Hubble's rescue, they began to depressurise the airlock. It had been hoped to deploy the HST on orbit 19 but even a release by orbit 21 was looking increasingly unlikely and battery power was slowly bleeding away.

With the limit switch off a third attempt was made with a command to again try deployment, a signal going out at 28hr 25min elapsed time, 4hr 14min since going on battery power. This time it worked and word went to McCandless and Sullivan to stop depressurising the airlock – they would not be needed after all! Nevertheless, there was a well-rehearsed set of contingency plans had anything seriously affected the ability of the crew to release the Telescope. These techniques had been worked out using the full-scale simulator at the Johnson Space Center, and the Shuttle crew had spent many hours evaluating procedures and the operation of tools for such a contingency.

Several manual overrides had been built into the HST and these included switching on electrical power by entering the payload bay and locating the astronaut control panel inside the trunnion bay of the SSM equipment section. This activity would be a priority, as thermal control and activation of the Telescope's internal power was crucial before deployment. Alternatively, the Orbiter could reorientate itself so as to use shadow to minimise sunlight going directly on to the HST structure until power was effected. If the HST could not stabilise its own internal temperature the crew would immediately attempt to reconnect the Orbiter umbilical, allowing the STOCC sufficient time to work out a procedure. The HST is extremely sensitive to temperature; the primary mirror could be permanently damaged if the mountings contracted in the cold.

Another possible anomaly could be if the Orbiter's RMS, which was to be used to lift the HST from the Flight Support Structure, failed to attach itself to the grapple fixture by remote control. EVA astronauts could unlatch the Telescope and literally push it gently out of the FSS, taking care not to impart too much energy so that it moved quickly away before it could be confirmed that the activation sequences had been successful.

Before release the solar arrays and the high-gain antennas were to deploy but if that failed astronauts could use a ratchet wrench in a fitting on the primary and secondary drive mechanisms to crank the drive by hand until the array deployed and the wing extended. There would be anti-torque power tools available for erecting the arrays quickly if necessary. Alternatively, if a solar array wing behaved erratically a crewmember could jettison the array to prevent damage to the Orbiter or injury to the EVA astronaut.

A wide range of contingencies had been worked through in order to provide a lexicon of manual or computer-commanded procedures to rescue the HST from a perilous state. These required a variety of different EVA procedures, and the crew worked many hours in the neutral-buoyancy simulator at the Johnson Space Center, Houston, familiarising themselves with procedures and learning the techniques that they hoped would never be required. In most eventualities where an off-nominal procedure was needed, time would be crucial. For instance, if power could not be applied to the internal batteries they would be irreparably damaged if they discharged below a certain level, and they are the only means of power when the Telescope is in the Earth's shadow.

Most arduous and engaging of all contingency operations would be recovery of the Telescope in the event it could not be made to operate, a condition where it would be recovered, re-attached to the FSS and returned to Earth. In the best of circumstances this would be conducted from inside the Orbiter using the RMS and by reconnecting the power umbilical to put Shuttle power into the Telescope for temperature control. If that was not possible it would require extensive EVA activity involving the astronauts either stowing or jettisoning all the appendages, grappling the HST on to the FSS and securing it back in the payload bay.

But none of these contingencies were necessary. The moment of truth came in a command to release the 23,981lb (10,878kg) Telescope to free flight and Hawley let go the grapple fixture and slowly withdrew the RMS arm back to a standby position above *Discovery*'s payload bay. The Hubble Space Telescope was released at 20:39 UTC on 25 April over the eastern Pacific Ocean at an elapsed time of 32hr 5min, fully 6hr 54min after going on internal power from its six batteries. At the time of release *Discovery* was flying with its underside facing the direction of flight, its nose pointed toward the Sun and its main engines pointing down toward the Earth.

Just 30 seconds after release the Orbiter's RCS thrusters were fired in a 0.6ft/sec^2 (0.18m/sec^2) burn which took *Discovery* 250ft (76m) away from the Telescope before arching 600ft (183m) above the HST as the two spacecraft crossed the southern coast of South America. Further tweaks to the orbit put the Shuttle in a station-holding position just 40 miles (65km) away from the Telescope, where it remained. The most likely reason for it being called back would have been if the aperture door on the Optical Telescope Assembly had failed to open, in which case a re-rendezvous would have been effected for McCandless and Sullivan to try a manual fix.

It had been a history-making mission which had successfully seen deployment and workaround procedures for a few heart-

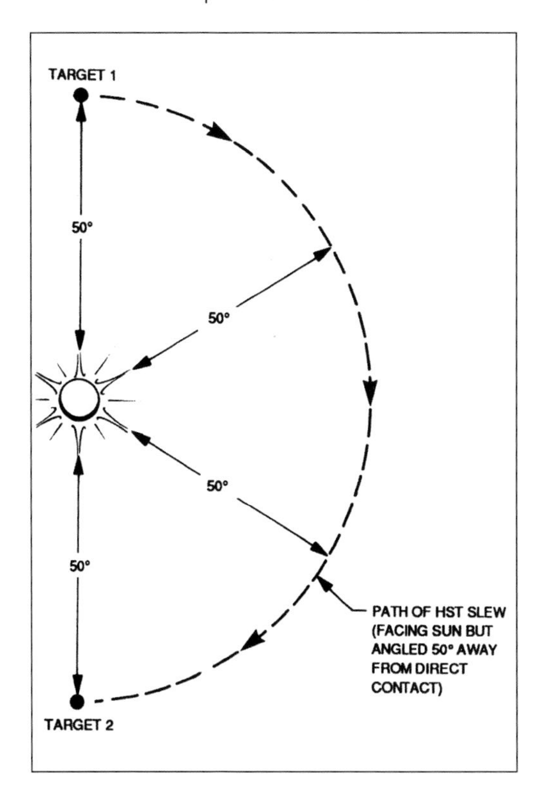

LEFT When viewing two targets 180° apart but close to the Sun exclusion zone, the HST follows an imaginary circle to prevent damage to the optics and locks on to the second target, avoiding a direct pitch up or down directly to the second objective. *(Lockheed)*

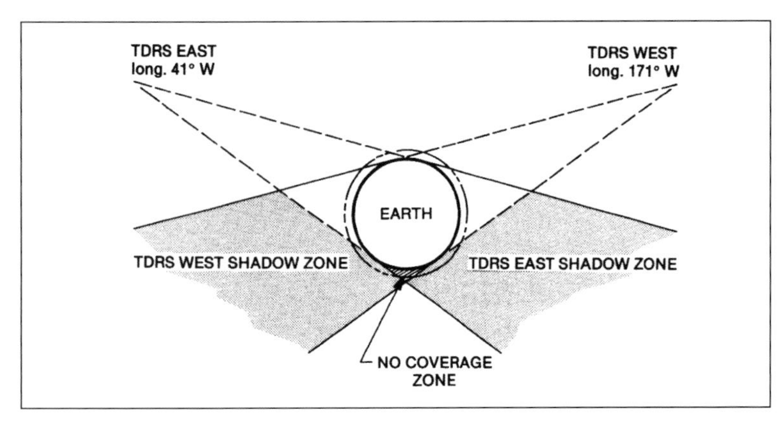

ABOVE Data transmitted to the ground can go via the Tracking and Data Relay Satellites, which at the time the HST was first deployed from STS-31 were located at 171°W and 41°W. *(Lockheed)*

BELOW Sullivan (left) and Shriver at work on the flight deck of *Discovery* with cameras and checklists. It was from the control stations behind Shriver that the RMS was commanded to raise and deploy the Telescope. Note the folded pilot's seat to Sullivan's left. *(NASA)*

stopping moments during deployment. Unfortunately this was only the start to a series of problems that would beset the Hubble Space Telescope almost from the outset. Immediately after deployment a series of minor anomalies began to appear, not least a problem with deployment of the No 2 High-Gain Antenna. It was a problem with this that initially caused concern because it proved impossible to move communications to the HGA system and increase the data transmission rate to begin checkout tests.

Initial communication was effected through the low-gain antennas, but when attempts were made to move data at the 500 bit/sec rate the system shut down and the STOCC was unable to regain communication for three hours. The complex but necessary fail-safe modes written into the Telescope's software made it difficult to overcome the natural tendency of the Telescope to save itself from anomalous commands which, due to the difficulties encountered, were valid but not understood as such by the electronics.

Because it was believed that the pre-programmed sequence of tests might compromise the operation of the Telescope the opening of the aperture door was moved up as a priority, just in case an EVA operation was necessary; *Discovery* had a limited time it could remain on station. Late in the evening of 26 April, with engineers and analysts working round the clock to solve the emerging problems, a plan was raised to open the door and the command went up at 05:00am on 27 April. During this process Goddard again lost contact and failed three times to establish links through the TDRSS satellites. Finally, the Telescope responded through the low-gain antennas and data showed that the Telescope had again entered a safe mode, but the door remained shut.

Controllers were concerned that the safe mode appeared to be triggered by excessive torque loads when the High-Gain Antennas slewed to find the TDRSS system. The controller decided to fool the Telescope into moving into a safe mode that preserved its ability to maintain contact if that should happen again, and this proved to be a fortunate decision. Unfortunately, when the command went up again to open the door the mechanical action caused two of the four primary attitude control gyroscopes to drop off line. This resulted in the software assuming that a major malfunction had occurred, and it commanded the Telescope to go into a 'software safe-mode' hold which was a much more complex problem to counter.

The earlier software command changes induced by Goddard allowed the spacecraft to manoeuvre to its earlier inertial safe mode when the software safe-mode commands had been issued automatically. Although the HST was still badly affected by HGA communications problems – none of which could be helped by the Shuttle remaining in the vicinity – it was at that point, with the aperture door open, that *Discovery* was released to go about the remainder of its mission and prepare for re-entry. The software fix required about ten hours of work to deliver that to the STScI, which then took a further 24 hours on the ground to integrate that data into a new computer command load for transmission to

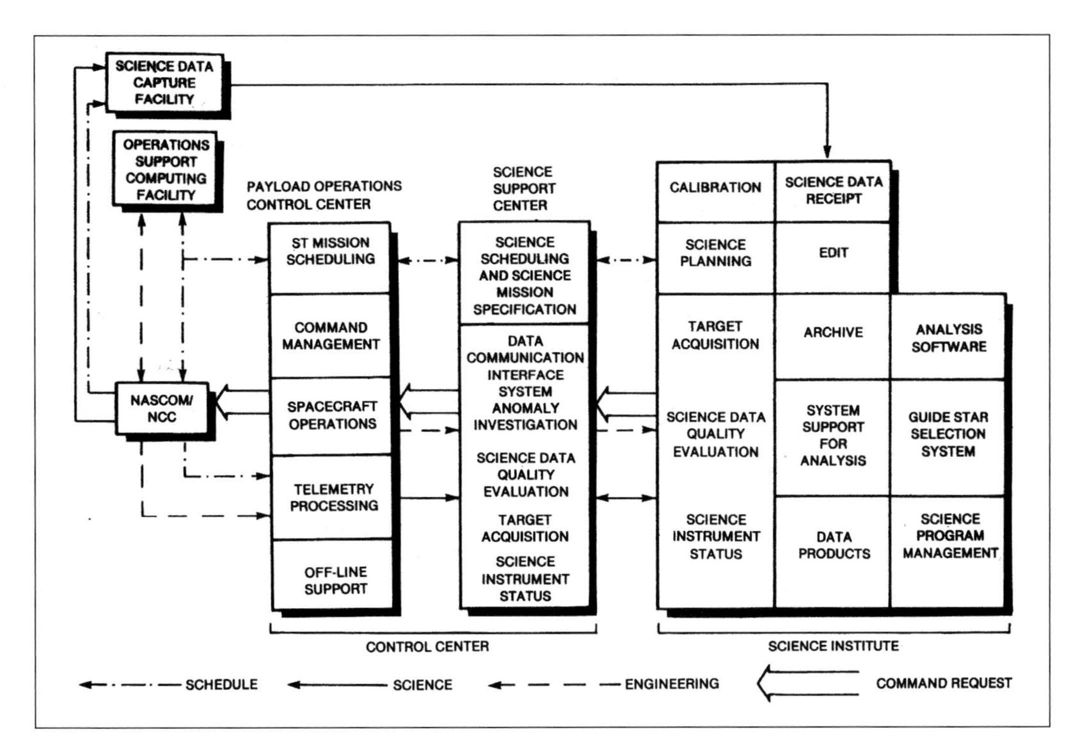

LEFT As the HST swung into action, processing the many requests for observer time and data capture went through a rigid process which began with a formal request for time and a slot and passed through a series of action-gates to the control centre. *(Lockheed)*

the Telescope. By late on 28 April the HST had been reconfigured and was out of its software safe-mode condition.

When the Shuttle returned home on 29 April it was precipitated by the longest OMS burn to date, a 4min 48sec firing that had been calculated to burn off 11,000lb (5,000kg) of propellant which had been on board in case a variety of rendezvous manoeuvres had been required under contingency circumstances. The burn took 573.3ft/sec (174.7m/sec) out of the Orbiter's velocity, bringing it down to Earth for a touchdown at Edwards Air Force Base in California on Runway 22 at an elapsed time of 5d 1hr 16min 6sec from a record altitude, which called for a slicing re-entry through the atmosphere carrying it 4,753 miles (7,648km) around the circumference of the Earth.

Meanwhile, on the HST, engineering data transmitted to the ground through the HGA indicated that there was a serious mechanical problem with getting it to point in the required direction. Photographs taken by the closeout crew at the Kennedy Space Center just before *Discovery* had been launched showed a wire bowed slightly out of place on the No 2 dish. A major engineering effort swung into action at the Goddard Space Flight Center and at the Huntsville Operations Control Center, at Lockheed and at Honeywell, as a solution was sought.

David Skillman of the GSFC built a model of the antenna to see if he could duplicate the problem. Using Tinkertoy construction kit pieces and a lamp cord, he worked with engineer John Decker to duplicate the problem, and the model was verified by computer graphic projections at Lockheed. This data was used to devise software that would move the antenna through most of its arc but avoid the wire that was rubbing against the counterbalance assembly. The software was written and transmitted on 30 April to reconfigure the way the antennas moved. It worked.

By 1 May normal operations were beginning to converge the modified test and verification plan with the script written, never in concrete, before the mission began – until a new problem emerged when, at about midnight UTC time, the aperture door closed unexpectedly, but as it is supposed to if the Telescope is orientated too close to the Sun. This triggered the same software safe-mode reaction as it had previously when a jitter from the No 2 HGA antenna had caused such delays and lengthy workaround procedures. Alerted this time by the same corrective procedure, controllers managed to get the door open again after about five hours. Had it not, the Telescope would have been inoperable until the first servicing mission, in 1991 or 1992 at the earliest.

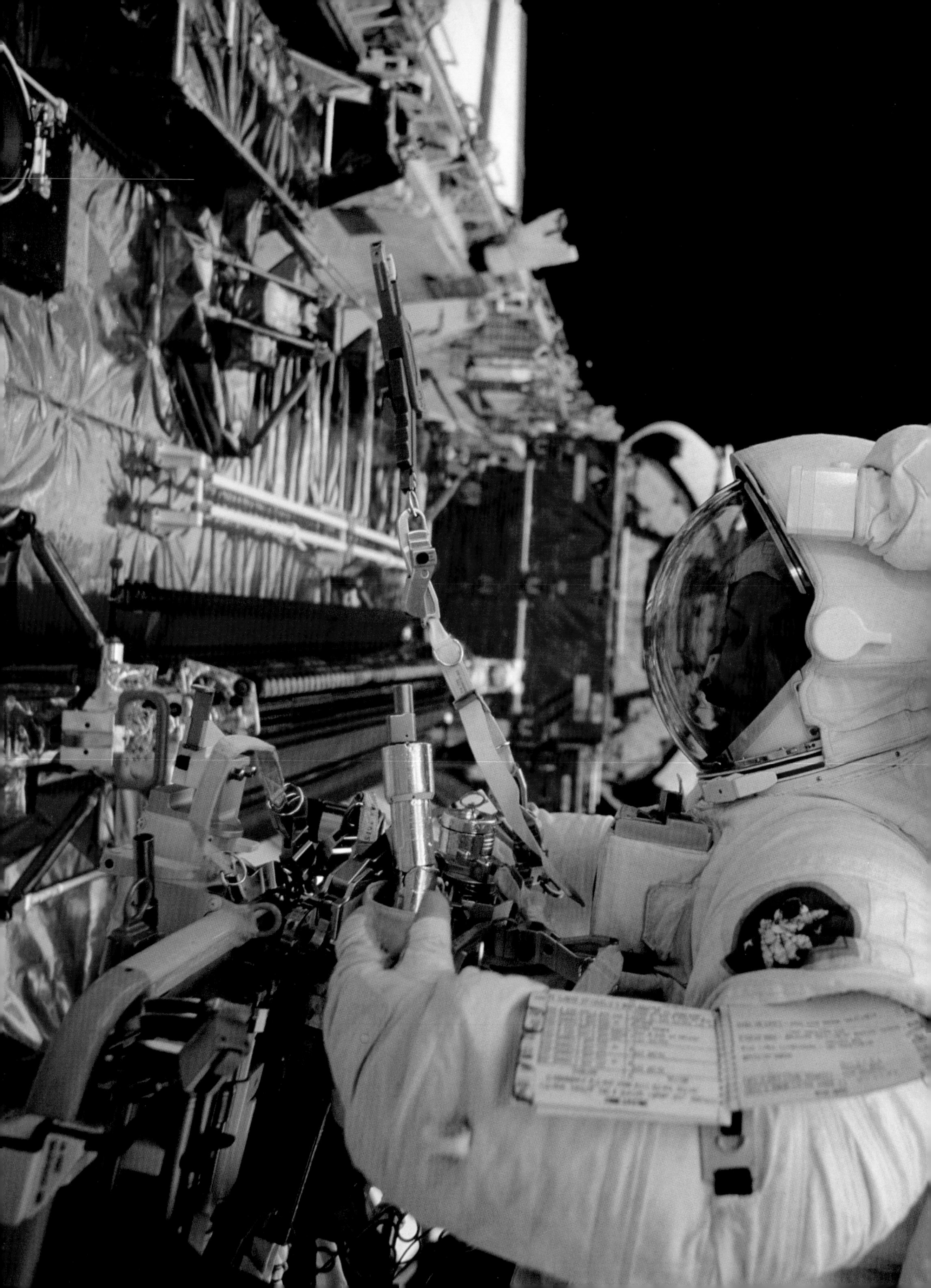

Chapter Seven

Corrective optics

When the Shuttle returned to Earth after deployment of the Hubble Space Telescope, hopes were high that a decade and more of development work and fabrication would usher in a new age for astronomy. However, teething troubles encountered in the early phase of the Orbital Verification phase were a foretaste of greater problems to come – issues which could be traced back to the fabrication and grinding of the primary mirror. This would change significantly the plan for getting the HST operational and delay the demonstration of full capability by more than three years.

OPPOSITE With checklist on the left sleeve, Kathy Thornton prepares the new solar arrays for installation on the Telescope. *(NASA)*

When the HST was launched the OV phase had been expected to last around three months, although with a new operating system in orbit that was always only a generalised estimate. The Science Verification period was expected to last a further five months, with the Telescope turned over to the astronomical community by the end of the year or early 1991. The latter stages of the SV phase would allow some of the Guaranteed Time Observer programme to begin, in which some degree of astronomer-participation was introduced.

On 20 May 1990, at 3:13pm UTC, the first image – known as 'first light' – was taken, a one-second exposure of the open-star cluster NGC 3532 from the Wide Field and Planetary Camera. A second exposure, lasting 30 seconds, was taken two minutes later. When received on the ground the images were showing a resolution of 0.7 arc-sec versus the 1.5 arc-sec required by the specification, but hopes were that an improvement over Earth-based observatories would be achieved by a factor of seven. Astronomers elated at the discovery that what appeared as a single point of light in Earth-based telescopes was now revealed to be two stars, realised that something was wrong.

First light for the Faint Object Camera came on 6 June with the first images on the 17th, and it became clear that what was blurring the image from the required specification was a classic case of spherical aberration. The specifications required the light to fall within a 0.2 arc-sec area and measurements were showing 1.4 arc-sec. About 15% covered 0.1 arc-sec with the remainder spread across 1 arc-sec, which disappointingly was only just equivalent to ground-based observations. The target of 0.1 arc-sec could only be achieved by computer processing for objects seen against a dark background. Low sky background at shorter wavelengths meant that ultraviolet observations were largely unaffected and the visible work suffered most.

Offering up its first images on 7 June, the WF/PC suffered worst, eventually taking only 10% of the observation time compared to the 40% planned; it was not worthwhile using it for many of the targeted observations planned. One of the primary objectives with the Hubble Space Telescope was to measure the expansion rate of the universe, and to that end observation of the Cepheid variables in distant galaxies was an essential requirement of the WF/PC; but serious improvement over ground-based measurements would have to await the installations of the WF/PC II (see 'STS-82' later in this chapter) in December 1993.

Further problems cascaded down on to the other instruments. Because of the lack of any improvement over ground-based telescopes, the Faint Object Camera was used only half the time it had been expected to operate, but the High Resolution Spectrometer and Faint Object Spectrograph operated well in the ultraviolet while the High Speed Photometer and Fine Guidance Sensors were largely unaffected. All four of these latter experiments, however, were unable to begin serious observations before the end of 1991 due to procedures worked out to enable some observations to be undertaken by the other instruments.

The state of affairs was far from satisfactory and on 21 June the project manager announced that the Telescope was unable to focus properly, negating the majority of its objectives. This took the astronomical world by storm and a fierce tirade of criticism and condemnation broke out. Hailed by its protagonists as the most important step forward in astronomical science since Galileo and by others as the greatest waste of money they could imagine, bitter disappointment ensued.

BELOW The honeycomb core of the Hubble Space Telescope backup mirror provided an analogy for the imperfections in the primary launched more than three years earlier. *(NASM)*

On 2 July 1990 an independent review body met under the chairmanship of Dr Lew Allen, director of the Jet Propulsion Laboratory, to find out what had happened. Over the next five months the board met and closely examined every minute piece of information from the design and assembly of the HST which might cast light on the matter, going over the records and the test data, examining the specification of the test equipment, evaluating procedures and investigating anomalies which appeared to have been minor and were disregarded, or to have been significant and were corrected.

It did not take long to discover that the primary mirror had been well figured and correctly ground by Perkin-Elmer but that it had been manufactured to the wrong shape. The Reflective Null Corrector laser interferometer that had been used to check the figure in 1981 was discovered to have had a lens 1.3mm out of position, and different testing which revealed that fact at the time was disregarded and ignored. It transpired during the review board's examination of the facts that contractual competitor Kodak had included an end-to-end test in its failed 1977 bid, for which there was no provision in the contract with Perkin-Elmer.

A second matter arose during the investigation which aired concerns from NASA senior managers that because Lockheed Martin had been working on highly classified reconnaissance and surveillance (spy) satellites only a small handful of NASA engineers and quality control inspectors were given clearance to enter the manufacturer's facility and maintain a check on the work and on the procedures. At least one manager told the board that he considered it a fundamental flaw in the overall management of the project, and that this was one of the principal reasons why the HST had the flawed mirror. The contractor implied in private that by seeming to move the blame to its industrial partner, this was just a way of NASA distancing itself from an embarrassing situation.

Whatever the truth in the detail, the principal problem would not go away by public explanation, words or with rewritten software. But there was an evolving series of workaround procedures which were gradually implemented to produce at least some valuable science as the astronomical world waited for the repairs that would be incorporated into the first servicing mission planned for late 1993. This servicing mission had been planned for that time since before the launch of the HST and was to have lifted the WF/PC II instrument to replace the original as well as install new solar arrays (about which more later).

As for astronomers, early observations of supernova SN87A and its ring of ejected matter using the FOC showed that resolution of 0.1 arc-sec could still be achieved under some circumstances, and further observation of the double stars R Aquarii (one a red giant, the other a white dwarf) confirmed that, together with much nearer targets such as Pluto and its moon Charon. Opportunistically, the WF/PC made significant contributions to studies of the planet Jupiter and the first new storm in its atmosphere since 1933. This disturbance first appeared in September 1990 and the inability to carry out some observations of distant targets afforded an opportunity to study this new phenomenon.

Satisfying too were the steps taken to carry out one of the Telescope's primary objectives: accurate measurement of the distance to the Large Magellanic Cloud. Previous measurements using ground-based telescopes gave a value of +/-30% but observation of SN87A provided a distance of 169,000 light years, +/-5%, a much more refined measurement.

Throughout 1991 measurements were made using workaround procedures for extracting as

BELOW Working out techniques for changing the Wide Field/Planetary Camera, the first major replacement instrument on the HST, using the NBF at the Johnson Space Center. A complete change-out took 4hr 15min in training. *(NASA)*

RIGHT The payload bay in *Endeavour* contained the Flight Support Structure at the aft end, the Spacelab (ORUC) pallet with the COSTAR instrument and the Wide Field/ Planetary Camera II forward, and the Solar Array Carrier and pallet at the front. *(NASA)*

much value as possible out of the Telescope. The High Resolution Spectrometer was able to carry out highly valuable work on measurement of the deuterium content of the universe, seeming to support the Big Bang theory, and it observed evidence of planetary formation around the star Beta Pictoris. The HRS also provided some puzzling evidence for large hydrogen clouds that should have evolved into galaxies billions of years ago. In satisfying another primary objective, that of searching for black holes, the WF/PC was able to show evidence for one 2.6 billion times the mass of the Sun in a region up to ten times that of the solar system, at the centre of galaxy M87.

One intriguing discovery during the early period of observation in 1992 was that many stars appeared to have planet-forming discs of accreting material. The WF/PC found evidence for at least 15 in the Orion Nebula alone. This prescient indication would result in a major search for such exoplanetary systems that would stimulate a wide and expanding search using dedicated spacecraft, and the confirmation that most stars – however bizarre their properties – have planets around them.

As in so many areas, the work conducted by the HST led to new branches of astronomical activity encouraged by questions posed through its results and the fresh questions they raised. While the Hubble Space Telescope was able to answer many questions and both pose and solve others, as with many branches of science it was knowing the right questions to ask that brought definitive results, and sometimes from new spacecraft developed specifically as a result of the research carried out by the Telescope itself. Even without correcting the deficiencies introduced in its manufacture, the HST was doing valuable work and achieving creditable results.

But as mathematical calculations permitted temporary corrections to get useful data from some instruments, the inevitable limitations caused by the spherical aberration built into the primary mirror at birth would require an engineering fix at the Telescope itself. Some form of corrective lens would have to be installed to realise the potential of the Telescope, and the anecdotal attribution of the solution is that it came to Jim Crocker, an engineer with Lockheed, while taking a shower during a visit to their ESA colleagues in Germany. The showerhead distributed the different jets of water in a broad spray that spread the single source equally to the full circumference of the outlet plate.

Between August and October 1990 a strategy for the recovery of the HST was formulated in a special study led by Robert Brown and Holland Ford that analysed the situation and summarised what would be required to put it right. It confirmed that the surface of the mirror 'is too low by an amount from the center to the edge that grows from zero to 0.002mm or four wavelengths of optical light'. It confirmed the true cause of the anomaly as explained above and that the 'light rays come to a focus at a different distance depending on the radius at which the rays strike the mirror…light from the edge of the primary mirror comes to a focus about 38mm beyond where the innermost rays converge'.

The limitation imposed by this meant that only about 15% of the image has a resolution of 0.1 arc-sec instead of the 70% expected. The remainder is cast about in a halo of about 3 arc-sec and, since aperture diffraction sets the size of the image core, the size is smaller at shorter wavelengths; but because aperture diffraction is set by geometrical optics the size of the halo remains constant.

At the time the strategy was being formulated NASA already had second generation instruments in development as part of the plan to periodically upgrade the HST with later and more sophisticated equipment. These were the Space Telescope Imaging Spectrograph (STIS), the Near-Infrared Camera and Multi-Object Spectrometer (NICMOS), and the Wide Field/Planetary Camera II. The WF/PC II, which had been under development since 1985, had been built as a clone of the WF/PC installed for launch and could be readily configured to compensate for the spherical aberration in the primary mirror.

The recovery plan scheduled the WF/PC II to be installed on the first servicing mission, leaving the STIS and NICMOS instruments to be carried on the second servicing mission. The WF/PC it would replace would be removed and returned to Earth in the Shuttle. It was the only instrument not aligned with the optical axis of the Telescope and was therefore incapable of benefitting from an engineering fix to the axial instruments.

To get all the other existing instruments back on specification, the panel recommended development of a new component, the Corrective Optics Space Telescope Axial Replacement (COSTAR), a device that would employ corrective optics. However, the COSTAR package would have to be installed in the bay used by one of the other axial science instruments, and the one chosen for removal was the High Speed Photometer, about the loss of which principal investigator, Dr R. Bless of the University of Wisconsin, lamented to Nancy Grace Roman, former head of scientific activities in NASA's astronomy programme: 'What wonderful results we could have obtained with the improved image quality!' As it was, the HSP was sacrificed to give the Hubble Space Telescope a corrective monocle!

The strategy panel's report was presented to Dr Riccardo Giacconi, Director of the Space Telescope Science Institute, on 18 October 1990, who promptly endorsed the recommendations and took the findings to NASA where a formal presentation was made on 26 October. NASA examined these recommendations closely and conducted a feasibility study of the engineering potential for the COSTAR and of the budget consequences of developing the corrective optics. By the end of the year it had formally approved development and sanctioned the first servicing mission (SM1) for December 1993.

Ball Aerospace began work on COSTAR in a contract dated February 1991 and finalised in October that year at a value of $30.4 million. The STAR component of the corrective optics package was in fact already in development for the HST to support updated axial instruments being considered for later servicing missions, so the development time was not as great as would have been the case with an all-new concept. Instead the existing device was adapted for the unexpected work now required as an essential prerequisite for normal operations. ESA had been considering requesting NASA for EVA time to do an in-orbit update to the Faint Object Camera with a new instrument, but at a cost of $80 million it proved too costly.

While strategy panels were working out what to do to get the HST back in focus, another problem appeared in the first few months of testing. During the Orbit Verification phase the fine pointing ability was marred by two thermal flexing motions revealing inherent design flaws with the original, albeit modified, set installed before launch.

In extensive tests and analyses using data from the sensors and the arrays (there were no cameras to show the precise effect), it was determined that one flexing was a 0.1Hz end-to-end motion lasting ten minutes after crossing the terminator leaving daylight for the night-side of the Earth. A transverse flexing of 0.6Hz, also lasting

BELOW COSTAR was developed over an extensive period from 1990, shortly after it was discovered that serious imperfections existed in the shape of the primary mirror. This cutaway shows the optical bench deployed with the pick-off mirrors in position to capture light from the secondary mirror and direct it to the four axial instruments. *(Lockheed)*

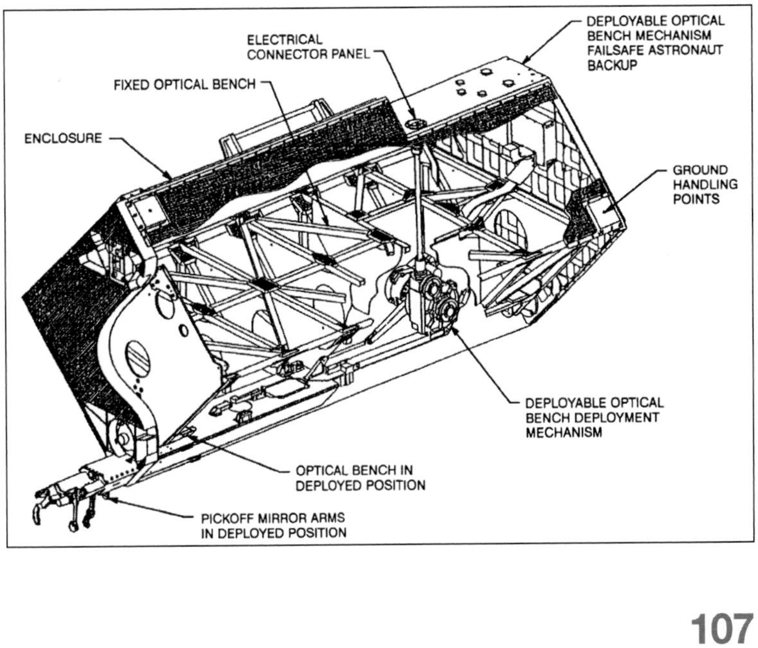

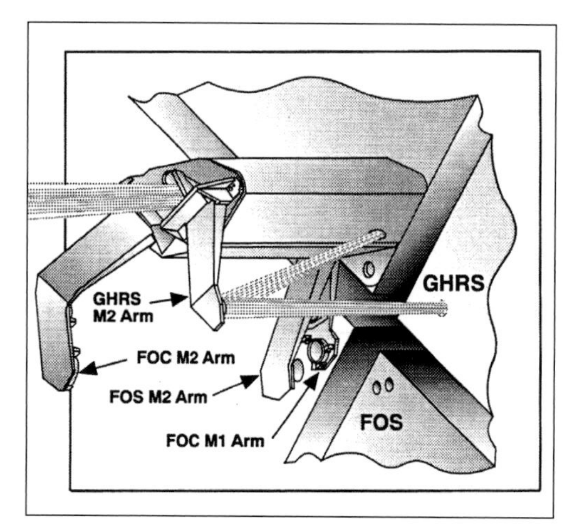

RIGHT The COSTAR mirror arms and light path for the Goddard High Resolution Spectrograph. The M1 mirrors for the Faint Object Spectrograph and the GHRS are inside the Deployable Optical Bench. *(Lockheed)*

up to ten minutes, occurred after crossing back into sunlight with the arrays flexing up to 3ft (1m) at the tips under worst conditions.

This vibration caused a twang effect that prevented long exposures and could only be partially compensated by software fixes which took up large amounts of memory. Replacement arrays were ordered with a critical design review by ESA held as early as January 1992.

On 3 December 1990 the No 6 rate gyroscope failed and its functions were taken up by one of the five remaining gyroscopes. Then No 4 failed in July 1991, which, like the first, was believed to have been due to high solar radiation. Soon after, No 5 gyroscope showed signs of wear, and this was attributed to dirt in its rotor mechanism. Only three operational gyroscopes are necessary but software was written for a two-gyro system as a contingency against further failure. It was decided to replace the two three-gyro packages during the first servicing mission.

Then, in July 1991, the Goddard High Resolution Spectrograph showed fluctuations in its power supply, and this was added to the inventory with a repair kit being developed so that an EVA astronaut could perform a manual repair job in space.

Added to the payload manifest for the SM1 mission were the Fairchild DF-224 coprocessor and a memory modification kit. One of the DF-224's six memory units had failed and another had partially failed, but as only three were required to operate the system it was not an urgent problem but one that had to be addressed by the first servicing mission. Also in *Endeavour*'s payload bay would be a 608lb (276kg) IMAX camera to film events during the EVAs, and a second IMAX, weighing 329lb (149kg) was carried inside the Shuttle for hand-held filming of events through the two aft-facing windows.

RIGHT COSTAR in the Payloads Hazardous Servicing Facility at the Kennedy Space Center on 13 September 1993. *(NASA)*

COSTAR

The design and development of the Corrective Optics Space Telescope Axial Replacement was a work of engineering art coupled to science craft and was the brainchild of Jim Crocker at the Space Telescope Science Institute. Developed and built by Ball Aerospace, it had been designed by optical scientist Dr Murk Bottema, who did not live to see his instrument installed on the Hubble Space Telescope in orbit. Over 30 years Dr Bottema had built more than 100 science instruments for NASA and Ball placed a small plaque on COSTAR in recognition of his many contributions.

The main structural components of COSTAR are contained within a protective enclosure 3ft x 3ft x 7ft (0.9m x 0.9m x 2.2m), exactly the same size as the High Speed Photometer case that it was built to replace, and weighing 660lb (299kg). Internally, it consists of a Fixed Optical Bench (FOB), a Deployable Optical Bench (DOB) and latches to attach the instrument case to the Telescope. The mechanical elements comprise the DOB, hinged arms that position the set of corrective mirrors, and actuators for fine adjustment for optical focusing. The optical subsystem comprises five pairs of corrective mirrors.

Light from the Telescope's secondary mirror falls on the first mirror in each pair of mirrors before it reaches the science instruments and is redirected to the second mirror in each pair (M2), which then directs it to the instrument's aperture. The mirrors were built to exacting tolerances with a roughness no greater than 10Å from the prescribed shape. The surfaces are highly polished, reflecting 78% of energy at the red end of the spectrum and 56% at the ultraviolet end. Each mirror is circular and approximately 0.787in (20mm) in diameter.

The M1 mirrors have a relatively concave spherical shape and the curvature of the M2 mirrors is steeper than the primary mirror to compensate for its spherical aberration. Because the primary mirror is 2 microns thick at the edges relative to the centre, light from the primary must map exactly on to the edges of M2 so that the extra 2 microns of path length from the primary mirror to the focal point are offset by the missing 2 microns between M2 and the focal point. In this way, light from every region on the primary mirror will come into focus at the same point and the aberration will be neutralised, even though the M2 mirror is 200 times smaller than the primary mirror.

The M2 shapes are known as anamorphic fourth-order aspheres on toroidal blanks and are mounted on small mechanical arms attached to the DOB, which emerges in an opening through the module. The optical bench acts as a robotic arm, extending the mirrors into the focal plane in front of the science instruments. Small actuators tilt or tip each M1 mirror to centre the image of the Telescope's primary mirror into the corrective M2 mirror. The Faint Object Camera and the Faint Object Spectrograph each have two sets of mirrors because each instrument has two operational modes with separate apertures for each. The Goddard High Resolution Spectrograph has only one set of mirrors.

The COSTAR instrument is a complex device, and packaging the four hinged arms, the 10 mirrors, the 11 actuators and heaters, together with the wiring and sensors, into a container with such a small cross-section was a major design challenge, and one met with considerable skill and ingenuity. The design is very modular. Individual parts were assembled in a hierarchy of subsystems and each small mirror cemented into a bezel that acts as a collar and holding fixture. The bezels were mounted on the small mechanisms that contain the fine adjustment motors and these mechanisms are attached to the Deployable Optical Bench, which in turn is installed in the Fixed Optical Bench. The fixed bench was mated to the enclosure and the latch system. In all, the instrument has approximately 5,300 parts and was managed by Paul Geithner of NASA's Goddard Space Flight Center.

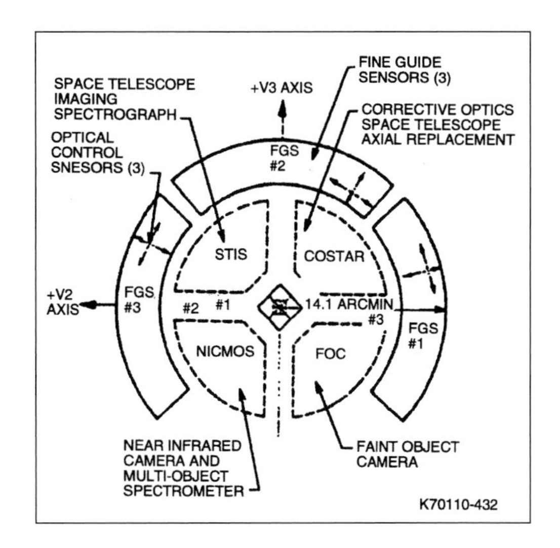

LEFT Acquisition instrument apertures for the four axial instruments and the Fine Guidance Sensors. *(Lockheed)*

BELOW COSTAR undergoes tests in Building AE at Cape Canaveral Air Force Station on 2 September 1993. *(NASA)*

BELOW The WF/PC II camera as viewed from one quadrant. *(Lockheed)*

BOTTOM Handrails and special handling links are attached in this view, which is rotated 180° to the previous view, hence the +V2 side seemingly being on the opposite side. *(Lockheed)*

Solar Arrays II (STSA-2)

Replacement solar arrays wings were developed by British Aerospace for the European Space Agency as part of its contribution to the Hubble Space Telescope. With dimensions identical to the first flight set installed for launch, STSA-2 addressed the jitter problem that had caused problems for focusing the Telescope. Each boom was covered with an aluminium layer supported by 900 aluminised fluorinated ethyl propylene/Teflon discs in an accordion-like structure, reducing the thermal gradient by a factor of 20.

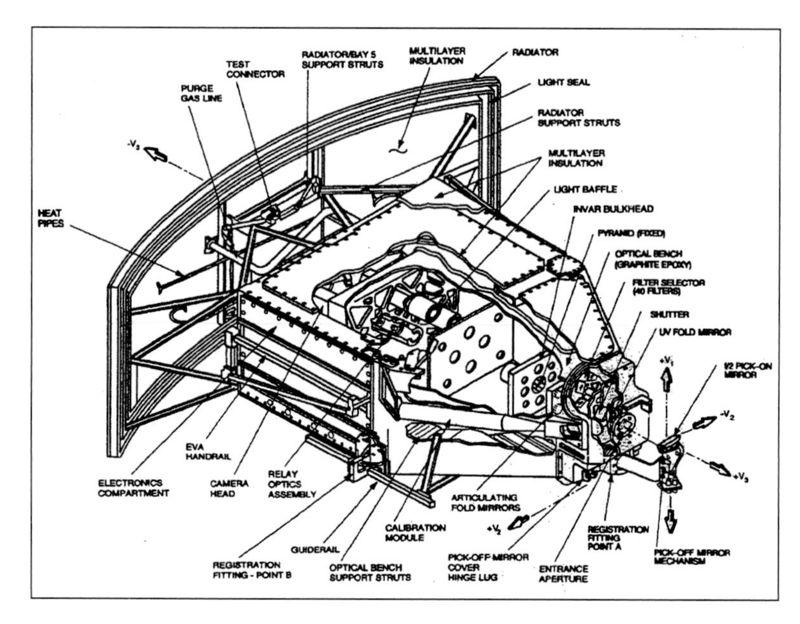

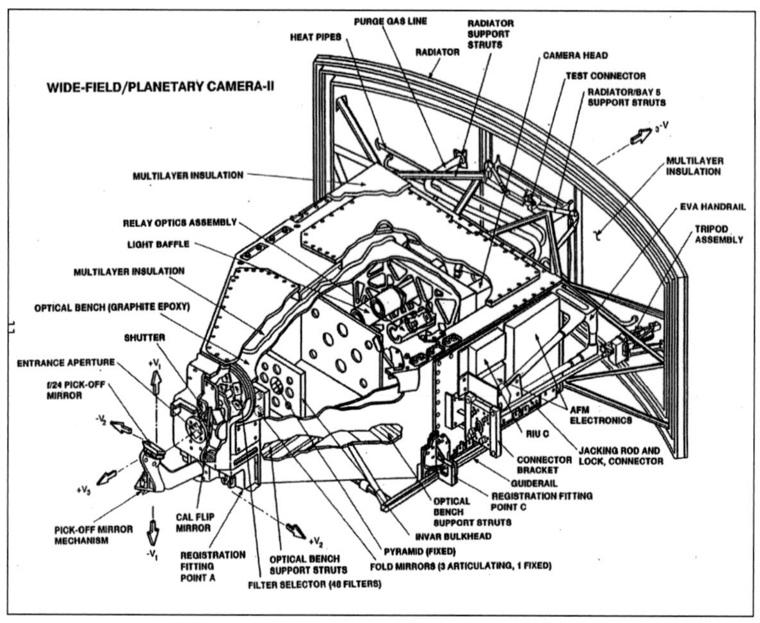

A second change substitutes a structure for the mechanisms that link the bistem booms to the arrays wings. The original pulley system had been replaced with an all-spring system with nine coil springs on each wing, with which there would be almost no likelihood of a stick-slip reaction causing vibration.

A system for countering expansion at the array tips was replaced by one of frictionless springs, and the storage drums were immobilised by an electric brake mounted on the outboard end of the drum. This would stop temperature changes inducing rotation movement of the drum in orbit. The drum could be switched on and off by ground command. In tests they were found to reduce the jitter effect by a factor of 20. A little heavier than the original solar array assemblies, the STSA-2 complement for the first servicing update weighed a total of 702lb (318kg). It was judged from additional tests that these arrays should be good until at least the year 2000.

During the redesign of the solar arrays and their deployment and stabilisation method, engineers at British Aerospace added a new handling system that aimed to make in-orbit replacement easier than it would have been with the old system. New inboard and outboard handles were fixed at each end of the drum in which the arrays had been stowed and a set of detachable handles was included to allow the astronauts to move the arrays around more easily.

A new Solar Array Drive Electronics (SADE) assembly was carried up on SM1, units designed to transmit positioning commands to the wing assembly. One of the drive units failed when transistors overheated and a replacement SADE, provided by ESA, restored that capability and provided better thermal control. Two SADE units were mounted on the inner door face to one of the service bays but only one was to be replaced on this mission. After opening the door, only six bolts needed to be loosened to free the old SADE unit and its quick-disconnect electrical fittings and replace it with SADE II.

Solar Array Carrier

New for the first servicing mission was the Solar Array Carrier (SAC), which was situated in the forward section of the Shuttle

RIGHT WF/PC II had three wide-field cameras and one planetary camera instead of the eight in WF/PC I. This was determined by the science team so that they could develop a system to align corrective mirrors on-orbit. Only one of four light trains is illustrated here. *(Lockheed)*

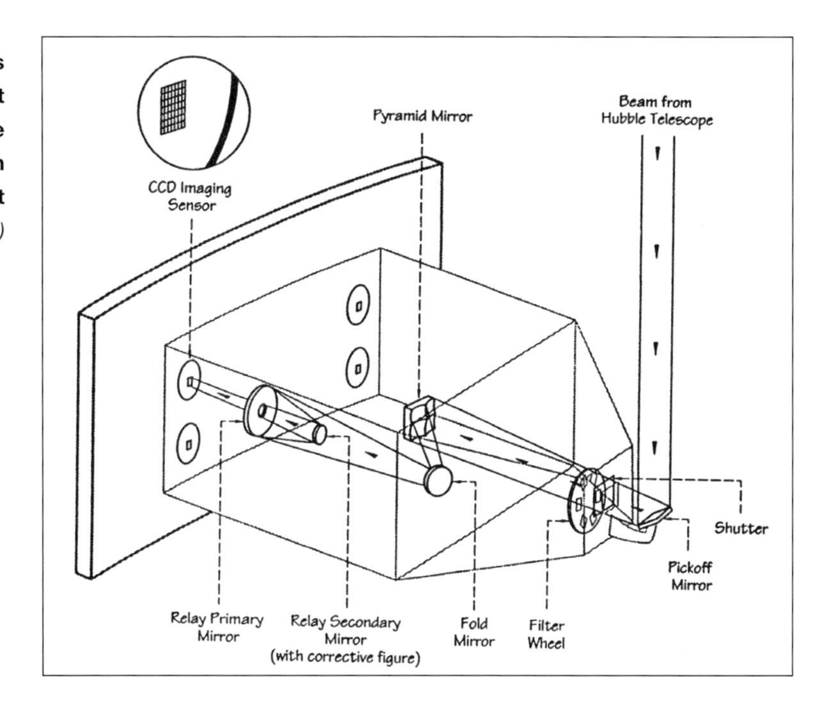

Orbiter payload bay. With a weight of 3,829lb (1,737kg), it was there to carry the STSA-2 arrays and to provide a support for the two existing arrays which were expected to be returned after removal from the HST. The SAC also had a temporary stowage location where an EVA astronaut could temporarily 'park' the old arrays before placing them in the fully stowed position prior to returning to Earth.

The SAC was constructed in the form of a flat pallet secured in the payload bay by two attachment pins at each side and a tripod keel with a central attachment pin securing it to the floor of the bay. A load isolator system held above the flat pallet supported the latching mechanisms for the solar arrays and an additional protective enclosure was attached to one side for the new SADE II on the way up and the original SADE unit, removed from the HST, on the way down.

Wide Field/Planetary Camera II

In development as an eventual replacement for the original WF/PC (see Chapter 4), the updated instrument incorporated enhancements such as upgraded filters, advanced detectors and improved ultraviolet performance. Although the original WF/PC had been the most used of all instruments on the HST, it suffered to some degree from the spherical aberration of the primary mirror and the replacement incorporated an integral compensation to further improve the resolution.

In WF/PC II four charge-couple device cameras are arranged to record simultaneous images in four separate fields of view at two magnifications. In three of the four fields, each detector occupies 0.1 arc-sec and each of the three detector arrays covers a square of 800 pixels on a side, as with the WF/PC. The Wide Field mode operates with a focal ratio of f/12.9, as does its predecessor, but the Planetary

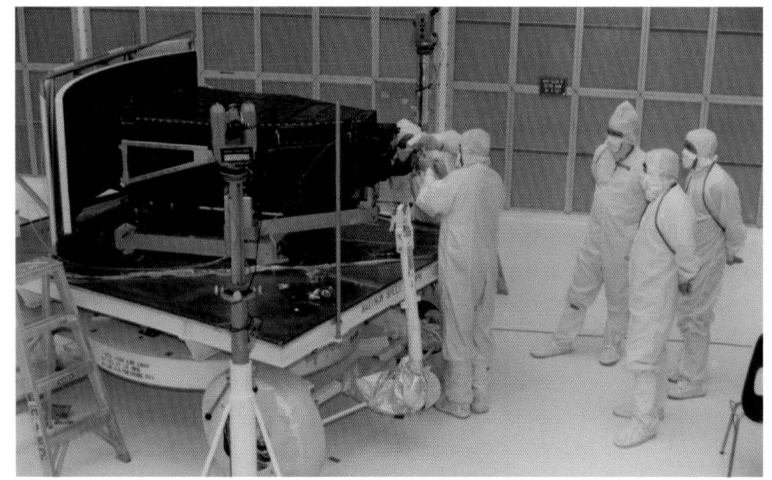

ABOVE WF/PC II in preparation for flight in Building AE at Cape Canaveral Air Force Station, 24 August 1993. *(NASA)*

LEFT Using a JPL handling fixture, the WF/PC II is checked out in the Payload Hazardous Servicing Facility at the Kennedy Space Center. *(NASA)*

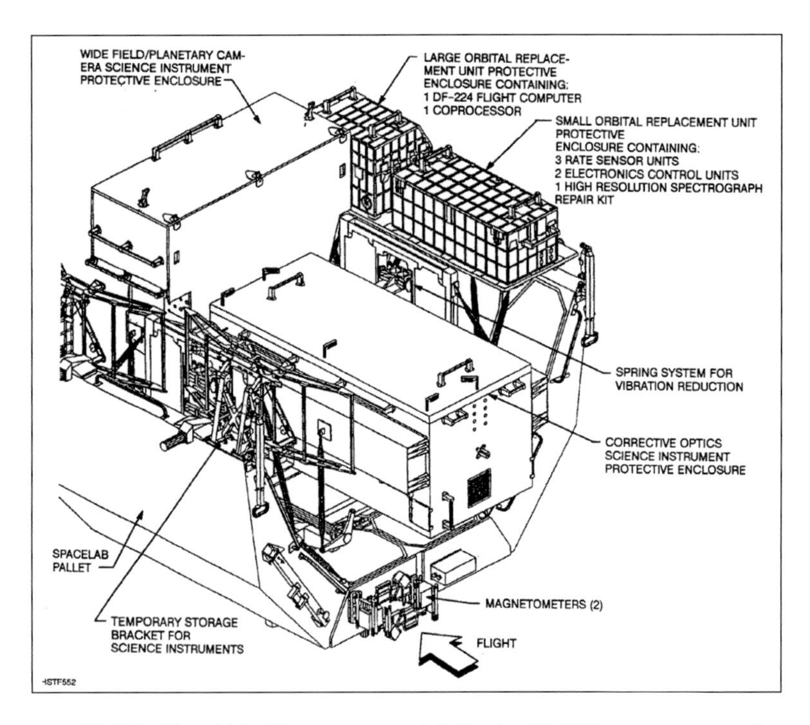

ABOVE The Orbital Replacement Unit Carrier (ORUC) was based on the European Spacelab pallet, which was itself used with and without the pressurised Spacelab module on Shuttle science missions. The ORUC was adaptable to a range of servicing missions and for the first of these it carried the COSTAR and Wide Field/Planetary Camera II in protective enclosures. *(Lockheed)*

BELOW The Solar Array Carrier was located in the forward section of the payload bay and supported the replacement solar arrays provided by the European Space Agency, as well as the Solar Array Drive Electronics Unit. *(Lockheed)*

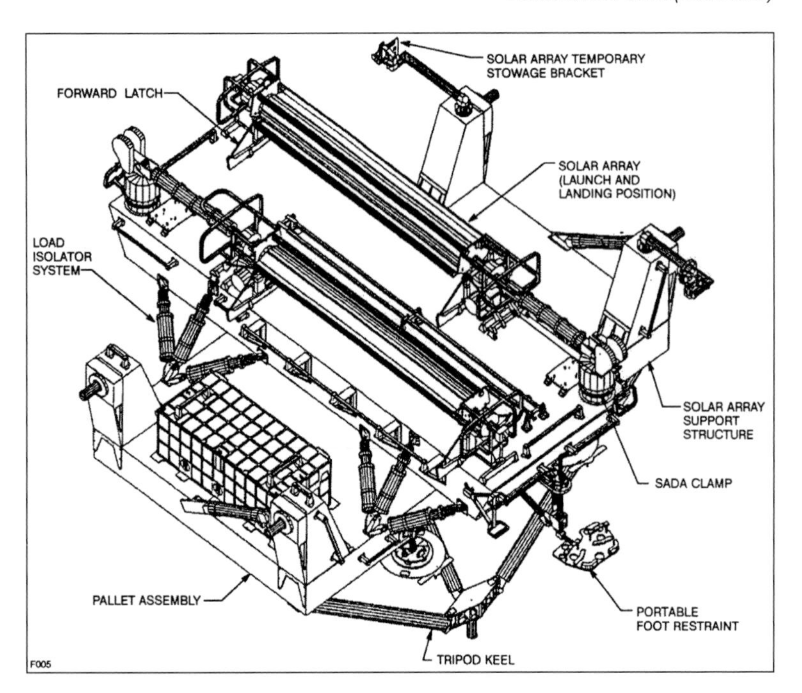

Camera – in which each pixel occupies 0.046 arc-sec and the square field of view is only 36.8 arc-sec – operates at a focal ratio of f/28.3.

The exterior dimensions of the WF/PC II are identical to those of the original instrument (which see in Chapter 4) but it is slightly heavier, with a weight of 619lb (281kg). The optical system consists of the pick-off mirror, an electrically operated shutter, a selectable optical filter assembly and a four-faceted reflecting pyramid mirror used to partition the focal plane to the four cameras. In the WF/PC II instrument three of the four old mirrors are equipped with actuators that will allow these mirrors to be controlled in two axes (tip and tilt) by remote control from the ground. These ensured that the newly provided spherical aberration correction built into the instrument will be accurately aligned relative to the Telescope in all four channels.

Orbital Replacement Unit Carrier (ORUC)

Located in a central position in the Shuttle payload bay, the ORUC was a Spacelab pallet of the type flown on the STS-31 mission when the HST had been delivered to orbit. It had a total weight in the payload bay of 6,369lb (2,889kg).

In the configuration for the first servicing mission it carried the WF/PC II and COSTAR stored in the Science Instrument Protective Enclosure (SIPE). The Rate Sensor Units, the Electronic Control Units, and the redundancy kit for the Goddard High Resolution Spectrograph were all contained inside the Small Orbital Replacement Unit Protective Enclosure (SOPE). The replacement DF-224 flight computer – which was carried as a contingency in case it was needed (which it was not) – and the DF-224 coprocessor were located in the Large Orbital Replacement Unit Protective Enclosure (LOPE) on the ORUC carrier shelf.

The protective enclosures contain heaters and thermal insulation that controls the temperatures inside the compartments. These enclosures are supported on struts and springs to isolate them from the shocks incurred during launch and during the flight into orbit. The vibration reduction structures were designed to provide a benign environment but some equipment was also for the temporary stowage of items removed from the HST for return to Earth, these including brackets and clips to secure them inside the payload bay. The two replacement magnetometers were located on the aft face of the U-shaped pallet.

STS-61 *Endeavour* (SM1)

2 December 1993
Mission duration: 10d 19hr 58min 33sec
EVA: 5
Commander: Richard O. Covey (Flt 4)
Pilot: Kenneth D. Bowersox (Flt 2)
MS1 (EV3): Kathryn C. Thornton (Flt 3)
MS2: Claude Nicollier (Flt 2)
MS3 (EV1): Jeffrey A. Hoffman (Flt 4)
MS4 (EV2): F. Story Musgrave (Flt 5)
MS5 (EV4): Thomas D. Akers (Flt 3)

The STS-61 mission, the 59th Shuttle flight and the fifth for *Endeavour*, had a planned flight duration of 10d 22hr 36min with a series of launch windows, the first of which opened on 1 December 1993 at 04:57am EST (09:47am UTC) and closed 1hr 7min later. Launch windows were constrained by potential abort landing times for the Shuttle Orbiter in the event of a problem during ascent to orbit, and by the orbit of the Hubble Space Telescope. Only one window was available for each successive day, the time marching further back into the night and eventually crossing into the hour before midnight by 10 December.

On 30 October technicians discovered that dust from sand-blasting on the adjacent LC-39B had blown over the Payload Changeout Room as the Shuttle and its payload stood on LC-39A, possibly contaminating the delicate

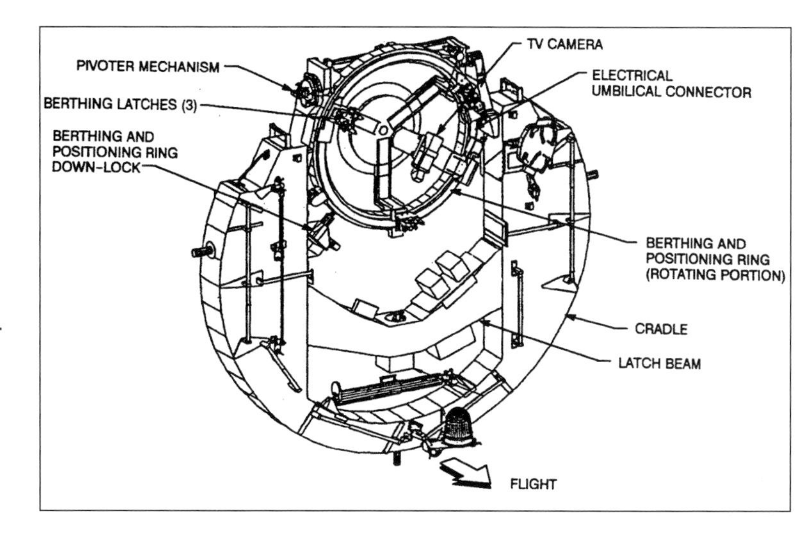

ABOVE The Flight Support Structure as specifically configured for the first servicing mission. It would be used a total of five times on each of the servicing missions between 1993 and 2009. *(Lockheed)*

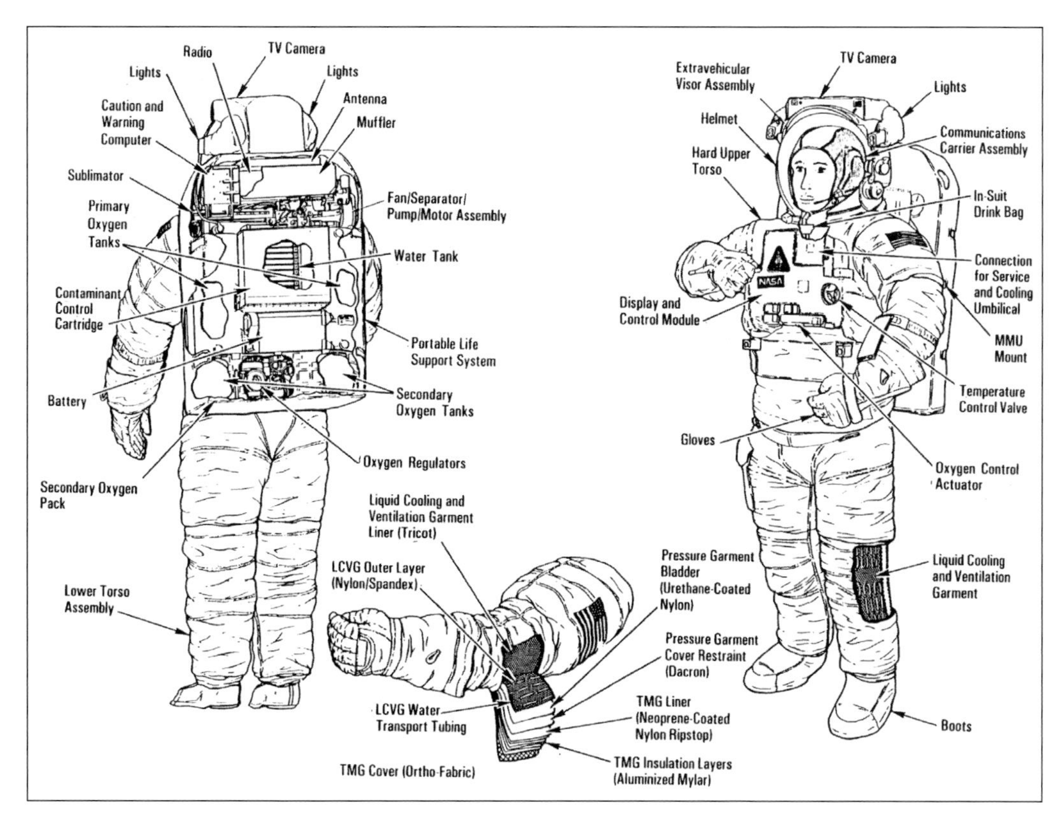

LEFT The ability of astronauts to conduct EVAs and replace ageing equipment on the HST depended entirely upon the effectiveness and reliability of their autonomous space suits – the Extravehicular Mobility Unit (EMU). Matured through several generations of spacesuit design, the EMU was developed specifically for use from the Shuttle and would be applied to spacewalks from the International Space Station. With integral life support equipment and a communications system carrying voice and biomedical data, the EMU has often been described as a spacecraft in its own right! *(NASA)*

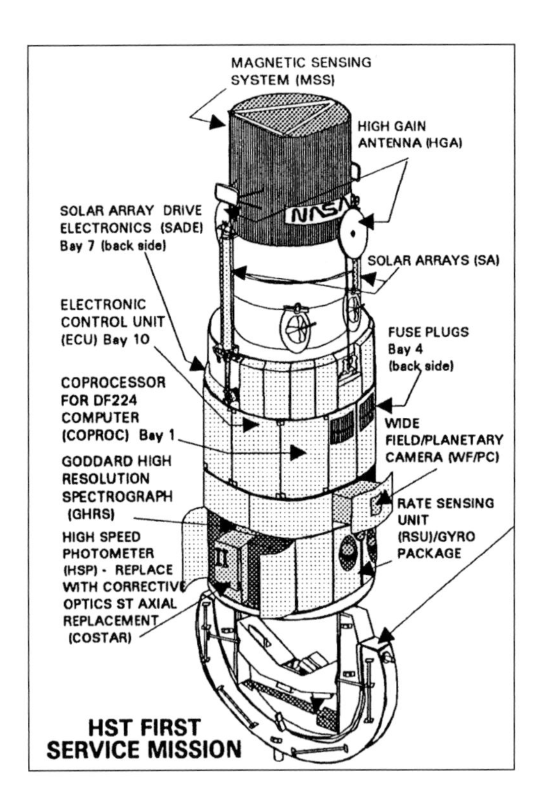

RIGHT The primary objectives of the first servicing missions are identified in this arrangement of tasks and equipment locations, essential to giving the HST the performance it was designed to achieve. *(Lockheed)*

STS-61 EXECUTE PACKAGE

FLIGHT DAY 2

CONTENTS:

PG	MSG NO.	TITLE
1-2	---	FD2 SUMMARY PAGES (NOT U/L)
3-5	005A	FD2 MISSION SUMMARY
6-9	004A	FD2 FLIGHT PLAN REV
10-14	003A	P/TV C/L UPDATES
15	006	FD2 PROCEDURE UPDATES
16	007	FD2 EARTH OBS
17	008	FD2 NOTE TO MS2
18	009	FD2 RELATIVE MOTION PLOT
19	010	SPOC AND PADM INFO

THE PLANNING SHIFT FINALLY GETS BANKERS' HOURS!!!

FAO: Gail Schneider

FAR RIGHT Replacing the time-consuming read-ups from Mission Control requiring crewmembers to write down updates or changes to their flight plans and data books, Shuttle crews received such material on teleprinter readouts direct from the ground. Known as Execute Packages, these were always prefaced by a lead sheet with humorous overtones! The Flight Day 2 package celebrates the relaxed pressure on planning shifts. *(David Baker)*

equipment in *Endeavour*'s payload bay. All the payload equipment had been sealed in double containers but a ten-day inspection was made beginning on 4 November when it was removed from the Shuttle and taken to Hangar AE at Cape Canaveral Air Force Station. As a precaution, *Endeavour* was moved from pad A to pad B, which had been recertified and guaranteed clean, from where it would be launched and to which the payload returned on 16 November after its inspection.

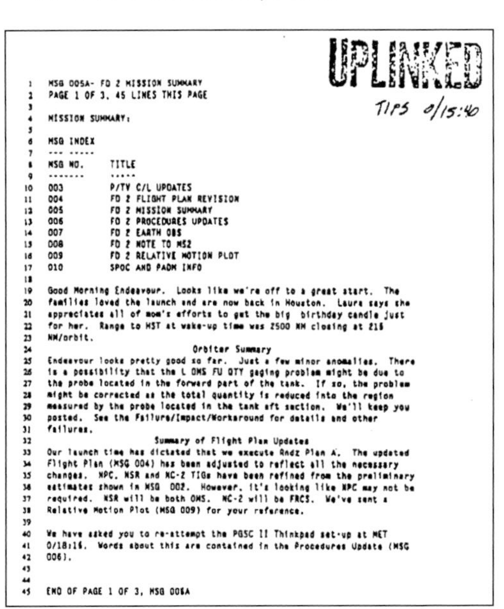
UPLINKED

TIPS 0/15:40

1 MSG 005A- FD 2 MISSION SUMMARY
2 PAGE 1 OF 3. 45 LINES THIS PAGE
3
4 MISSION SUMMARY:
5
6 MSG INDEX
7 --- -----
8 MSG NO. TITLE
9 ------- -----
10 003 P/TV C/L UPDATES
11 004 FD 2 FLIGHT PLAN REVISION
12 005 FD 2 MISSION SUMMARY
13 006 FD 2 PROCEDURES UPDATES
14 007 FD 2 EARTH OBS
15 008 FD 2 NOTE TO MS2
16 009 FD 2 RELATIVE MOTION PLOT
17 010 SPOC AND PADM INFO
18
19 Good Morning Endeavour. Looks like we're off to a great start. The
20 families loved the launch and are now back in Houston. Laura says she
21 appreciates all of mom's efforts to get the big birthday candle just
22 for her. Range to HST at wake-up time was 2500 NM closing at 215
23 NM/orbit.
24 Orbiter Summary
25 Endeavour looks pretty good so far. Just a few minor anomalies. There
26 is a possibility that the L OMS FU QTY gaging problem might be due to
27 the probe located in the forward part of the tank. If so, the problem
28 might be corrected as the total quantity is reduced into the region
29 measured by the probe located in the tank aft section. We'll keep you
30 posted. See the Failure/Impact/Workaround for details and other
31 failures.
32 Summary of Flight Plan Updates
33 Our launch time has dictated that we execute Rndz Plan A. The updated
34 Flight Plan (MSG 004) has been adjusted to reflect all the necessary
35 changes. NPC, NSR and NC-2 TIGs have been refined from the preliminary
36 estimates shown in MSG 002. However, it's looking like NPC may not be
37 required. NSR will be both OMS. NC-2 will be FRCS. We've sent a
38 Relative Motion Plot (MSG 009) for your reference.
39
40 We have asked you to re-attempt the PGSC II Thinkpad set-up at MET
41 0/18:15. Words about this are contained in the Procedures Update (MSG
42 006).
43
44
45 END OF PAGE 1 OF 3, MSG 005A

RIGHT The first sheet of the Flight Day 2 Execute Package provides the crew with a brief summary of the events on launch day and a review of upcoming activities. *(David Baker)*

Due to bad weather, launch took place a day late when the window opened at 4:27am EST (9:27am UTC). This was the most complex Shuttle mission to date and would involve five EVAs and a difficult sequence of tasks that had been fully rehearsed over the previous 18 months. Between March and April 1992 eleven astronauts conducted 16 tests in the Neutral Buoyancy Simulator developing procedures and EVA plans for what was the most arduous sequence of activities yet conducted in space.

The ascent to orbit was normal and proceeded through a series of post-insertion rendezvous manoeuvres incurring five firings of the OMS engines, a series of sequentially higher orbits narrowing the distance from 6,790 miles (10,925km) and gradually reducing the rate of closure until a rendezvous was achieved. ESA astronaut Claude Nicollier secured the RMS arm on to the HST's grapple fixture at an elapsed time of 1d 23hr 19min 56sec over the south-west Pacific Ocean and had it berthed in the FSS just 37min 34sec later. The three latches on the FSS were closed and the RMS arm withdrawn.

During the approach phase the deformed solar arrays became noticeable – the bend on the upper outboard bistem of the +V array was the most severe. 'One side of it is bent way over', radioed Jeff Hoffman from *Endeavour*.

EVA-1

Preparations for EVA really began after the first sleep period, at 19hr 17min, with a reduction in pressure within *Endeavour* to 10lb/in^2 (68.95kPa) for pre-breathing. The first EVA began earlier than scheduled at a mission time of 2d 18hr 17min, with Musgrave and Hoffman replacing two of the rate sensor units (RSUs), each of which contain two gyroscopes, to replace the two that had failed, possibly due to bad electronics.

Hoffman worked off a foot restraint attached to the RMS arm controlled by Nicollier inside *Endeavour* and used a programmable power tool operating with pre-set speed, torque and turns, to open five latches to two doors in the Aft Shroud. Musgrave set up a foot restraint inside Hubble and with Hoffman's help eased himself inside the open spacecraft. He disconnected two electrical plugs and removed three bolts to remove RSU No 2, installed the new box and, after repeating the process for RSU No 3, went to the aft end of the payload bay to prepare for removing the solar arrays the following day.

While he was doing this, Musgrave was closing up the doors to the Aft Shroud but found they did not align and let a small gap through which light could leak. Leaving it as an unsolved problem, a solution to be worked out on the ground so as not to waste EVA time, the two went halfway up the Telescope and to the System Support Module to replace two Electronic Control Units for the rate sensors. ECUs Nos 1 and 3 had each experienced failures in one of their two channels. Each box is connected by four bolts and a cable.

After this, Hoffman on the RMS arm began closing the door while Musgrave returned to the Solar Array Carrier, after which they started on replacing eight of the HST's 12 fuses, a task that took a mere 12 minutes. Then it was back to the troublesome doors on the Aft Shroud. By carefully pulling and adjusting and using a strap

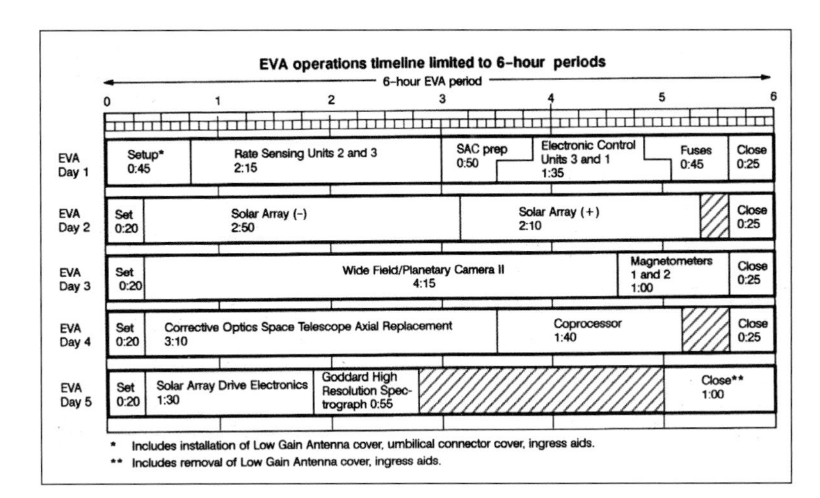

ABOVE Planning for the five scheduled EVAs included simulated workouts in the Neutral Buoyancy Facility and required contingency margins to accommodate any delays caused by unexpected problems. While this plan was not adhered to rigidly it represents the best-efforts planning developed from simulations. *(Lockheed)*

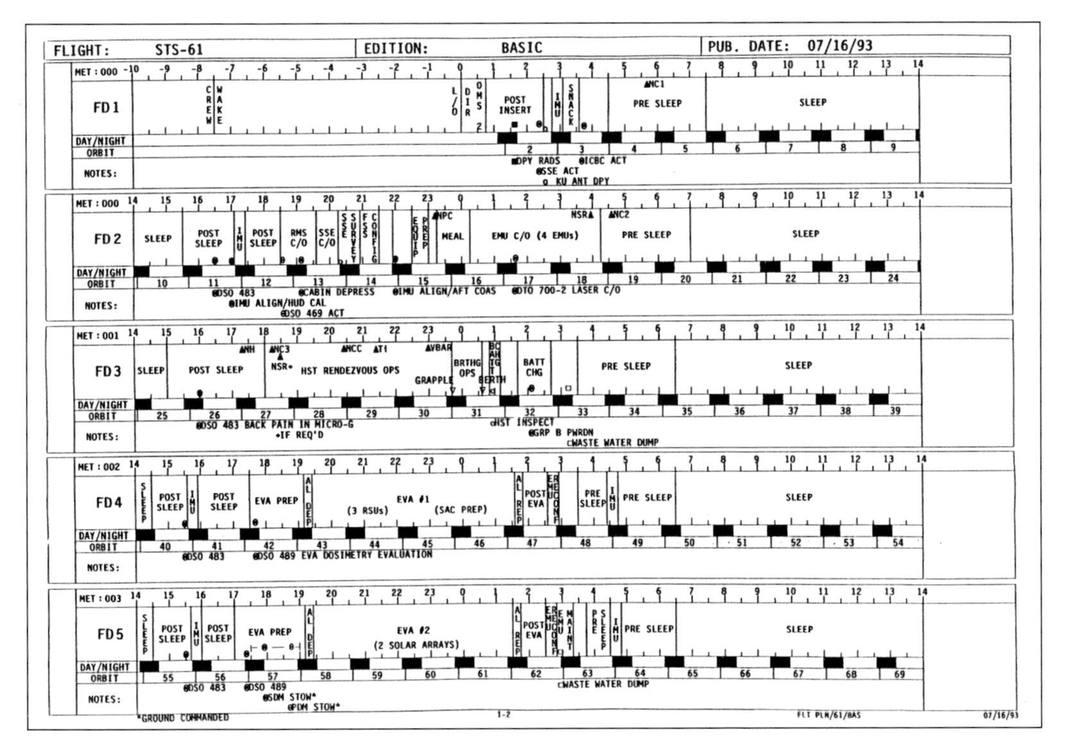

LEFT This page from the flight plan displays the balance of time slots on successive mission days up to and including the second EVA on Flight Day 5. *(David Baker)*

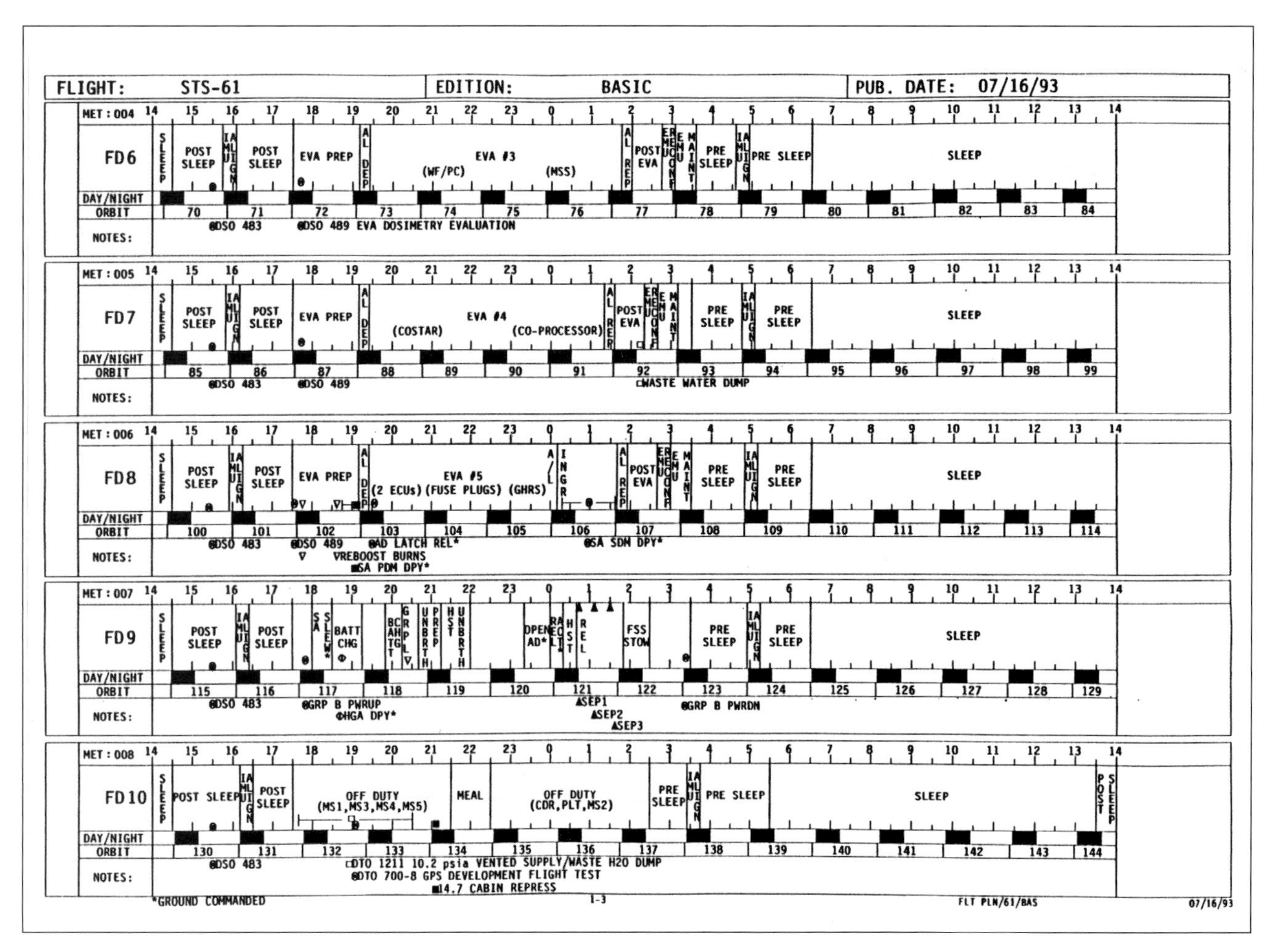

ABOVE Flight plan details for Flight Days 6 to 10, with the last three EVAs and deployment of the HST. *(David Baker)*

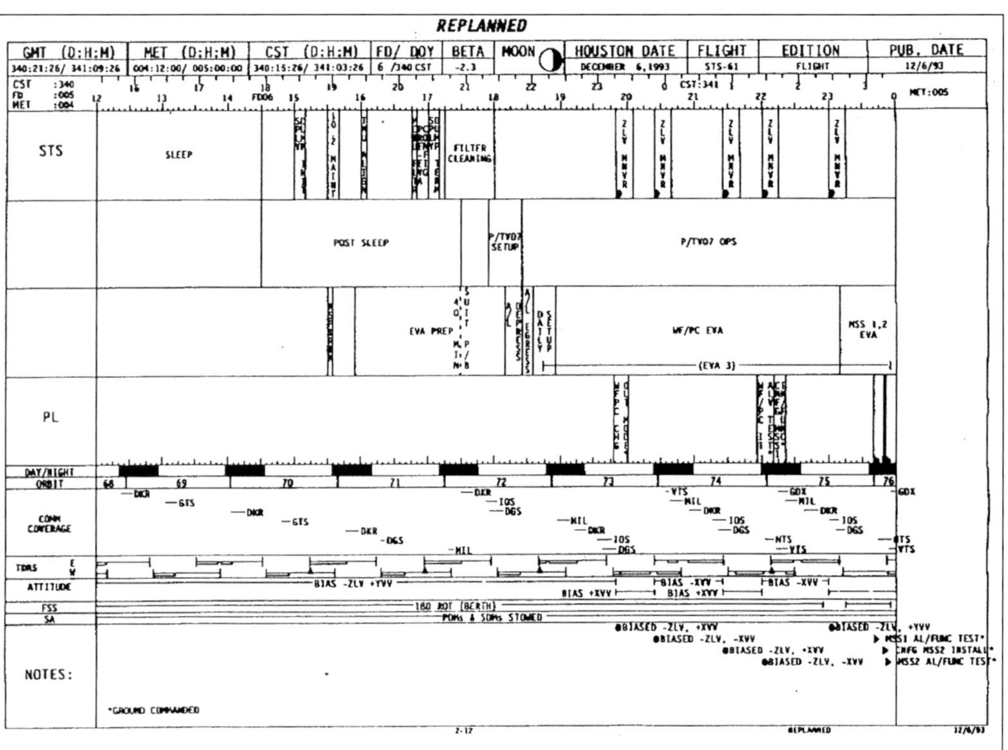

RIGHT Uplinked by teleprinter, Flight Day 6 activities included in the Execute Package for that day are typical for the changes reflected through real-time amendments and small timeline changes made necessary by actual activities. *(David Baker)*

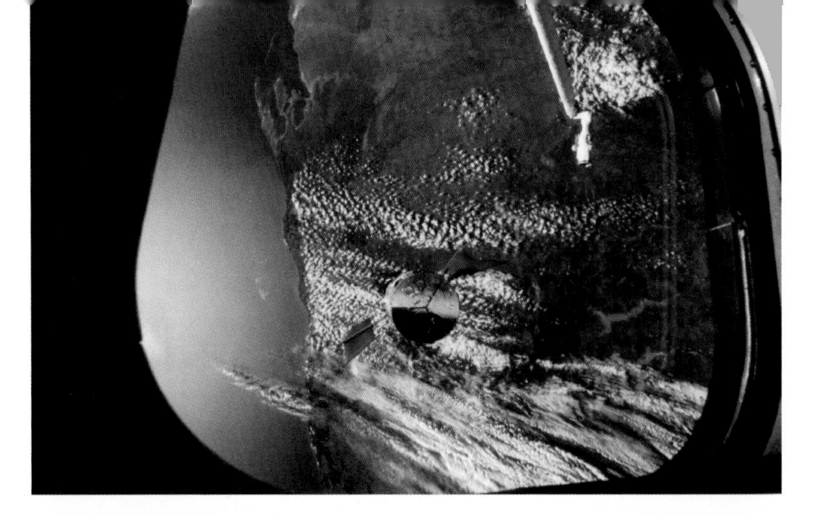

RIGHT With the RMS arm deployed (top), ***Discovery*** closes in on the HST. *(NASA)*

CENTRE Secured on the Flight Support Structure, the HST is ready for installation of new solar arrays (seen on the SAC in foreground), the COSTAR corrective instrument and a new Wide Field/Planetary Camera. *(NASA)*

to pull the two handles together they finally managed to get the doors closed and bolted back up. The EVA lasted 7hr 53min 58sec.

After the crew were reunited back in *Endeavour*, the angle of the HST was pivoted down so the solar array wings would clear the Orbiter tail, and rotated 90° prior to beginning the process of rolling the arrays back up and then rotating the arms back in 90° against the side of the Telescope ready for removing and stowing them on the second EVA. The first array tripped a microswitch a few times but was eventually furled up and was stowed as planned. The second array was so deformed that it could not be retracted and the decision was made to jettison it as the first order of business on EVA-2.

EVA-2

The next spacewalk was performed by Akers and Thornton, who began EVA-2 at an elapsed time of 3d 18hr 2min with the first task being to disconnect the solar array that the previous day had refused to furl back up due to its distorted shape. Thornton had difficulty with getting a communications link to the Orbiter, so for some periods messages to her had to be relayed by Akers.

Working from the end of the RMS arm, Akers manually released the solar array arm still in its deployed configuration and, grasping the weightless 352lb (160kg) assembly in a special yoke-like tool, held it above her head as Nicollier manoeuvred the arm from inside *Endeavour*

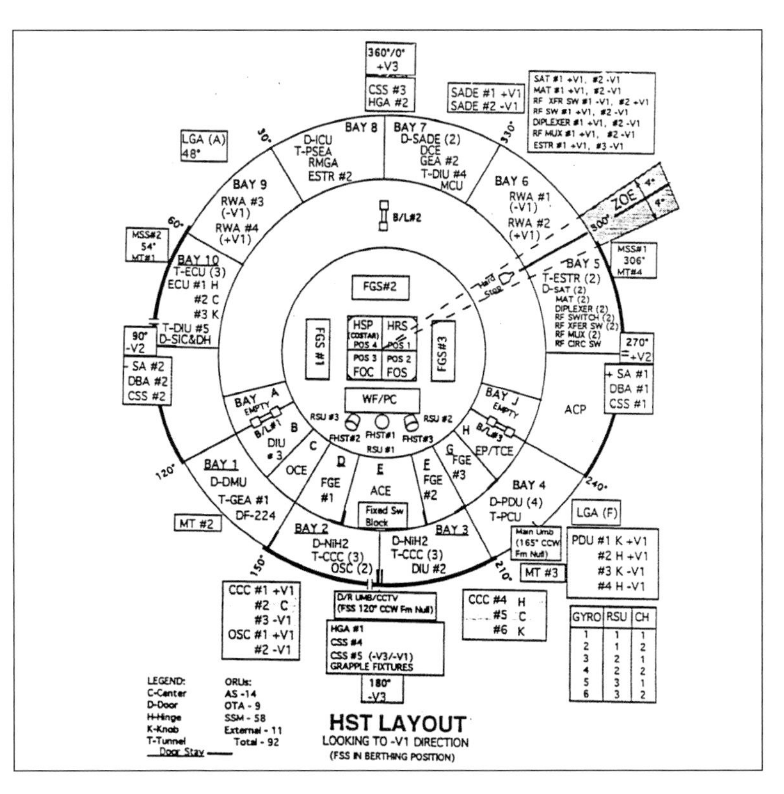

RIGHT The EVA checklist and a flight support document inside the Orbiter carried a full map of the layout of the HST and its systems and equipment for orientation purposes during the mission. This view shows the HST looking down from the top, as it would be secured to the FSS. *(NASA-JSC)*

ABOVE A fisheye-lens view from high up on the HST, displaying grapple fixture, solar arrays and the prolific array of handholds. *(NASA)*

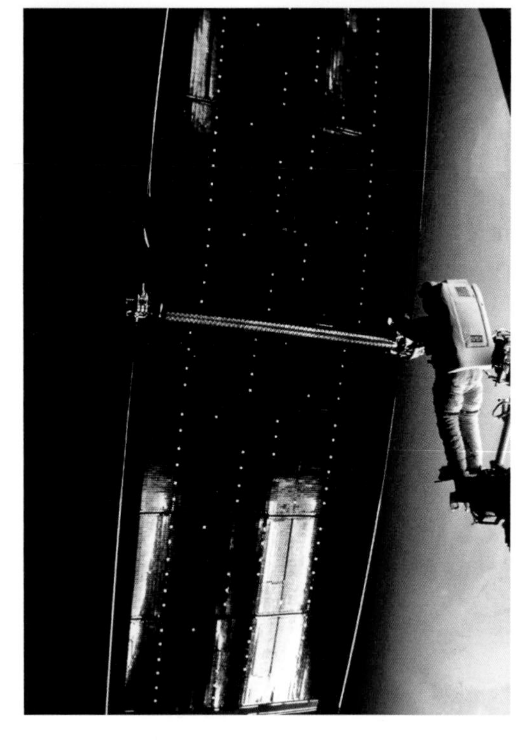

RIGHT Kathy Thornton prepares to release the failed solar array that, unable to be rolled back up, could not be returned. *(NASA)*

RIGHT Uplinked on the daily Execute Package, an amateur artist at Mission Control likens the gently wafting arrays to a prehistoric Pterodactyl! *(NASA)*

away from the Telescope. With a call of 'Here we go. No hands!' she let go of the array and watched briefly as it drifted slowly away. In *Endeavour*, Covey fired the Orbiter's thrusters to secure a 1ft/sec (0.3m/sec) separation rate. The array was struck by thruster jets and it fluttered and tumbled, the new boost increasing the separation rate to 4ft/sec (1.22m/sec) so that on each orbit it would separate by 12.5 miles (20km) and would re-enter the atmosphere within two years.

The next task was to detach the remaining (port side) solar boom and place it down on the Solar Array Carrier in the payload bay. Periodically, inside *Endeavour* Nicollier swapped control of the RMS arm with Bowersox; it was an exacting and precise task which while critical to the activities assigned to the EVA carried the potential for harm to the astronaut riding the foot restraint at the end of the arm's end effector. The work was meticulous but carried on apace and the two spacewalkers soon had the two new solar arrays installed, giving the HST a more efficient, and effective, power system.

The EVA ended at an elapsed time of 6h 35min 30sec, but re-pressurisation in the airlock was slowed when Thornton complained of pressure in her ears, the more gradual increase in pressure avoiding painful earache or potential damage. Both astronauts would be required for EVA-4 two days hence.

EVA-3

Having got the engineering aspects of the HST sorted out by installing the new and improved jitter-free solar arrays and the two RSUs, the next two EVAs were assigned to getting the HST back on track for scientific work, replacing the WF/PC camera with the new one and installing the COSTAR corrective lens set in place of the High Speed Photometer.

The EVA officially began at 4d 18hr 9min with Musgrave and Hoffman first installing a handhold on the existing WF/PC so that they could withdraw it from its stowed position on the side of the SSM Aft Shroud. With Claude Nicollier operating the RMS from *Endeavour* and Hoffman on the arm, Musgrave moved around freely to work from a foot restraint on the HST. After disconnecting the electrical connector drive umbilical to the WF/PC and

ABOVE **European astronaut, Swiss-born Claude Nicollier, works the RMS arm from inside *Discovery*.** *(NASA)*

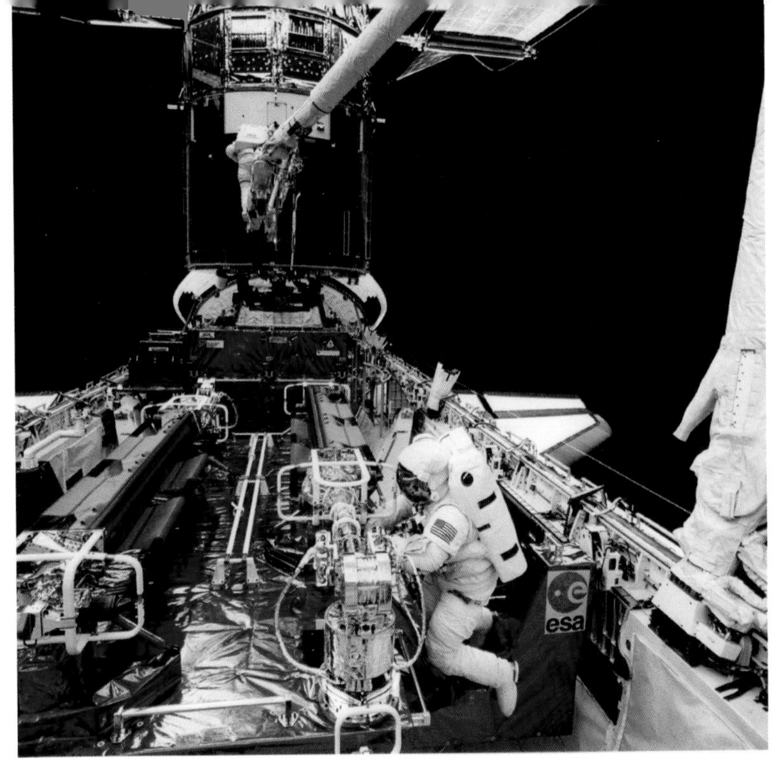

RIGHT **One astronaut rides the RMS arm up to the Aft Shroud on the SSM while a second spacewalker attends to the European Space Agency's solar arrays installed on the SAC.** *(NASA)*

unlocking it from its restraining latches, they slowly withdrew the old camera and then tested its ease of motion by reinserting it to check clearances for its replacement.

Clear of its place within the shroud, the WF/PC was moved down on to a temporary location on the port side of the payload bay from where it would later be attached to the ORU Carrier for return to Earth. The handle was then removed and attached to the WF/PC II camera for an enactment of a reverse procedure, Musgrave carefully removing a protective cover over the pick-off mirror. 'It's in!' said Hoffman as they confirmed that the self-corrected instrument was securely mounted back in the bay. Musgrave then checked on an indicator light to confirm it was latched and the handhold was removed and used to lock the old unit back down to the ORU Carrier.

On the ground, the STOCC performed an 'aliveness' test to check that it was responding as expected to engineering commands and that the ground had dialogue with it. Full evaluation of its effectiveness required several weeks of testing.

Meanwhile, the two spacewalkers were moved up the side of the HST on the RMS arm almost 44ft (13.4m) above the payload bay to replace the two magnetometers which had been giving problems for some time (MSS-1 and MSS-2). These items were not designed to be replaced in orbit and the replacements, with better insulation and new electronics, were to be bolted on to the top of the existing boxes. Hoffman noted that one of the existing magnetometers was pulling away and had almost come off, threatening to contaminate the optical path.

Before the spacewalk ended, the two men tried out some 'get-ahead' tasks to ease the schedule for the next two spacewalkers. The EVA had lasted 6hr 47min 21sec.

BELOW **The Wide Field/Planetary Camera is held temporarily on a holding fixture awaiting relocation.** *(NASA)*

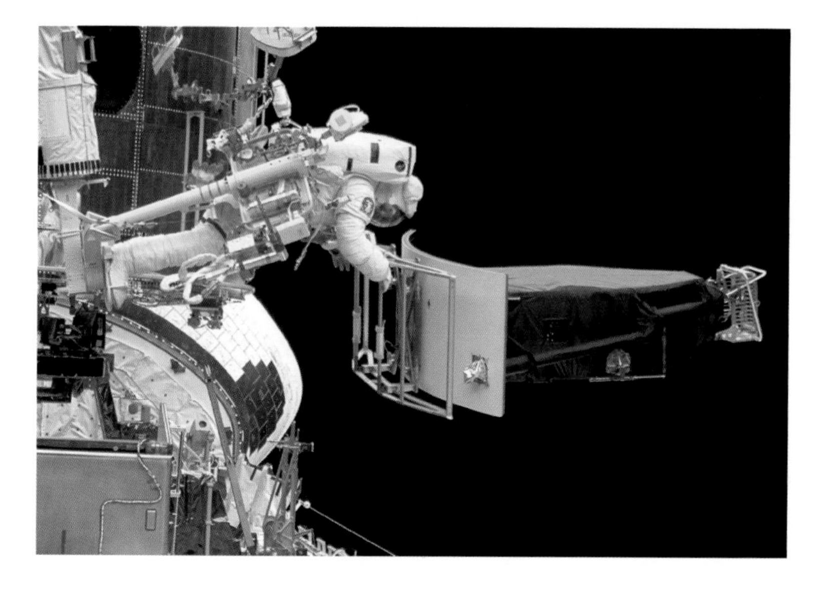

LEFT Hoffman manoeuvres the WF/PC II with a temporary handhold attachment that allows handling from the Scientific Instrument Protective Enclosure and for the previous camera to be installed for return to Earth. *(NASA)*

EVA-4

Akers and Thornton began their second EVA at a mission elapsed time of 5d 17hr 46min when they started on a spacewalking day that would see successful installation of the much awaited COSTAR corrective instrument. With Thornton on the RMS arm, the two astronauts opened a pair of doors on the Aft Shroud and removed four connectors and a grounding strap for the High Speed Photometer. Just as had been the procedure on the previous day, the ability to insert a replacement module was first tested by sliding back the one being removed. After checking that there was a smooth movement in as well as out, and that nothing was present to snag its successor, the HSP was removed and COSTAR was installed.

BELOW A fine study of spacewalkers at work, the WF/PC parked on its holding fixture. *(NASA)*

The pair then turned their attention to installing an Intel 386-architecture coprocessor on the DF-224 main computer. The existing unit had to be powered down during this task, which was only nervously accepted as necessary by the Goddard engineers. It was, said David Leckrone, Goddard's HST project scientist, like putting the HST through brain surgery and eye surgery at the same time. After removing a handle from the DF-224 and bolting the coprocessor on top and connecting the two together, the main computer was brought back up and demonstrated that it was working just fine; a similar 'liveness' test had been completed on the COSTAR instrument, although it would be some weeks before the full benefit of the corrective lenses was apparent.

Before getting back inside, Akers removed some multi-layer insulation from the old WF/PC handhold and its carrier box to bring inside so that covers could be fashioned for the magnetometers, a task carried out by Bowersox and Nicollier and allocated to the last EVA. In the 6hr 50min 52sec spent outside, this spacewalk made Akers the top-scoring NASA astronaut for EVA time, having now accumulated 29hr 40min, some 5hr 28min longer than previous record holder Eugene Cernan, the last Apollo astronaut on the Moon. It also made Thornton the woman with the most EVA time: 14hr 26min 22sec.

EVA-5

Before the final spacewalk, the Hubble Space Telescope was tilted downwards 45° for a re-boost manoeuvre. The two pilots had been conservative with propellant consumption and it was judged that the forward firing thrusters could be used to raise the orbit to extend the orbital life of the Telescope. Orbital decay is notoriously difficult to predict several months in advance. Any minor change in solar intensity can significantly influence the outer atmosphere and 'stiffen' it up so that the trace molecules have a greater effect on slowing large spacecraft with broad surface areas that can act as drag brakes; any change in velocity will result in a change to that orbit and a reduction in speed means that it will begin to decay out of orbit at a much faster rate.

Four nose thrusters fired for 6sec at an elapsed time of 6d 16hr 59min, consuming 4,981lb (2,259kg) of propellant and exacting a change of 12ft/sec (3.6m/sec). At rendezvous, the HST was in an orbit of 370 x 364 miles (596 x 586km) and when it was released again the Shuttle had boosted that back up to an orbit of 373.5 x 367.9 miles (601 x 592km), the small increase in altitude preserving it in a safe orbit until the SM2 mission more than three years later.

The final EVA of STS-61/SM1 was conducted by Hoffman and Musgrave and began at 6d 18hr 0min 33sec. It started with installation of the new SADE unit that transmits commands to the array wings so that they track the Sun. One had failed due to transistor overheating but their replacement proved more difficult than expected. Located in Bay 7 of the Support System Module, as they were unscrewing the original device one of the screws went adrift but Nicollier was nimble with the RMS arm and manoeuvred Musgrave to catch it. Then the connectors proved particularly difficult to manipulate, but installation was eventually completed.

Fortunately for the timing, controllers at STOCC reported that the first new solar array would not deploy from its stowed position and the astronauts were called upon to provide some leverage, to which it responded. Next up was the redundancy package for the Goddard

ABOVE As the Shuttle and the docked HST begin a pass over the southern Indian Ocean, flight controllers in Mission Control keep a continuous televised watch on the spacewalkers attending to the Telescope. *(NASA-JSC)*

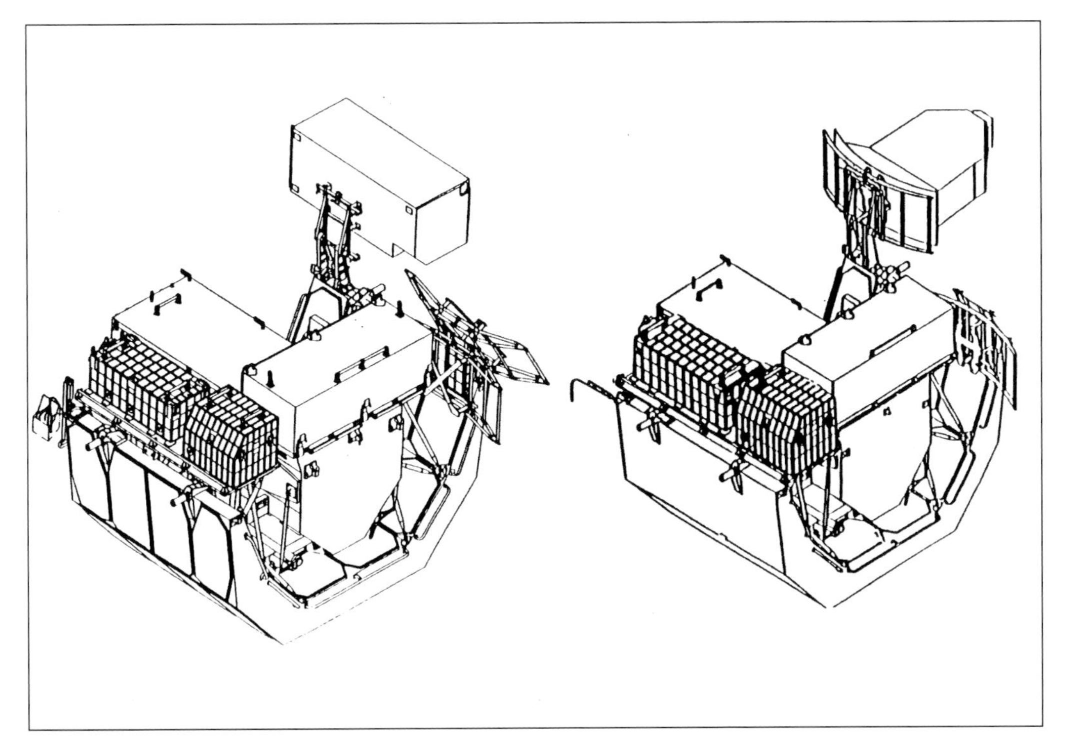

LEFT Special holding fixtures on the ORUC for the COSTAR (left) and the Wide Field/ Planetary Camera II to position replacement and replaced instruments as they are passed between their enclosures and the HST. *(Lockheed)*

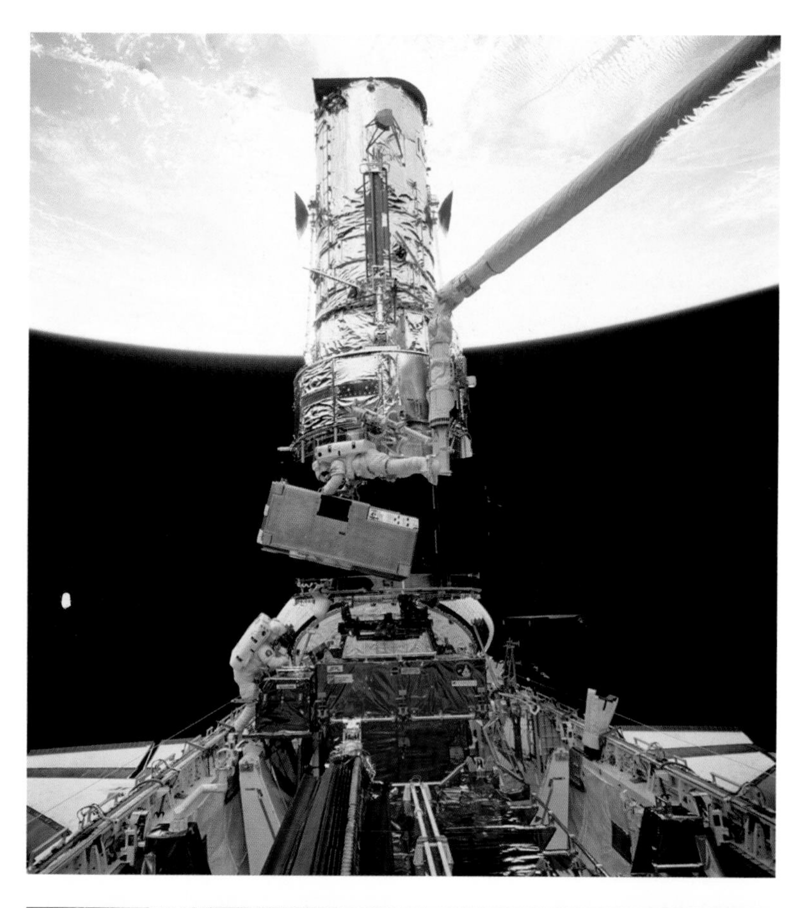

LEFT The COSTAR instrument is manhandled across to the HST from its enclosure on the ORUC. *(NASA)*

High Resolution Spectrometer, which involved four cables and a relay box designed to bypass the erratic power supply for one detector. This work was carried out in the Aft Shroud. The final task for the spacewalkers was for them to go up to the top of the Telescope once more and install the makeshift insulation covers for the magnetometers, securing them with plastic ties. The EVA had lasted 7hr 21min 0sec.

Verification

The five EVAs had been an outstanding success, with minor anomalies corrected along the way. Some 200 tools were carried on this mission, of which around 40 were used at some point or other during the accumulated EVA time. Only two tools had issues: a power tool developed a switch problem while another had a problem with its speed setting. But that was all.

The RMS arm was used to reattach the end effector to the HST at 7d 22hr 17min 14sec and the Telescope was unberthed from the Flight Support Structure 8min 57sec later. It was released back to orbit at an elapsed time of 8d 0hr 59min 20sec. Having been reduced by about one-third for EVA pre-breathing, the pressure in *Endeavour* was brought back up to normal about 20 hours later and the crew prepared for landing, touching down in the dark at the Kennedy Space Center just after midnight on 13 December.

Initial evaluation of the performance of the Telescope indicated that the corrective optical COSTAR instrument was effective and that the WF/PC II was also performing according to expectations. One of the most beneficial improvements was to reduce the jitter caused by thermal expansion and contraction in the solar arrays. New software was written to replace the previous SAGA (Solar Array Guidance Augmentation) program written to help observers compensate for the 'twang' effect with the first

LEFT Working inside the Aft Shroud, astronauts replace the HSP with the COSTAR. *(NASA)*

RIGHT Musgrave rides the arm while Hoffman works in the payload bay. *(NASA)*

set of arrays, which was now reduced to a level which was no greater than that induced by tape recorders and momentum wheels.

The completion of the first servicing mission also drew to a close the hardware development phase for the European Space Agency. For some 16 years, ESA had been actively involved in not only developing a scientific participation but also in engineering the hardware for the Hubble Space Telescope itself. With the installation of the STSA-2 wings with SM1, that engineering development came to a close. But the science lived on, and continues today, the hardware investment by countries across Europe and the UK bringing an agreed 15% of observation time to European astronomers and physicists. Most of the time it is more, and that activity involved scientists at the STScI in Baltimore, USA, and at the European Southern Observatory at Garching, Germany.

RIGHT Suitably upgraded and the optical corrective instrument inserted, the HST is set for full specification performance. It will be 38 months before it receives another visit. *(NASA)*

BELOW Part of the uplinked Execute Package for the final day of the mission produced this summary of events and an advisory about landing conditions. *(David Baker)*

```
 9                    Summary of Flight Plan Updates
10  As we told you last night, the second landing opportunity to KSC looks
11  questionable.  The weather is predicted to deteriorate due to low
12  pressure over the northeastern seaboard.  To ensure two KSC
13  opportunities for nominal EOM, we would like to try for an earlier
14  opportunity on rev 162.  This will require another one hour sleep shift
15  tonight.  Bottom line:  We will end up with two opportunities to KSC
16  (TIG's on rev 162 and 163) and two to EDW (TIG's on rev 163 and 164) on
17  the nominal EOM day.  See MSG 080 for an overview of this plan.
18  The PAO event is currently scheduled at MET 9/20:35.  The duration has
19  been extended to 40 min.  We suggest you review the PADM
20  (MESSAGES/NEWS) before the event for the latest headlines.  We've also
21  provided Claude the translations of anticipated questions from the
22  European press (MSG 079).
23  IMAX has found one more target where they think the weather has
24  improved since yesterday.  The target is shown in the Flight Plan
25  revision at MET 10/00:01.
26
27                         Payload Summary
28  HST remains in a sunpoint attitude.  All subsystems are nominal.  Real-
29  time trending of Gyro and MSS data shows normal operating conditions.
30  In the last 24 hour period both high and low gyro biases have been
31  uplinked.  An attitude reference update was also uplinked to reduce the
32  vehicle attitude RSS error to approximately one third of a degree.
33  There is no evidence of uncompensated gyro bias drift.
34  Preliminary results indicate that the tracker to tracker alignments
35  have not shifted relative to pre-servicing alignments.
36  HST tip-off rates at release were 0.054 degrees/sec.  Our goal was less
37  than 0.2 degrees/sec.  That's 73% better than our goal. Great job on
38  the release!  Coarse Sun Sensor data indicated an attitude error of
39  approximately 3 degrees in both V1 and V2 axes.  HST software sunpoint
40  capture occurred within 1 minute 20 seconds of release.
41  The power system is functioning properly with the new solar arrays
42  delivering a peak current of 154 amps.  This is 25 amps greater than
43  the old arrays.
```

Chapter Eight

New servicing missions

With the Hubble Space Telescope now performing as intended, and even in some aspects better than expected when it was designed, it was time to begin the process of harvesting additional data with better and improved instruments, equipment which would be delivered over four servicing missions that, in 1996, were scheduled for early 1997, mid-1999 and mid-2002.

OPPOSITE Key to understanding how to provide suitable materials for the resilience required in the HST, incidental to the Telescope programme but highly relevant was the Long Duration Exposure Facility (LDEF), carrying 57 different materials and coatings to evaluate their decay and erosion in space. Looking deceptively small in this view of the LDEF about to be released from the RMS of Shuttle Orbiter *Challenger* in April 1984, LDEF was 30ft (9.14m) long, had a width of 14ft (4.27m) and weighed 21,396lb (9,705kg). *(NASA)*

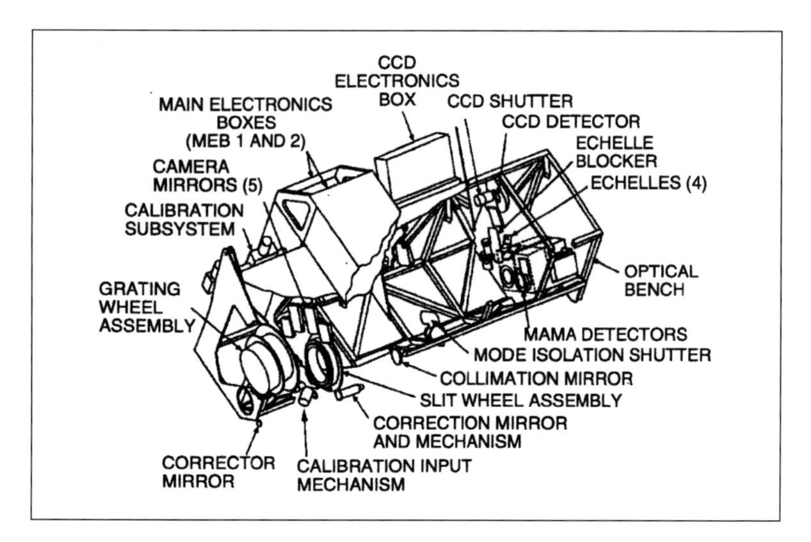

ABOVE The Space Telescope Imaging Spectrograph was designed by Ball Aerospace to provide two-dimensional capability to HST spectroscopy and to replace the Goddard High Resolution Spectrograph. *(Lockheed)*

The first of these (SM2) was planned as a near ten-day mission carrying a crew of seven supporting at least four EVAs and replacing two ageing science instruments with new and refined equipment capable of a broader range of objectives.

The two instruments that were to be removed and returned to Earth were the Goddard High Resolution Spectrograph and the Faint Object Spectrograph. In their place would go the Space Telescope Imaging Spectrograph (STIS) and the Near Infrared Camera and Multi-Object Spectrometer (NICMOS). The astronauts would also replace some items and upgrade others, including a refurbished Fine Guidance Sensor, a Solid State Recorder to replace one of the existing reel-to-reel recorders, and a replacement for one of the HST's four Reaction Wheel Assemblies with a refurbished spare.

A Solar Array Data Electronics (SADE) device was replaced during the first servicing mission. That device after return to Earth was refurbished and was assigned to SM2 as a replacement for the second unit, SADE-2, which would similarly be returned to Earth. The SADE units were contributed by the European Space Agency. One of the three Engineering Science Tape Recorder (ESTR) devices failed and was assigned a replacement on SM2. There are three ESTRs on board with two assigned to storing scientific data and the other for engineering data on a periodic basis. The failed ESTR was to be replaced with a new one and a second ESTR already on board was to be replaced with a Solid State Recorder (SSR).

Space Telescope Imaging Spectrograph (STIS)

Developed under the direction of Dr Bruce E. Woodgate and Ball Aerospace, the STIS was regarded at the time of launch as the most complex science instrument sent into space. It was designed to be a versatile and efficient spectrograph using the most modern technology to provide a two-dimensional capability used either for long-slit spectroscopy, where spectra of many different points across an object are obtained simultaneously, or in an echelle mode to gain greater wavelength coverage in a single exposure. It can take both ultraviolet and visible images through a limited filter set.

The instrument was designed to replace many of the functions of the GHRS and some of those in the Faint Object Spectrograph and carried its own corrective optics that did not require the use of the COSTAR installed on SM1. The STIS was built to fit in the axial bay behind the primary mirror and had a comparable size to the GHRS it replaced. The instrument measures 7.1ft x 2.9ft x 2.9ft (2.2m x 0.98m x 0.98m) and weighs 825lb (374kg). It consists of

LEFT The components and detectors of the STIS are featured in this diagrammatic representation of the internal arrangement. *(Lockheed)*

Mode	Band Wavelength Range (nm) Detector	Band 1 115-170 Det #1 (MAMA/CsI)	Band 2 165-310 Det #2 (MAMA/Cs_2Te)	Band 3 305-555 Det #3 (CCD)	Band 4 550-1000 Det #3 (CCD)
Low resolution spectral imaging (first order)	Mode number Resolving power (l/Dl) Slit length (arcsec) Exposures/band	1.1 770-1,130 24.9 1	2.1 415-730 24.9 1	3.1 445-770 51.1 1	4.1 425-680 51.1 1
Medium resolution spectral imaging (first order scanning)	Mode number Resolving power (l/Dl) Slit length (arcsec) Exposures/band	1.2 8,600-12,800 29.7 11	2.2 7,500-13,900 29.7 18	3.2 4,340-7,730 51.1 10	4.2 3,760-6,220 51.1 9
Medium resolution echelle	Mode number Resolving power (l/Dl) Exposures/band	1.3 37,000 1	2.3 23,900-23,100 2	—	—
High resolution echelle	Mode number Resolving power (l/Dl) Exposures/band	1.4 100,000 3	2.4 100,000 6	—	—
Objective spectroscopy (prism)	Mode number Resolving power (l/Dl) Field of view (arcsec)	—	2.5 (115-310 nm) 930-26 (at 120-310) 29.7 x 29.7	—	—

K70110-409

LEFT STIS spectroscopic modes in the five resolution settings available. *(Lockheed)*

a carbon fibre optical bench that supports the dispersing optic and three detectors.

Although designed to work in three different wavelengths with their individual detectors, there was overlap in detector response and backup spectral modes. The Mode Selection Mechanism (MSM) allows selection of a wavelength region and contains 16 first-order gratings, an objective prism and four mirrors. An optical bench supports the input corrector optics, focusing and controlling tip/tilt motions, the input slit and filter wheels, and the MSM.

The MSM consists of a rotating wheel, its axis being a shaft with two inclined outer sleeves, one fitting inside the other. These are constructed so that the rotation of one sleeve turns a wheel to align the appropriate optic with the beam. Rotation of the second sleeve changes the inclination of the wheel's axis or the tilt of the optic to select the wavelength range and point the dispersed beam to the corresponding detector. One of the three mirrors can be selected to take an image of the object. Of the 16 gratings, six are cross-dispersers that direct dispersed light to one of the four echelle gratings for medium and high resolution modes.

In operation, light from the primary mirror is corrected and brought to focus at the slit wheel and after passing through it is collimated by another mirror on to one of the MSM's optical elements, at which point a computer selects the appropriate mode and wavelength. The MSM both rotates and nutates to select the correct optical element, grating, mirror or prism and points the beam along the appropriate optical path to the correct detector.

In the case of first-order spectra, a first-order grating is selected for the wavelength and dispersion and the beam is pointed to a camera mirror that focuses the spectrum on to the detector, or goes directly to the detector itself. For an echelle spectrum, an order-sorting grating that directs the light to one of the four fixed echelle gratings is first selected and the dispersion echellogram is focused via a camera mirror to the appropriate detector. The detectors are housed at the rear of the optical bench so that they can readily dissipate heat through an outer panel and are thermally controlled, with both detectors and mechanisms controlled by an on-board computer.

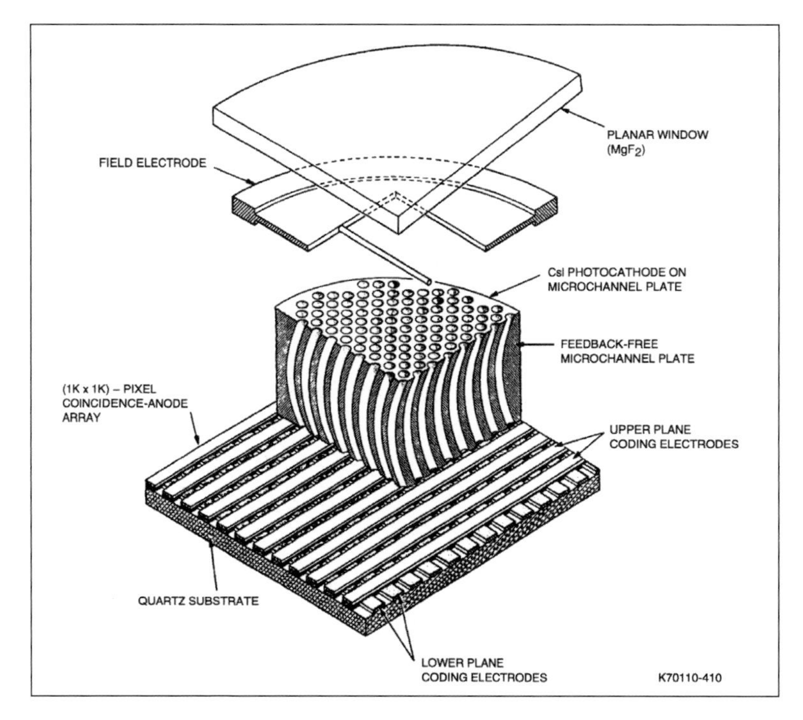

BELOW The Multi-Anode Microchannel Plate Array (MAMA) detector in the Band 1 tube admitting ultraviolet photons. *(Lockheed)*

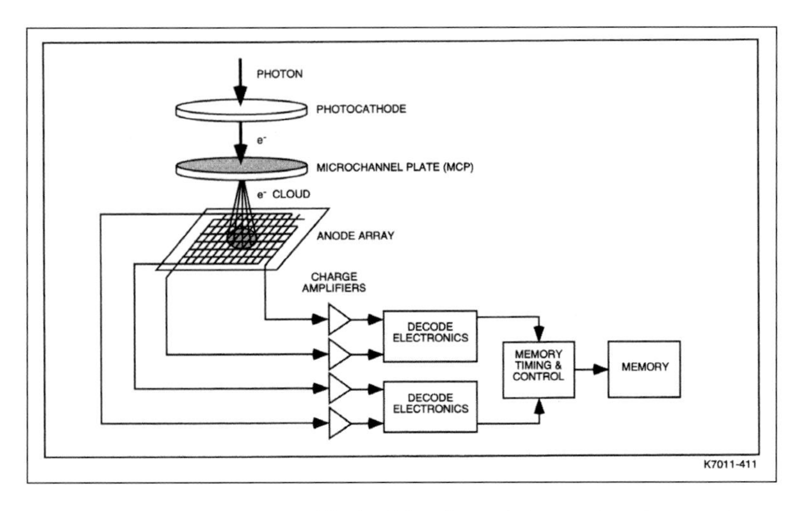

ABOVE A simplified schematic of the MAMA system in the Space Telescope Imaging Spectrograph. *(Lockheed)*

BELOW Ball Aerospace engineers fitting the STIS into its shipping container. *(Ball Aerospace)*

Each of the three detectors has been set for a specific wavelength region. Band 1 (115–170nm) uses a Multi-Anode Microchannel Plate Array (MAMA) with a caesium iodide (CsI) photocathode; Band 2 (165–310nm) also uses a MAMA but with a caesium telluride (CsTe) photocathode; Bands 3 (305–555nm) and 4 (550–1,000nm) use the same CCD detector.

The MAMA is built round a micro-channel plate that consists of a thin disc of glass about 0.06in (1.5mm) thick and 2in (50mm) in diameter honeycombed with 12.5-micron holes. Both sides are metal coated and when a voltage is applied across the plate an electron entering any hole is accelerated by the electric field. The electron eventually collides with the wall of the hole, its kinetic energy giving up two or more secondary electrons. These continue down the hole and collide with the wall to emit more electrons, and so on. This produces a cascade of a million electrons at the end of the hole.

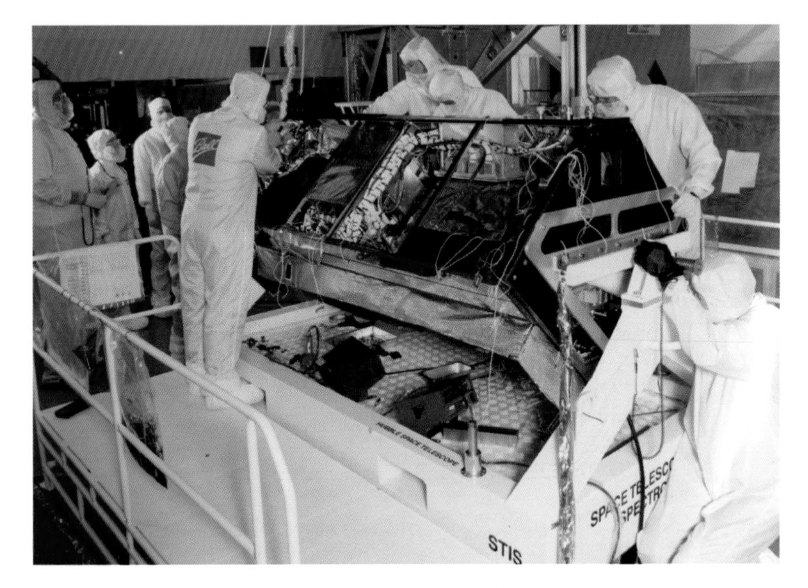

In the Band 1 tube, ultraviolet photons hit the CsI photocathode deposited on the front of the micro-channel plate to produce an electron when a photon hits it, which is accelerated into its holes which in turn amplifies the number of electrons which cascade out as a shower on to the anode array. This has been designed so that only 132 circuits are required to read out all 1,024 x 1,024 pixels. As the MAMA records the arrival of each photon it can provide a time sequence so that any periodicity such as that presented by a pulsar will be determined.

In the Band 2 tube, the CsTe photocathode is deposited on the inner surface of the front window as a semi-transparent film. As the photons pass this window they are stopped in the cathode film, at which point they generate electrons. These are amplified and detected in the same manner as those in Band 1. Each detector has 1,024 x 1,024 pixels, each 25 x 25 microns square, but data from the anode array can be interpolated to give a higher resolution, each pixel splitting into four 12.5 x 12.5 micron pixels. In this high-resolution mode observers can examine the fine structural details of an object that broadens out the sampling of optical images and of spectra.

The principal advantages of the STIS are that it provides two-dimensional detectors allowing both long-slit spectroscopy and an echelle format in ultraviolet light, provides lower background by a factor of 100 to 200 over the GHRS, affords use of a CCD detector for improving visible efficiency and wavelength coverage out to 1,000nm (compared with 700nm for the Faint Object Spectrograph), and gives coronographic capability for spectroscopy and imaging while enabling wide-field spectroscopy of many objects at once.

Near Infrared Camera and Multi-Object Spectrometer (NICMOS)

Designed as a second-generation instrument, NICMOS extended observations into the near-infrared and allowed limited spectroscopic observations between 1.5 and 2.5 microns for detailed analysis of

RIGHT The NICMOS instrument replaced the Faint Object Spectrograph and was designed to extend the HST's spectroscopic capabilities into the infrared using a cryogenic dewar with a planned life of five years. *(Lockheed)*

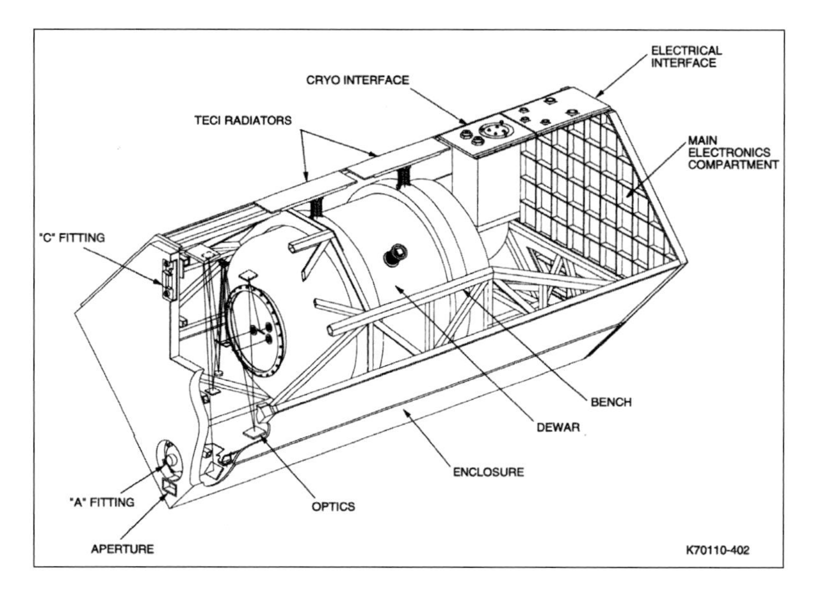

young star clusters and protostar clouds as well as brown dwarfs, obscured active galactic nuclei, and the changes over time in planetary atmospheres. It did this through the application of a cryogenic dewar system with a five-year life.

With principal investigator Dr Rodger U.I. Thompson of the University of Arizona, the NICMOS was built by Ball Aerospace and occupied a space of 7.1ft x 2.8ft x 2.8ft (2.2m x 0.85m x 0.85m) with a weight of 861lb (391kg).

The instrument is an all-reflective imaging system with foreoptics that relay images to three focal plane cameras contained in an insulated cryogenic dewar. Each camera has the same spectral band of 0.8–2.5 microns with a different magnification and an independent filter wheel and each camera views a different segment of the Telescope's field of view at the same time.

Light enters the entrance aperture and falls on a flat folding mirror, from where it is redirected to a spherical mirror mounted to an offset pointing mechanism. This corrects the spherical aberration in the Telescope's primary mirror and incorporates a cylindrical deformation to correct for astigmatism in the optical path. Thus corrected, the image is relayed to a three-mirror field-dividing assembly that separates the light into three second-stage optical paths. In addition, each second-stage optic uses a two-mirror relay set and a folding flat mirror.

The field-dividing mirrors are tipped to divide the light rays by almost 4.5° which allows physical separation for the two-mirror relay set for each camera and its field of view. The curvature of each mirror permits the required degree of freedom to set the exit pupil at the cold mask placed in front of the filter wheel of each camera. A corrected image is produced in the centre of the Camera 1 field mirror, and the remaining mirrors for this camera are confocal parabolas with offset axes to relay the image into the dewar with the correct magnification and minimal aberration.

Cameras 2 and 3 have different amounts of astigmatism because their fields are at different

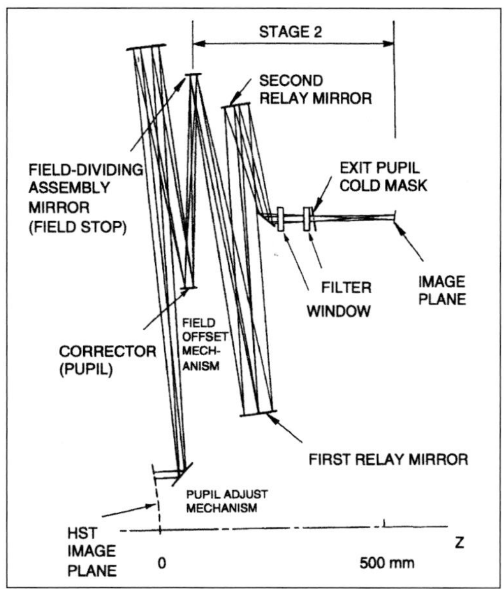

LEFT The light path of the NICMOS system, which divides incoming rays into three second-stage optical paths. *(Lockheed)*

BELOW A schematic of the NICMOS cryogenic dewar with solid sublimating nitrogen. *(Lockheed)*

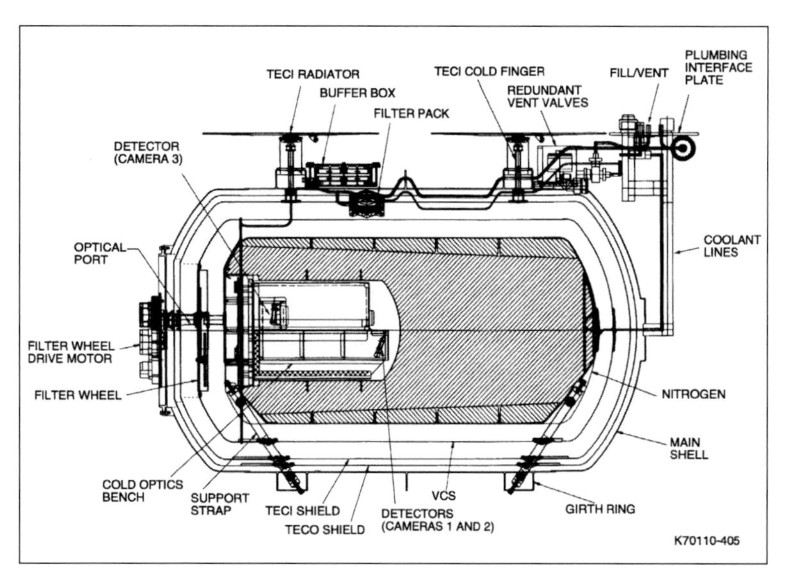

BELOW **The Second Axial Carrier for the second servicing mission was an adaptation of the Solar Array Carrier on the first servicing mission and was a design based on the Goddard Space Flight Center's Pallet Assembly. It contained the NICMOS instrument among a wide range of ORU items.** *(Lockheed)*

off-axis points; one of the off-axis relay mirrors in Camera 3 is a hyperbola and one of the relay mirrors in Camera 2 is an oblate ellipsoid. Camera 2 also allows a coronographic mode by positioning a dark spot in its field-dividing mirror. In this mode, the Telescope is manoeuvred so that the light from the observed star falls within the Camera 2 field-dividing mirrors and is occulted for coronographic measurements.

Camera 1 has a total field of view of 11 arc-sec with a pixel size of 0.043 arc-sec and a magnification of 3.33 at f/80; Camera 2 has a 19.2 arc-sec field with a pixel size of 0.075 arc-sec at a magnification of 1.91 at f/45.7; Camera 3 has a field of 51.2 arc-sec for a pixel size of 0.20 arc-sec providing a magnification of 0.716 at f/17.2. All the detectors are 256 x 256 pixel arrays of mercury cadmium telluride (HgCdTe) with 40-micron spacing of pixels. An independent cold filter wheel is positioned in front of each camera and is rotated by motors at room temperature at the external access point of the dewar.

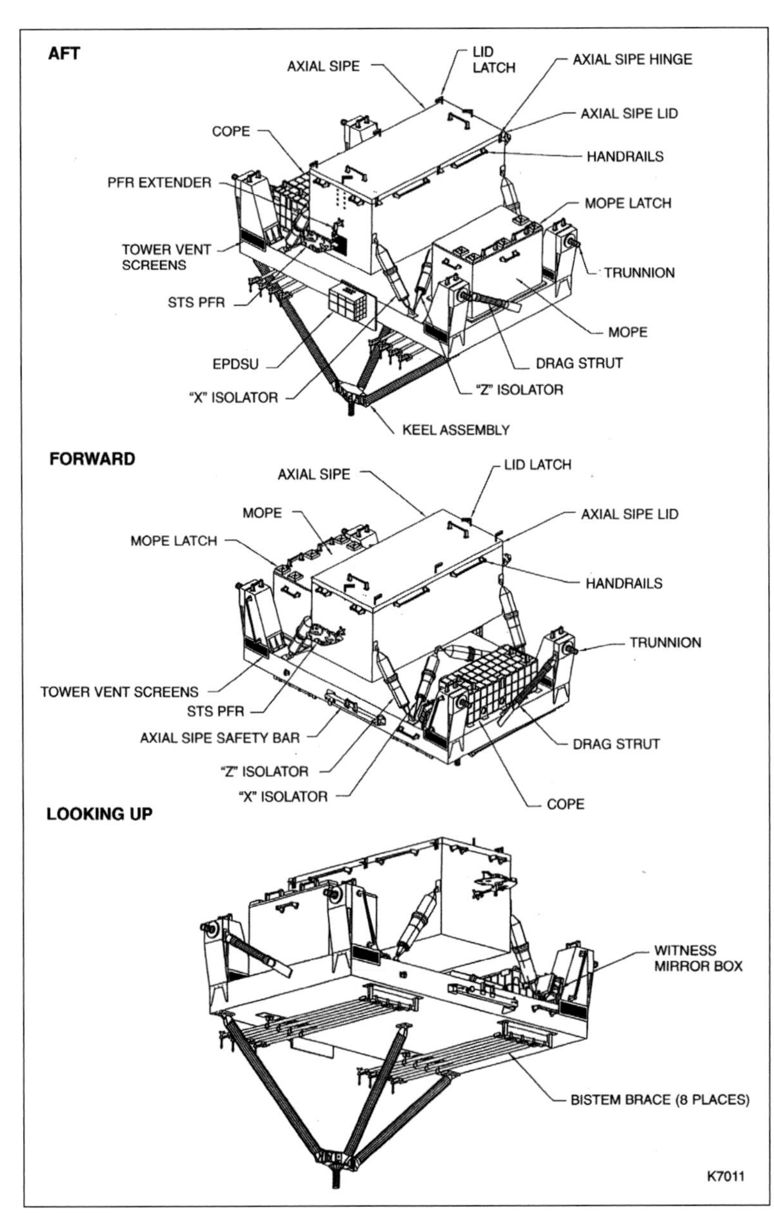

The NICMOS enclosure, designed so as to be the same size as the Faint Object Spectrograph that it replaced, incorporated two separate compartments. The optics compartment contained a graphite-epoxy optical bench kinematically mounted within the enclosure and physically separating the two compartments. The foreoptics and the cryogenic dewar are mounted to the bench and maintained at a constant 270°K (-3°C). The electronics compartment has a thermally controlled environment utilising radiators on the outboard enclosure panels, reflective surface coatings and multiplayer insulation.

The cryogenic dewar contains a 225lb (102kg) mass of solid-sublimating nitrogen maintaining the three cameras on a cold bench at a constant 58°K (-215°C) encased within three shields interspersed with multi-layer insulation. The innermost shield is a vapour-cooled jacket with vented nitrogen. The two thermoelectric cooled outer shields reject heat to the external radiators and are isolated to prevent thermal soakback into the dewar. In all, the dewar system uses 124 layers of insulation consisting of double-aluminised mylar with polyester net spacer material placed in varying quantities between the three shields and external to the outer thermoelectric cooled shield, as well as between the innermost shield and the solid nitrogen tank.

The tank itself is suspended by six fibreglass/epoxy straps attached to girth rings on the vacuum shell, and the three shields are structurally attached and thermally neutralised at intermediate points along the straps. The dewar itself has three plumbing lines and is filled with liquid nitrogen through a single vent and fill line, used on orbit as the vent path for the nitrogen vapour. The remaining two loops form a coolant ring where gaseous helium is circulated to pre-cool the dewar before loading, freezing the liquid nitrogen to a solid and periodically supercooling the solid nitrogen. The optical paths penetrate the dewar at three locations and each port has vacuum shells and cold masks to prevent the detectors from seeing the warm structure.

Second Axial Carrier (SAC)

The SAC is the second mission for the pallet-carrier designed by the Goddard Space Flight Center. When used on the first servicing mission it was known as the Solar Array Carrier (see Chapter 7). For SM2, the modified SAC transported the NICMOS instrument inside an Axial Scientific Instrument Protective Enclosure (ASIPE) replacing the Solar Array Support Structure carried on STS-61. Some Orbital Replacement Units were also carried on the SAC in two separate box-shaped containers. The SAC was located forward of the ORUC within the payload bay.

For SM2 the Second Axial Carrier weighed 5,275lb (2,393kg) and spans the payload bay, with structural ties at five points: four at the two longerons and one on a keel trunnion. Active isolators reduce launch vibration loads with each consisting of a spring and a magnetic damper that converts mechanical vibration energy from the launch events into heat energy, which reduces the loads on the NICMOS instrument. The SAC has eight isolators arranged in four sets of two with each one attached to the pallet assembly and to the ASIPE at its base or end plate.

Orbital Replacement Unit Carrier (ORUC)

The ORUC for the second servicing mission was identical structurally to that flown for the first servicing flight but in this configuration weighed 6,978lb (3,165kg) and carried Fine-Guidance Sensor No 1 in a box-shaped Axial Scientific Instrument Protective Enclosure lying transversely across the U-shaped pallet. The STIS instrument was contained in a similar box-shaped enclosure lying along the long axis at the bottom of the pallet. Two smaller ORU enclosures were carried on the port side of the pallet as installed in the payload bay. The

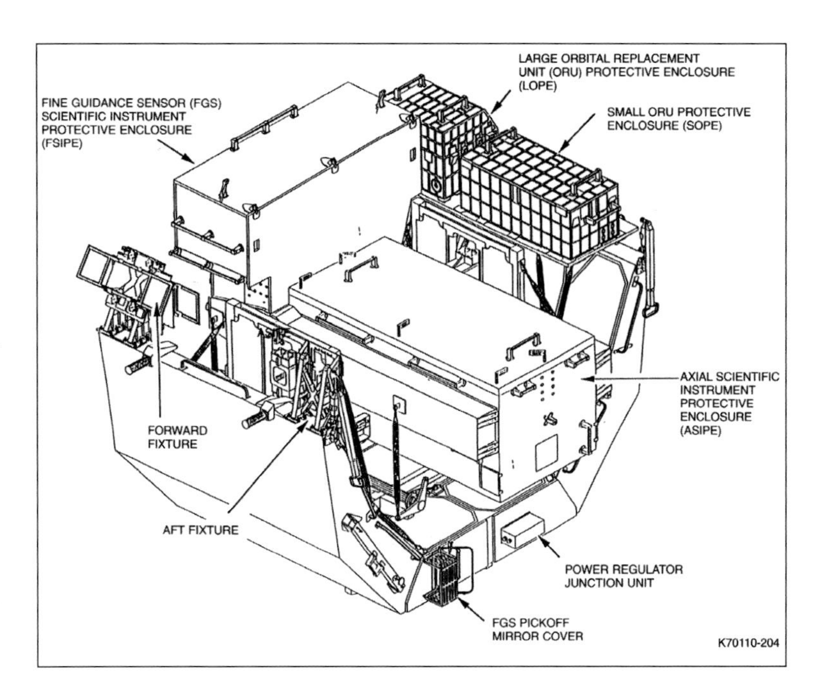

ABOVE The ORUC for the second servicing mission housed the STIS instrument and the Fine Guidance Sensor No 1. *(Lockheed)*

BELOW The Flight Support Structure for SM-1 showing the berthing latches and other support items. *(Lockheed)*

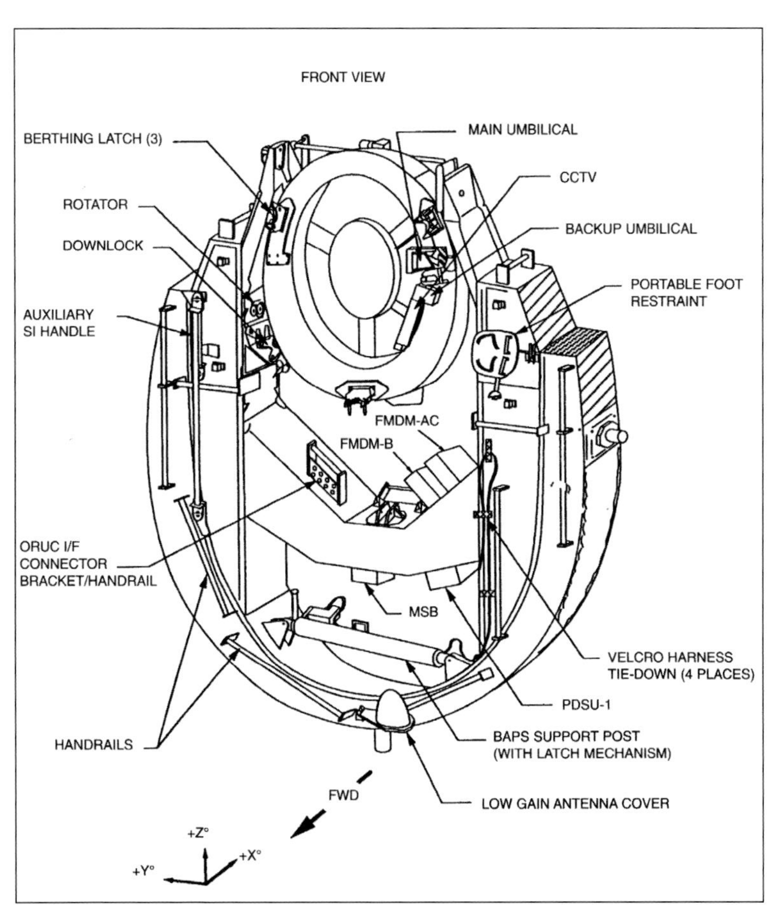

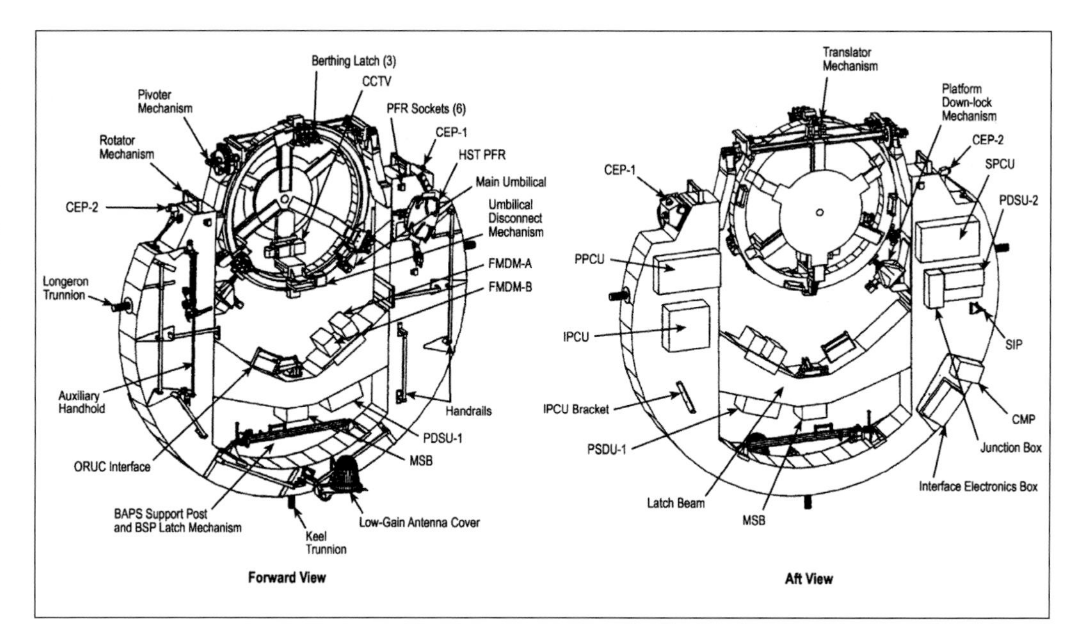

RIGHT Forward and aft views of the FSS show the complexity of design and the multiple interfaces and pick-off points for attachments and support mechanisms. *(Lockheed)*

BELOW The arrangement of servicing support equipment on the SAC and the ORUC together with the position of the HST when berthed on the FSS and locked down for access by spacewalking astronauts. *(Rockwell International)*

LARGE ORU PROTECTIVE ENCLOSURE (LOPE)
SMALL ORU PROTECTIVE ENCLOSURE (SOPE)
ORBITAL REPLACEMENT UNIT CARRIER (ORUC)
HST PORTABLE FOOT RESTRAINTS (PFR)
HUBBLE SPACE TELESCOPE (HST)
MULTI-MISSION ORU PROTECTIVE ENCLOSURE (MOPE)
ORBITER DOCKING SYSTEM (ODS)
ASIPE
FSIPE
ASIPE
TOOLS STOWAGE ASSEMBLY (TSA)
SECOND AXIAL CARRIER (SAC)
CONTINGENCY ORU PROTECTIVE ENCLOSURE (COPE)
FORWARD FIXTURE
AFT FIXTURE
MANIPULATOR FOOT RESTRAINT (MFR)
FLIGHT SUPPORT SYSTEM (FSS) OUTLINE
SOLAR ARRAY DEPLOYED 0°
INCHES
0 50 100 150 200 250 300 350
ASIPE – AXIAL SCIENTIFIC INSTRUMENT PROTECTIVE ENCLOSURE
FSIPE – FINE GUIDANCE SENSOR SCIENTIFIC INSTRUMENT PROTECTIVE ENCLOSURE
K70110-202

ORUC was located between the SAC and the Flight Support Structure to which the Hubble Space Telescope would be docked for in-flight servicing.

STS-82 Discovery (SM2)

11 February 1997
Mission duration: 9d 23hr 37min 7sec
EVA: 5
Commander: Kenneth W. Bowersox (Flt 4)
Pilot: Scott J. Horowitz (Flt 2)
MS1 (EV4): Joseph R. Tanner (Flt 2)
MS2: Steven A. Hawley (Flt 4)
MS3 (EV3): Gregory J. Harbaugh (Flt 4)
MS4 (EV1): Mark C. Lee (Flt 4)
MS5 (EV2): Steven L. Smith (Flt 2)

Preparations for the launch of STS-82 ran surprisingly well since this was the first flight for *Discovery* since a major programme of extensive modifications had been carried out following its last flight in July 1995. A noticeable modification was the replacement of an internal middeck airlock with an external airlock, which, instead of protruding into the pressurised middeck area where the crew lived, occupied a location on the forward face of the payload bay. The countdown began on 8 February but in the final determination of the precise position of the HST in orbit, the launch time was delayed by 1min 17sec, lift-off occurring at 3:55:17am local time (08:55:17am UTC) from LC-39A on 11 February 1997.

Because both payload bay doors had been removed for the upgrade, after reaching orbit the crew opened the doors partially before reclosing them to check the alignment fit. While it was vital to get the doors open as soon as possible to use the radiators on their inner face, it would be catastrophic if the doors could not be closed for re-entry. They were finally driven to their fully open position at 1hr 41min elapsed time. While rendezvous manoeuvres were proceeding, at 19hr 8min the cabin was partially depressurised to 10.2lb/in² (70kPa) in preparation for EVAs and at 19hr 34min the RMS arm was checked out.

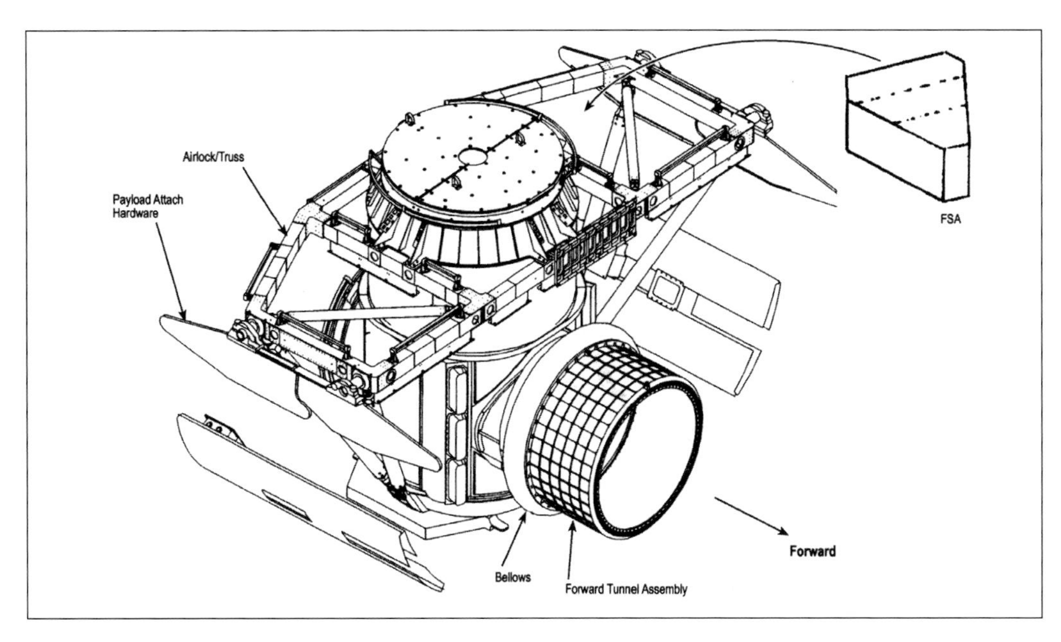

LEFT Developed from the docking unit and airlock module built for the Shuttle Orbiter *Atlantis* when that vehicle was used to dock with the Mir space station, the external Orbiter Docking System was built for use with the International Space Station. The original airlock module was located internally within the middeck area of the pressurised crew compartment and the ODS provided the first mechanism for attaching the Orbiter to another structure via a docking mechanism. It was redundant for use with this HST servicing mission but it did eliminate the bulky airlock module inside the Orbiter and was entered through the hatch in the forward bulkhead of the payload bay. *(Rockwell International)*

In the airlock module, the three EMU suits were checked and final preparations were made for approaching the HST.

The manipulator arm was used to grapple the Telescope at 1d 23hr 38min and it was berthed to the Flight Support Structure 22min later. While preparing for EVA, the venting atmosphere from the airlock module being lowered down to 5lb/in² (34.5kPa) at 2d 17hr 48min struck the solar arrays. This caused the +V2 HST solar array (on the port side of the Orbiter) to rotate about 80° to the hard-stop position and then rebound back 40° before stopping its wayward motion. The airlock was evacuated to a vacuum at 2d 19hr 19min.

EVA-1

The first EVA of the second servicing mission began at 2d 19hr 34min 40sec and would last 6hr 42min 21sec, during which time Lee and Smith worked on the Aft Shroud area, opening the doors and removing the Goddard High Resolution Spectrograph and the Faint Object Spectrograph and replacing them with the Space Telescope Imaging Spectrograph and the Near Infrared Camera and Multi-Object Spectrometer.

EVA-2

Harbaugh and Tanner got a fast start and left the airlock module a little earlier than planned, their EVA officially beginning at 3d 18hr 27min 54sec and lasting 7hr 27min 31sec. First order of business was to remove and replace Fine Guidance Sensor No 1, with Tanner on the end of the RMS arm and Harbaugh working down in the payload bay passing equipment coming and going from the HST. The data recorder was also replaced and the Optical Control Electronics Enhancement Kit was installed, increasing the capability of the FGS.

There was also time to examine several areas of cracking on thermal insulation on the HST in areas that most commonly faced the Sun. There was concern at Mission Control, as well as at the STOCC, that this could worsen and begin to affect the thermal stability of

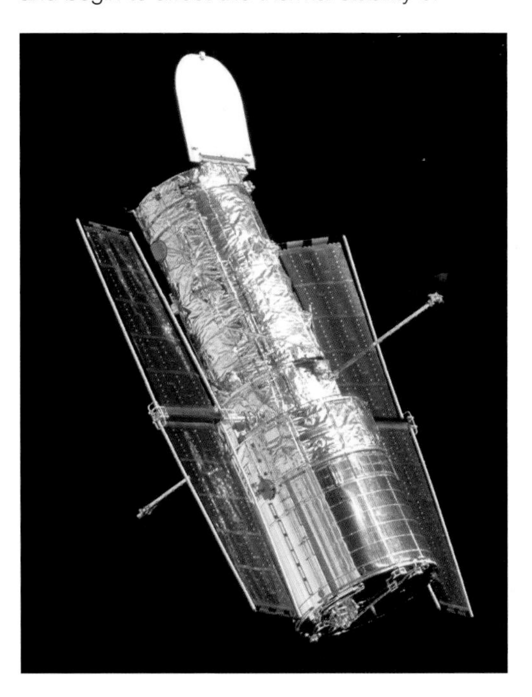

LEFT The target for this second servicing mission – Hubble in quiescent mode with the Aperture Door open. *(NASA)*

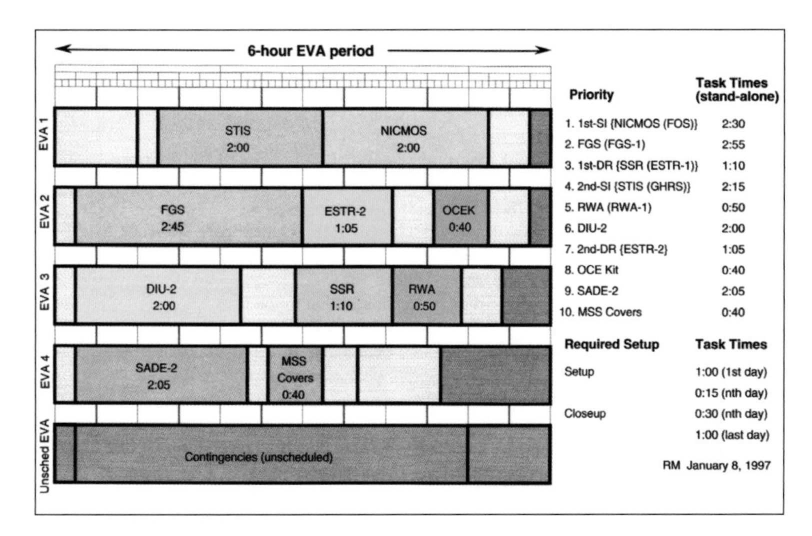

ABOVE The pre-flight timeline for the four planned spacewalks together with activity times as judged from simulated tasks in the NBF. *(Lockheed)*

the Telescope unless it was repaired on this mission. This was something that would be worked on as the next several days progressed, leading to the addition of a fifth EVA, which was always an option in pre-flight planning.

During the second EVA a modest propulsive boost was given to the docked configuration at 4d 1hr 9min 28sec when the RCS Vernier thrusters were fired for 20min 41.9sec to raise altitude by two miles (1.3km) to 372.5 x 367.3 miles (599.3 x 591km). This imparted an almost negligible acceleration and did nothing to interfere with crew activities. A second reboost manoeuvre, this time to avoid debris from an old satellite, began at 4d 6hr 7min 4sec with an RCS burn lasting 10min 12.6sec producing an orbit of 374.5 x 368.3 miles (602.6 x 592.6km).

Between spacewalks, a further reboost manoeuvre using the RCS thrusters was carried out beginning at 5d 1hr 15min 3sec and lasting 19min 46.9sec, delivering an orbit of 385.6 x 369.4 miles (620.4 x 594.4km).

EVA-3

Lee and Smith began their second EVA of the mission at 4d 17hr 55min 39sec, during which they removed and replaced a Data Interface Unit, removed a reel-to-reel recorder and installed the SSDR and removed and replaced one of the four Reaction Wheel Assemblies. As previously, yet another reboost manoeuvre was conducted during this spacewalk when the RCS Vernier thrusters fired up at 5d 1hr 15min 3sec for 19min 46.9sec, changing the orbit to 378.5 x 368.8 miles (609 x 593.4km). The EVA lasted 7hr 11min 0sec. It was during the closeout activities for this that Mission Control decided to add a fifth EVA for repair work on the degraded thermal insulation and to reschedule activities on EVA-4 to accommodate that.

EVA-4

Harbaugh and Tanner began their second spacewalk of the mission at 5d 18hr 48min 7sec and worked to replace the SADE package to the solar arrays, retrieved, repaired and returned from the preceding mission. They then went to the top of the HST to add additional covers to the two magnetometers and then set about starting on insulation repair, beginning high up on the light shield. They worked on two areas damaged by solar radiation and used pieces of multi-layer insulation to prevent further decay and patch the affected surfaces.

The ability to inspect and repair spacecraft in orbit was new to the space programme and very little was known about the long-term effects of exposure to the space environment over extended periods. Much of what was known was extrapolated from ground tests and also from the direct examination of many different types of materials, surface coatings and paints carried into orbit on the massive Long Duration Exposure Facility (LDEF) lifted

BELOW The replacement of the Fine Guidance Sensor as depicted in a planning graphic, new computer tools only just coming into use for astronaut training and familiarisation. The crude low-definition image is difficult to reconcile with 21st-century graphics tools. *(NASA)*

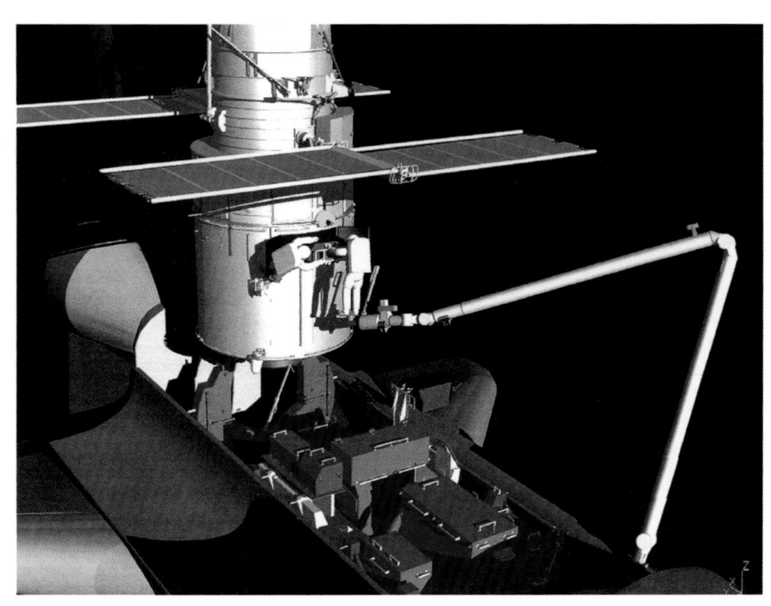

by Shuttle in April 1984. It was retrieved by another Shuttle in December 1989 only four months before the launch of the Hubble Space Telescope, its return delayed several years by the *Challenger* disaster in January 1986.

What was learned from examining the effects of the space environment on LDEF samples, and from periodic inspection of the HST, fed across to the International Space Station which would start assembling hardware in orbit less than two years after the return of this second servicing mission.

The EVA ended at an elapsed time of 6hr 34min 30sec.

EVA-5

Lee and Smith began their third spacewalk of the mission at 6d 18hr 17min 25sec and set about adding repair patches of multi-layer insulation to the HST. Thermal blankets were added in three areas of the Telescope before they returned to the area of the airlock module. Some concern had been expressed about the condition of one of the three Reaction Wheel Assembly units that had been installed prior to launch, not the one changed out on EVA-3. Ground controllers worked a potential problem that could have required the EVA astronauts to retrieve a replacement from the ORUC pallet and install that, but the problem was solved and the EVA ended at an elapsed duration of 5hr 17min 21sec.

Checking out

Shortly after the conclusion of the final EVA, the last reboost manoeuvre was made with the RCS Vernier thrusters at 7d 1hr 32min 58sec, a continuous burn for 31min 53.5sec placing the docked spacecraft in an orbit of 385.5 x 369.4 miles (620.3 x 594.4km). This was a record altitude for the Shuttle. The RMS arm grappled the HST at 7d 18hr 25min and released it to free flight 3hr 20min later. After reconfiguring the Orbiter, packing away flight items and preparing *Discovery* for return to Earth, the landing was delayed for one orbit by unacceptable weather at the Kennedy Space Center. The payload bay doors were closed at 9d 18hr 13min 43sec and the de-orbit burn was made with ignition by the Orbital Manoeuvring System motors at 9d 22hr 26min 37.7sec, changing velocity by 504.5ft/sec (153.8m/sec). Landing occurred just after 03:32am (08:32 UTC) on 21 February on runway 15 at the Kennedy Space Center, for a mission duration of 9d 23hr 37min 7sec.

At first the new instruments checked out well with excellent images from NICMOS Cameras 1 and 2 but with the focus on Camera 3 out of range of the internal adjustment mechanism. It was believed that unexpected thermal contact in the dewar provided a path for excess heat to travel to the outer structure. It was hoped that this contact might release at some point but very shortly after the SM2 mission this short caused early exhaustion of the solid nitrogen, reducing its anticipated life from five to two years.

The other instruments checked out well, the

ABOVE The GHRS is exchanged for the STIS on orbit during the first EVA. *(NASA)*

BELOW The Solid State Recorder replacement in Bay 5, as digitally depicted for astronaut training. *(NASA)*

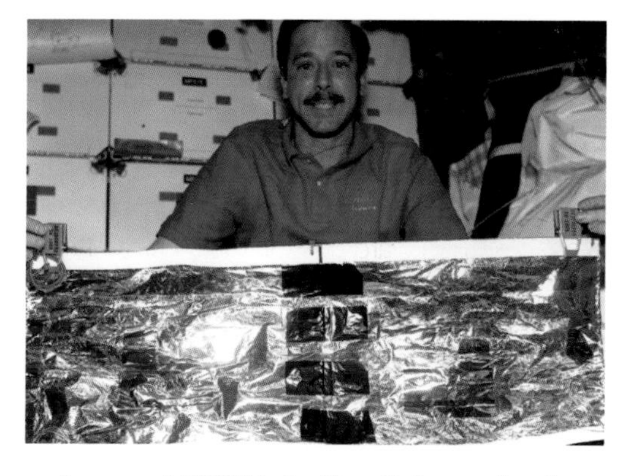

RIGHT **Insulation prepared inside *Discovery* would be used on a fifth, unscheduled, EVA to patch up areas which had suffered from exposure to the space environment.** *(NASA)*

replacement FGS1R significantly increasing the astrometric performance of the original Fine-Guidance Sensor, allowing 1R to be assigned as the prime instrument for this science. The STIS too lived up to expectations and engineering tests went well allowing an early start on making the instrument available to investigators.

An orbital test

Another milestone of note was achieved with the successful test of Telescope equipment destined for installation on SM-3A with the STS-95 mission launched on 29 October 1998. The STS-95 mission was a long-duration science and technology flight that also carried John Glenn back into space for the second time after more than 36 years. On 20 February 1962 Glenn had been the first American to orbit the Earth and was allowed to make the flight for several reasons, not least because he had pushed hard to become accepted for a return to space but also because it was of medical interest to study the physiological reactions of a 77-year-old person.

BELOW **An image of the HST raised just above the FSS, showing the latch pins.** *(NASA)*

The Hubble Space Telescope Orbital Systems Test Platform (HOST) weighed 2,800lb (1,270kg) and was carried on STS-95 to demonstrate that electronic and thermodynamic equipment scheduled for installation on the HST during the third servicing mission would work satisfactorily in the radiation and microgravity environment of space at the altitude occupied by the Telescope. One of the items of equipment was a new cooling system for the NICMOS instrument installed during SM2. Comprising a reverse turbo-Brayton cycle cooler and capillary pump loop, it represented a significant advance that offered the possibility to eliminate cooling systems that depend on dewars to supply super-cold liquids for cooling sensors.

Another piece of equipment tested on STS-95 and destined for the Telescope was the Intel 486-technology computer being developed to replace the troublesome DF-224, still presenting operating problems in the HST. The HOST test allowed engineers to determine the level of heavy ions that cause single-event upsets in similar electronic devices. By measuring the radiation levels and the effect those have on the components and equipment, the final configuration of the replacement computer could be either verified or changed.

HOST also carried on test a Solid State Recorder (SSR) to determine whether it would be as robust within the space environment as engineers believed. Errors had been noted with the single-event upsets and transients to the equipment on the HST and this was a persistent problem. It was not known whether the transients were unique to the equipment on the Telescope or whether it was common to SSRs of this kind.

The platform for the HOST equipment was the airborne support equipment cradle that had first flown on the STS-48 mission when it carried the Upper Atmosphere Research Satellite into orbit. The cradle itself was reworked to accommodate the HOST equipment and was left installed in the Orbiter's payload bay, exposed to the space environment throughout the period the payload bay doors were open.

Failure rates

By March 1999 it had become apparent that plans for a single SM3 servicing mission were unrealistic and that it would have to be split into two components, with a launch late in the year to replace ailing gyroscopes, without which the Telescope would be unusable. Moreover, the NICMOS instrument was unusable from January 1999 when the coolant ran out, but any repairs or refurbishment would have to wait until the following mission; NASA was already working a plan to install a cooling system, but that would not be ready for at least a year or 18 months.

The plan was to launch as soon as possible to keep the HST operational, and to do that the two RSUs and their six gyroscopes would have to be replaced. One had failed in 1997, another in 1998, a third in 1999 and the fourth on 13 December. Because two gyroscopes were not sufficient for science operations the HST was placed in safe mode. Analysis had shown that as the gyroscopes spin at 19,200rpm on gas bearings mounted in a sealed canister floating in a thick viscous fluid, pressurised air used to force the fluid into the gyroscope cavity reacted with it to form a corrosive material that degraded some of its thin electrical wires. As a precaution the replacement gyroscopes in the two RSUs had pressurised nitrogen instead of air.

Planning for the third servicing mission assigned that role to STS-103, which at the beginning of 1999 was scheduled for June 2000. That got changed when the problems with the HST began to show concern about its operating life, and the third servicing flight was moved to October 1999. When technical concerns held back the availability of Shuttle Orbiters while inspections were made, the date slipped to December. Originally designated for ten days with four EVAs, as the flight drifted toward the end of the year it was shortened to eight days with three spacewalks so as not to be in space on 31 December.

The crucial factor was the uncertainties surrounding the computer world that the effect of the new century on clock systems – which had not been built to understand the turn of the year from 99 to 00 in the numerical registration – would cause global meltdown in computer systems and software programmes. This was known as the Year 2000 question, or Y2K. Fears grew that there could be widespread failures, and so as not to incur risk unnecessarily NASA was determined not to have a Shuttle flight in space over the New Year.

As defined, the refurbishment mission would be designated SM3A as a rescue flight to keep the Telescope working, followed by SM3B in the second quarter of 2001 to remove the Faint Object Camera and install the Advanced Camera for Surveys (ACS). Plans for SM3B included replacement of the STSA II solar arrays built by British Aerospace with a pair of STSA III arrays. There were also plans for a special cooling system for the ailing NICMOS instrument and a new cooling system for the Aft Shroud.

Although preliminary, there were tentative plans for a fourth servicing mission in 2003 to carry up a Cosmic Origins Spectrograph, a Wide Field Camera III to replace the WF/PC II, and a third Fine-Guidance Sensor to bring the Telescope fully up to service. There were thoughts about ending the programme in 2010, either by boosting the HST to a very much higher orbit where it would remain almost indefinitely, or returning it intact aboard a Shuttle.

Advanced Computer (AC)

Based on a successful orbital test on STS-95, the Goddard Space Flight Center cleared the Advanced Computer for SM3A as a replacement for the aged DF-224. Based on the Intel 80486 microchip, it has 20 times the operating speed and six times the memory capacity of the DF-224. Designed around existing commercially available components, the AC is configured as three separate and autonomous computers in one, each single-board computer (SBC) possessing 2MB of fast static RAM and 1MB of non-volatile memory. It is built to communicate with the HST through direct memory access on each SBC through the Data Management Unit, although only one SBC can control the Telescope at any one time.

On power-up, each SBC executes a self-test and copies operating software from the slower non-volatile memory to the RAM, and this is capable of diagnosing problems with the AC and reporting that to the ground. The AC measures 18.8in x 18in x 13in (47.7cm x 45.7cm x 33cm), weighs 70.5lb (32kg), and is located in Bay 1 of the System Support Module.

The Data Management Unit (DMU) is a set of printed-circuit boards interconnected through a backplate and external connectors. It measures 26in x 30in x 7in (60cm x 70cm x 17cm) and weighs 83lb (37.7kg). It encodes and sends messages to selected HST units and all Data Management System units, powers the oscillators and is the primary timing source. It also receives and decodes incoming commands, transmitting each processed command to be executed. The DMU receives science data from the SI C&DH while engineering data comes from each subsystem. The data can be stored in the on-board recorders if direct transmission cannot go via TDRSS.

Orbital Replacement Unit Carrier (ORUC)

The ORUC for the third servicing mission was identical structurally to that flown for the first servicing flight but in this configuration carried the Advanced Computer and two Voltage Improvement Kits in one container and a Solid State Recorder, an S-Band Single Access transmitter, RSUs, and associated flight harnesses in a second container. A third container houses the Fine-Guidance Sensor and a fourth container encloses the spare Advanced Computer and RSUs should they be needed.

STS-103 *Discovery* (SM3A)

19 December 1999
Mission Duration: 7d 23hr 10min 47sec
EVA: 3
Commander: Curtis L. Brown (Flt 6)
Pilot: Scott J. Kelly (1)
MS1 (EV1): Steven L. Smith (Flt 3)
MS2: Jean-Francois Clervoy (Flt 3)

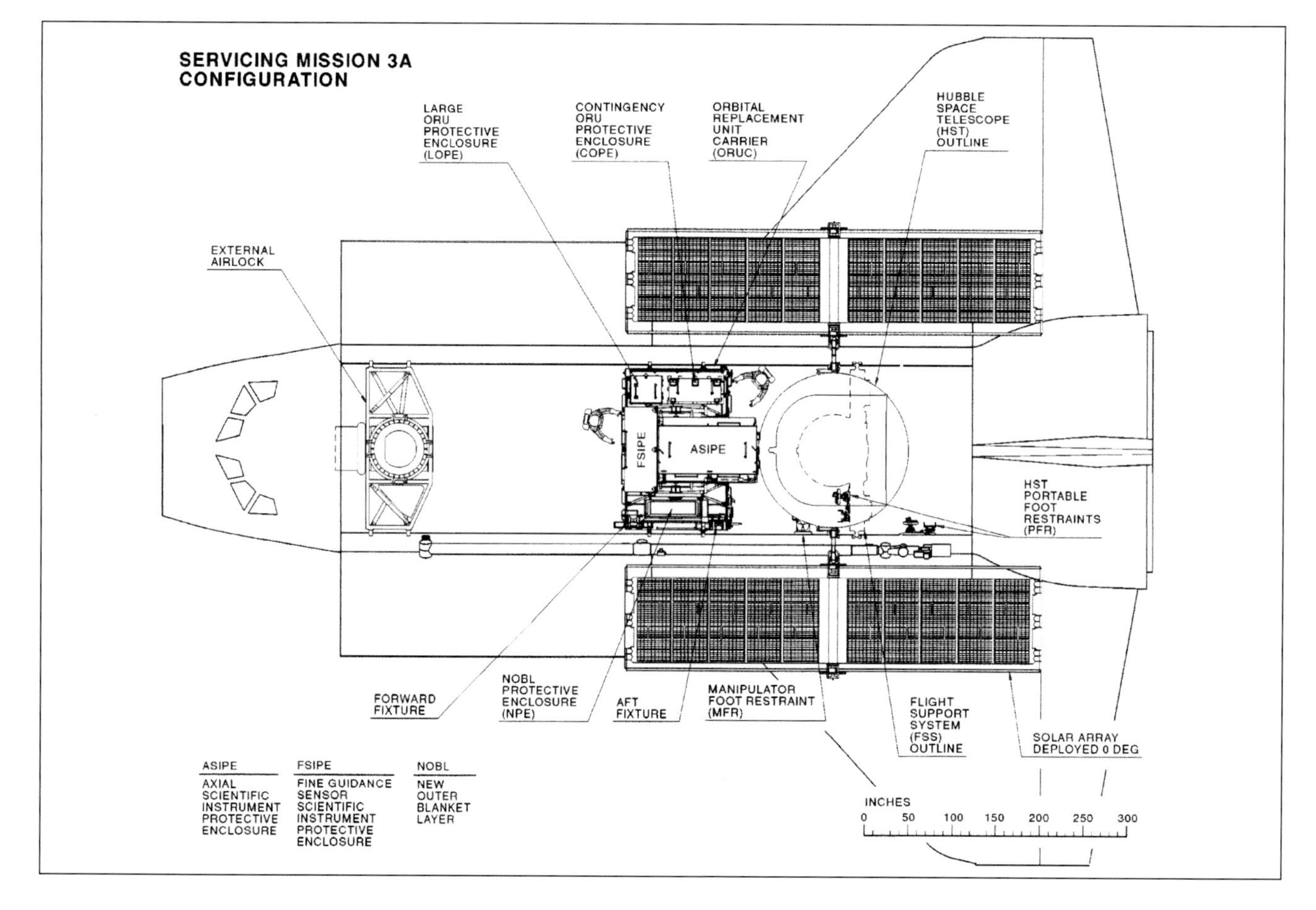

BELOW The general arrangement of equipment in the Shuttle Orbiter payload bay for the STS-103 mission, less crowded than on the first two servicing missions, with the position of the HST when berthed at the FSS for SM3A. *(Rockwell International)*

RIGHT The ORUC carrier for the SM3A mission, with containers for the Fine Guidance Sensor and other items destined for installation on the HST. *(Lockheed)*

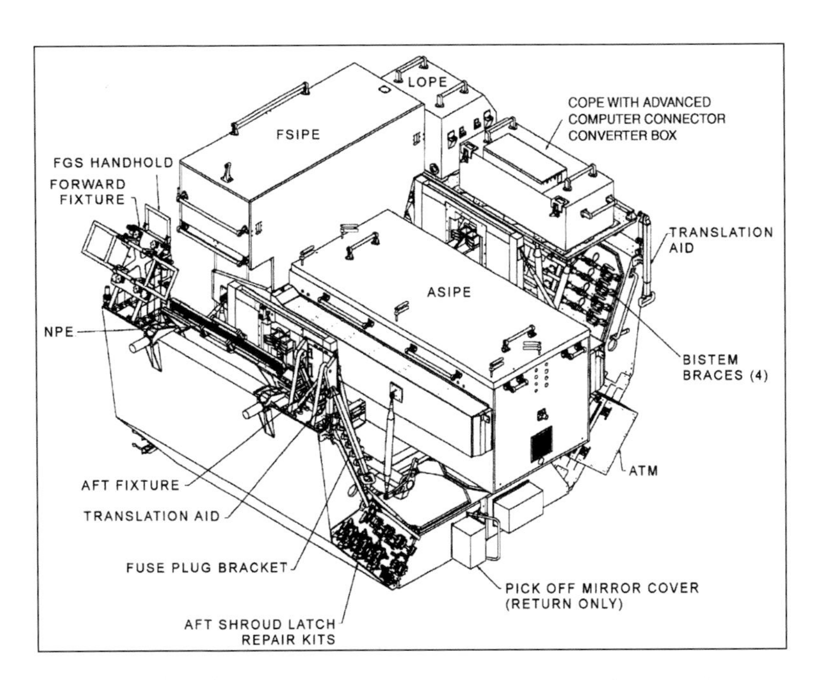

MS3 (EV2): John M. Grunsfeld (Flt 3)
MS4 (EV3): Michael Foale (Flt 5)
MS5 (EV4): Claude Nicollier (Flt 4)

A long and complicated sequence of delays and postponements from 28 October as the original flight date preceded the launch of STS-103/SM3A at 7:50:00pm local time (00:50 UTC, 20 December) from LC-39B at NASA's Kennedy Space Center. The usual sequence of events with a visit to the Telescope preceded rendezvous and RMS arm grapple attached at 1d 23hr 44min 1sec and berthing to the Flight Support System 1hr 8min later.

EVA-1

After depressurising the airlock using a new procedure so as not to cause the solar arrays on the HST to move (as they had on the previous flight), Smith and Grunsfeld began their first EVA of the mission at 2d 51min 37sec, exiting the airlock to begin what would become the second longest spacewalk in Shuttle missions to date. The EVA in general was plagued with small but irritating anomalies, problems with tools, foot restraints and EMU battery chargers, making it far from fault free.

Nevertheless, they replaced the three Rate Sensor Units and were manoeuvred by the RMS arm to the ailing NICMOS instrument, where they removed caps to purge nitrogen coolant and allow residual coolant in the line to vent to the vacuum of space in readiness for a new coolant jacket on the next servicing mission. The spacewalkers then installed six Voltage/Temperature Improvement Kits in Bays 2 and 3, to increase efficiency and reduce the

CENTRE Astronauts Smith and Grunsfeld study the detail on an electrical panel that they will address on orbit during the third servicing mission. *(NASA)*

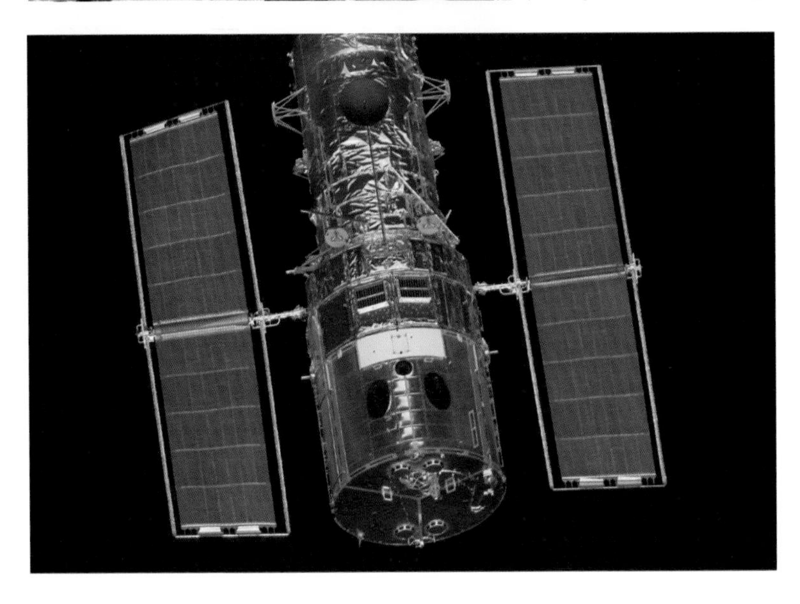

RIGHT Rendezvous as viewed from the flight deck of *Discovery* making a return visit to the Telescope. *(NASA)*

ABOVE Just before the RMS arm grappled the HST, a view of the forward section of the Telescope. *(NASA)*

ABOVE At the point of contact with the FSS, the Telescope is still in the grip of the RMS arm. *(NASA)*

RIGHT A view of the Telescope from the aft-facing windows on the flight deck of *Discovery*, with the latches and engagement tongues visible on top of the Orbiter Docking System. *(NASA)*

likelihood of over-charging, a condition to which older batteries are vulnerable.

A general slippage in the timeline resulted in permission from Mission Control for an extension, their EVA ending at a duration of 8hr 15min 30sec, with sundry items postponed to the next EVA.

EVA-2

British-born, now a US citizen, Michael Foale and Swiss astronaut Claude Nicollier began the second mission spacewalk at 3d 17hr 42min 33sec and set about the business of replacing the DF-224 with the Advanced Computer in Bay 1, and followed the door closure with a fresh blanket of multi-layer insulation. Foale was an old hand, having conducted spacewalks on a previous Shuttle mission and also from the Russian Mir space station.

Next up was a traverse via the RMS arm to +V3 Fine-Guidance Sensor, and after opening the doors to the existing FGS-2 the connectors were removed and the latches released. The old unit was temporarily stowed on the ORUC and the replacement inserted and connected before closing the latches. Video imagery was taken of the installation and the doors were closed. The

LEFT A close-up view of the FSS and HST interface with detail on the HST and the support structure. *(NASA)*

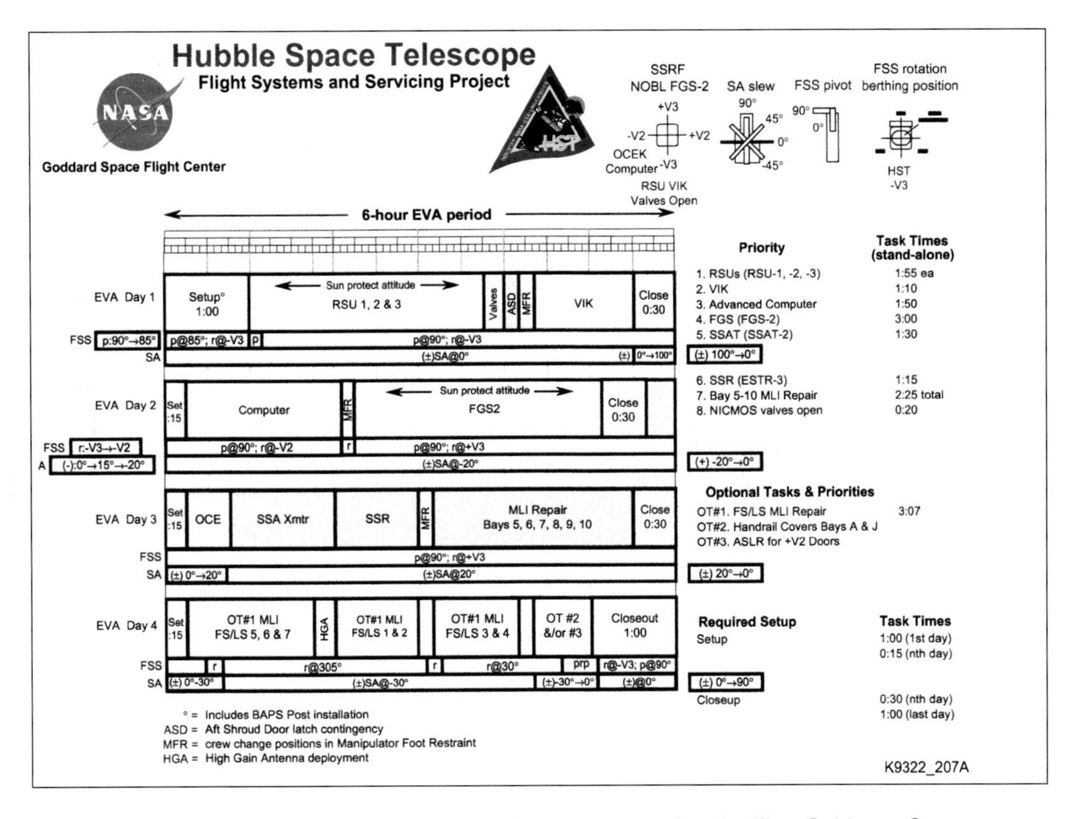

LEFT The pre-mission flight timeline for the three planned EVA periods with information on the attitude of the HST on the Flight Support Structure, the assessed duration of individual tasks and a broad range of information about the events scheduled for each spacewalk. *(Rockwell International)*

EVA lasted 8r 10min, the third longest mounted from the Shuttle to date.

EVA-3

Smith and Grunsfeld began the final spacewalk of this mission at 4d 18hr 59min 51sec. Earlier, a problem with one of the EMU suits caused a switch to the third (spare) suit prior to suiting up and depressurisation. The first task was to Bay 3, where the doors were opened and the additional Optical Control Electronics Enhancement Kit (OCE-EK) cable connectors were mated to the Fine-Guidance Sensor, after which the doors were closed and the astronauts moved to Bay 5. Here the S-Band Single Access Transmitter was removed and the SSAT-2R replacement inserted in its place. Then the old reel-to-reel tape recorder was removed and the SSR-3 digital recorder installed as a replacement.

The next task was to apply new insulation to the various bays, but after Bays 9 and 10 had been addressed Bays 5, 6, 7 and 8 were left as they were, due to time constraints on the

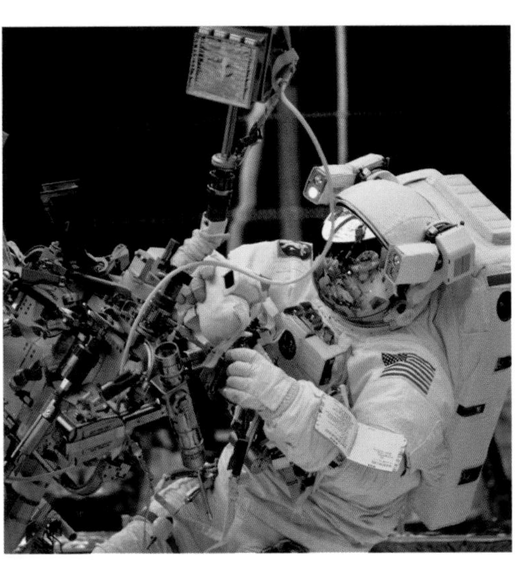

FAR LEFT Grunsfeld works on replacing one of the radio transmitters. *(NASA)*

LEFT Smith tangles with leads, wires, tethers, power tools and aids for carrying out the EVA tasks. *(NASA)*

LEFT Smith and Grunsfeld at the Aft Shroud of the SSM in a view that displays the extreme distortion in the solar arrays. *(NASA)*

ABOVE ESA astronaut Claude Nicollier works on a storage enclosure with a power tool. *(NASA)*

BELOW Smith (right) and Grunsfeld working inside the Telescope. *(NASA)*

BELOW Smith gets a grip on an HST handrail while Grunsfeld closes up inside. *(NASA)*

RIGHT British-born astronaut Michael Foale (left), veteran of the Russian space station Mir, and Swiss-born Claude Nicollier change out the Fine Guidance Sensor. *(NASA)*

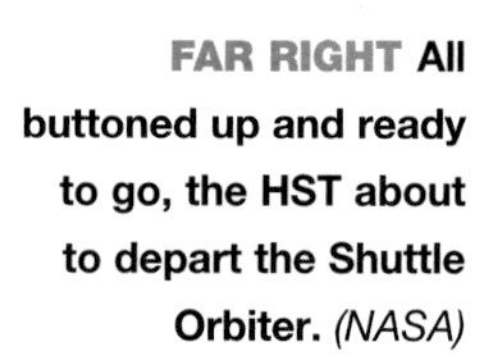

FAR RIGHT All buttoned up and ready to go, the HST about to depart the Shuttle Orbiter. *(NASA)*

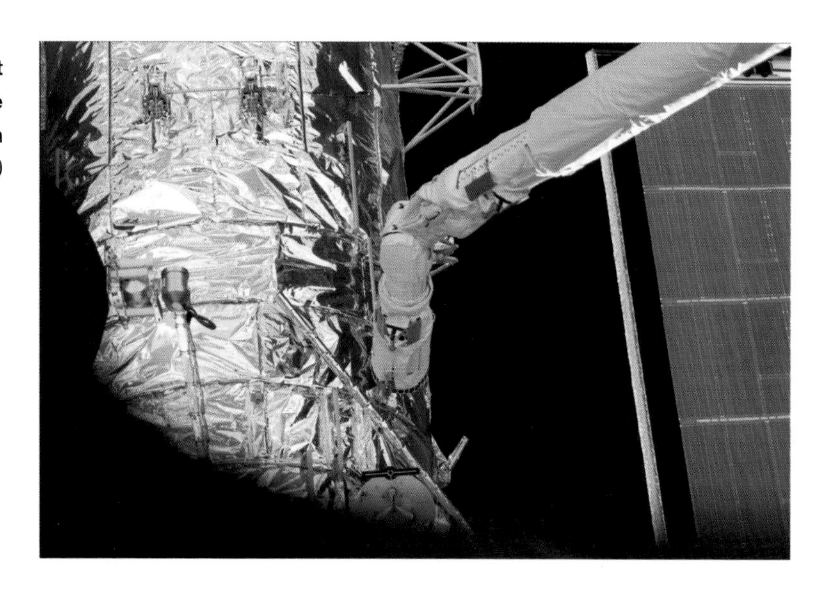

RIGHT About to be released for independent flight and another tranche of observations, the HST will benefit from its engineering upgrade on this third servicing mission. *(NASA)*

spacewalk. Again the EVA was compromised to some degree by failed tools and some minor anomalies but overall there were no seriously important objectives left unaddressed. When it was time to get back in, the Airlock Power Supply would not power EMU-3 with the servicing and cooling umbilical, and so repressurisation was performed with the suit on battery power. The EVA lasted 8hr 8min 30sec.

Adaptation

With the three spacewalks completed there was an urgency to return to Earth and avoid any looming catastrophes at the stroke of midnight on 31 December but this was inevitably – given the launch date – the first time a NASA crew had been in space over Christmas since the Skylab flight in 1973.

The HST was unberthed by the RMS arm at 5d 20hr 28min 41sec and released back into orbit 1hr 44min 20sec later. No reboost burns had taken place on this flight but two Orbiter separation manoeuvres occurred to carry the Shuttle away from the Telescope. Excessive crosswinds at the Kennedy Space Center delayed the return by one orbit and *Discovery* landed on 27 December at an elapsed time of 7d 23hr 10min 47sec.

With the third servicing flight a distinct success and upgrades to major engineering systems, the observation programmes proceeded on a surer guarantee of reliability and success. The popularity of the Telescope was undiminished, with many times the available observation slots requested by astronomers. But the overall programme had to fit within the parameters of NASA's ambitious science and astrophysics research effort, and an evaluation of the Telescope, together with much soul-searching at the STScI, resulted in a decision to fund the HST through to the end of 2010, its fate beyond that date uncertain.

However, the second component of the third servicing mission (SM3B) was scheduled for late 2001 or early 2002, with a fifth and final

BELOW The HST drifts away from the Orbiter *Discovery* as the Shuttle puts on a gentle separation manoeuvre at the end of the third servicing mission. *(NASA)*

servicing mission (SM4) in 2004. Three new science instruments were being developed for those flights to replace existing instruments on the Telescope. SM3 would carry the Advanced Camera for Surveys and the NICMOS Cooling System, restoring observers' time on the near-infrared instrument, while SM4 would transport the Cosmic Origins Spectrograph and the Wide Field Camera III.

Much had changed in the ten years since the Telescope had been placed in orbit and only the Faint Object Camera remained of the original five science instruments installed when it had been launched. This had become obsolete and had been decommissioned, and would be replaced by the Advanced Camera for Surveys, an instrument which would transform the optical capabilities of the HST.

The performance of the Reaction Wheel Assemblies so crucial to the effective operation of the Telescope became a cause for concern when RWA-1 experienced a seven-minute dropout in telemetry on 10 November 2001, preventing flight controllers getting data on the wheel's rotation speed but noting two torque disturbances in the attitude of the HST. The performance resumed as normal but managers decided to replace this wheel on the fourth servicing mission.

Advanced Camera for Surveys (ACS)

Conceived, designed and developed in a collaborative effort between the Johns Hopkins University, NASA's Goddard Space Flight Center, Ball Aerospace and the Space Telescope Science Institute, the purpose of this third-generation instrument was to fit the Telescope with a combination of detector area and quantum efficiency surpassing that available from extant instruments by a factor of ten, measured in an almost surreal balance of the product of the imaging area and the instrument 'throughput'.

The ACS consists of three completely separate and independent channels for unique imaging roles with wide-field, high-resolution and ultraviolet imaging capability together with various filters. The ACS is five times more

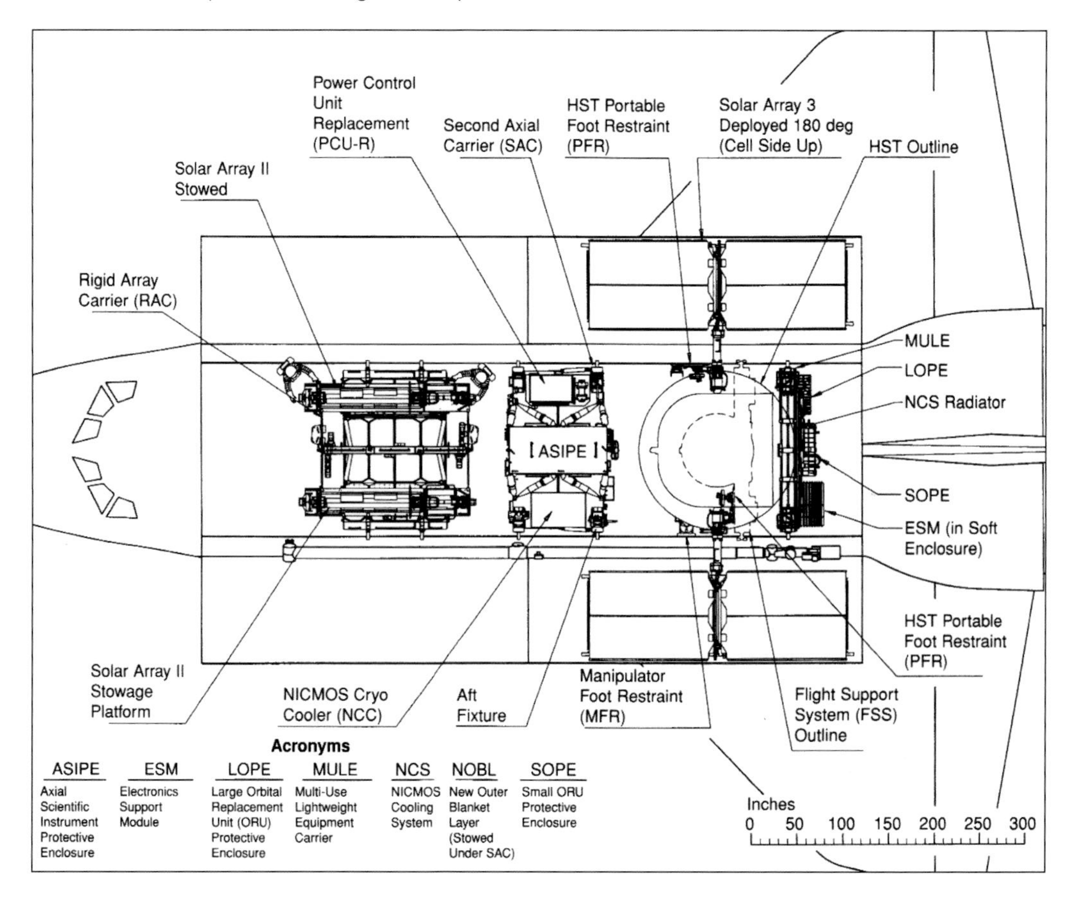

RIGHT **The primary objectives for SM3B were to install the SA-3 arrays and the new NICMOS cooling system, replace the Faint Object Camera with the Advanced Camera for Surveys, and provide upgrades to the spacecraft's power control unit. The payload bay was full for this fourth servicing mission.** *(Rockwell International)*

sensitive than the WF/PC II and boasts more than twice its field of view. With a weight of 875lb (397kg), it is 7ft x 3ft x 3ft (2.2m x 0.9m x 0.9m) in size and has a total wavelength range of 115–1,050nm. It sits in the axial bay behind the primary mirror, replacing the Faint Object Camera. The three operating channels are Wide Field (WFC), High Resolution (HRC) and Solar Blind (SBC).

The WFC in the visible and red wavelengths is the instrument of choice for imaging. It has a spectral range of 350–1,050nm, and a detector array of 4,096 x 4,096 pixels, at 15 x 15 microns per pixel, with a field of view of 200 x 204 arc-sec. This channel would be preferable for wide sky surveys for studying the nature and distribution of galaxies, while its red light sensitivity lends it toward studies on old and very distant galaxies with red-shifted spectra resulting from the expansion of the universe.

The HRC is designed to provide astronomers with highly detailed surveys of the inner regions of galaxies and to search neighbouring stars for planets and protoplanetary disks. It has a coronograph to suppress light from bright objects and allows astronomers to study the interiors of galaxies where massive black holes reside, but it can also be used for high-precision photometry. The HRC has a spectral range of 200–1,050nm, a detector array of 1,024 x 1,024 pixels at 21 x 21 microns per pixel and a 26 x 29 arc-sec field of view.

The SBC allows observers to block visible light to enhance the ultraviolet portion and allows study of emission lines that indicate the presence of certain molecules which can only be observed in this region of the spectrum. It has a spectral range of 115–180nm, a detector array of 1,024 x 1,024 pixels, a 25 x 25 micron pixel size and 26 x 29 arc-sec field of view. This channel employs a highly sensitive photo-counting detector to enhance visibility of certain molecules.

In addition to observing in the optical portion of the spectrum, the ACS can perform prism spectroscopy, low-resolution, wide-field spectroscopy in the spectral range 5,500–11,000Å in both the HRC and the WFC. The ACS offers objective prism spectroscopy at low resolution in the near-ultraviolet at 2,000–4,000Å in the HRC only. It can also conduct objective prism spectroscopy, low-resolution far-ultraviolet spectroscopy in the range 1,150–1,700Å available only in the SDBC. Aberrated beam coronography in the HRC at 2,000–11,000Å is provided with optional occluding spots of 1.8 arc-sec and 3 arc-sec. Imaging photopolarimetry in the HRC and WFC channels provides polarisation angles of 0°, 60° and 120°.

To serve the three operating channels, the ACS has two main optical channels, one for the WFC and one shared between the HRC and the SBC, each channel possessing independent corrective optics to compensate for the spherical aberration in the primary mirror. The

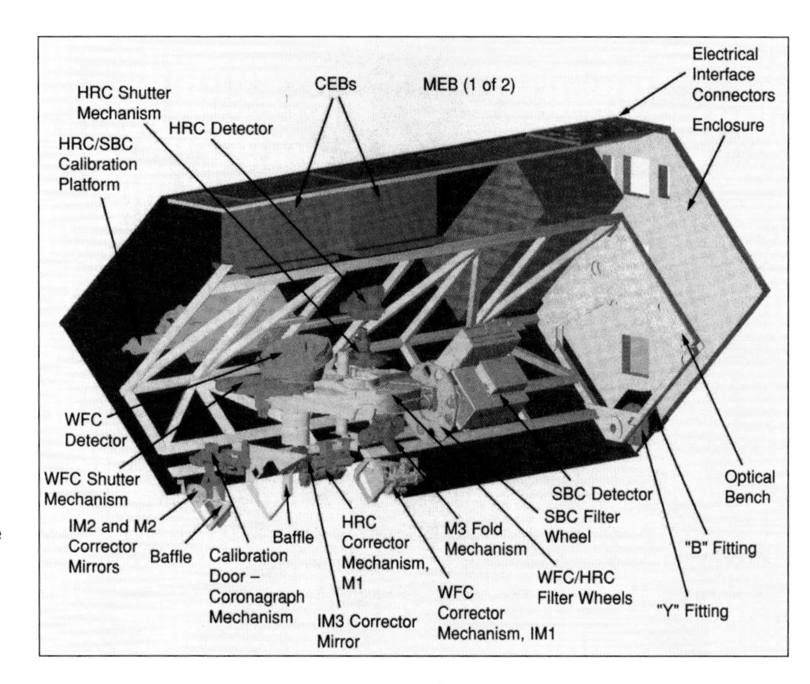

ABOVE Technology and the science of instruments had moved on in the two decades since the original suite of science equipment had been developed for the HST. The Advanced Camera for Surveys (ACS) would replace the FOC and provide a quantum step in efficiency and a major improvement in detector area. *(Ball Aerospace)*

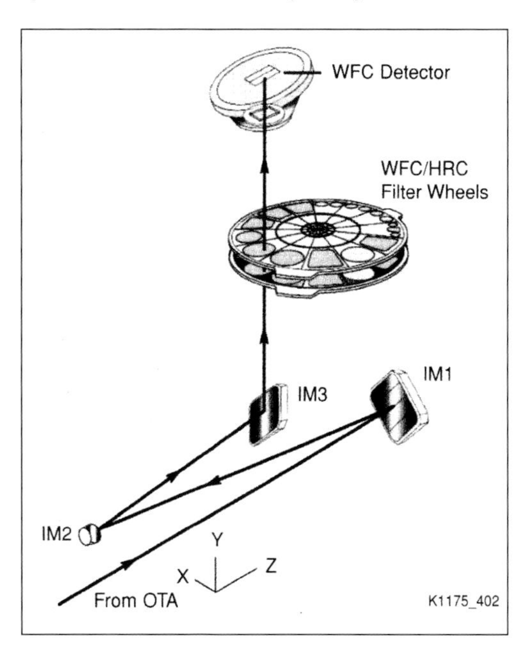

LEFT The design of the wide-field channel in the ACS was new to the HST and this diagram shows the optical path. *(Ball Aerospace)*

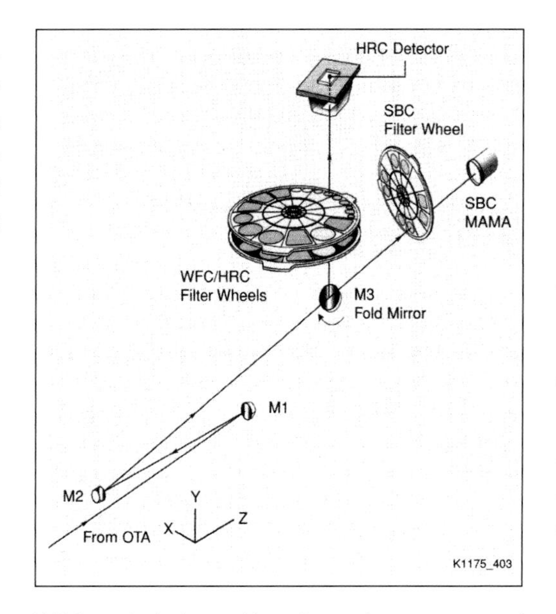

RIGHT The high-resolution and solar blind channels of the ACS are selected by means of a plane fold mirror. *(Ball Aerospace)*

WPC optical channel has three elements coated with silver, which cuts off wavelengths short of 3,700Å to optimise observations of visible light. With two filter wheels shared with the HRC it can provide internal parallel viewing for some filter combinations.

The HRC and SBC are selected by means of a fold mirror, in one position to move it so that the image is directed to the HRC through the WFC and HRC filter wheels, and to the SBC the mirror is moved out of the beam to a two-mirror optical chain that passes images through the SBC filter wheel to the SBC detector. To obtain an aberrated coronograph a mechanism is inserted into the optical chain of the HRC, and this places a substrate with two occulting spots at the aberrated telescope focal plane and an apodiser at the re-imaged exit pupil.

The ASCS has three filter wheels of which two are shared by the WFC and the HRC and one is dedicated to the SBC. Each wheel contains one clear aperture for each of the WFC and HRC applications. Because of this it is possible to conduct parallel observations with some filter combinations but since the wheels are shared it is not possible to independently select the filter for the WFC and HRC separately in parallel observations.

BELOW A cutaway of the NICMOS coolant system designed to restore functionality to an instrument that had exhausted its cryogenic nitrogen more than three years earlier in January 1999. *(Lockheed)*

NICMOS Cooling System (NCS)

Designed by the Goddard Space Flight Center as an experimental cooling system for the NICMOS instrument which failed in late 1998, the NCS was an attempt to revive operations by using a primary coolant loop to carry heat away from the NICMOS cryostat to keep the detectors at their operating temperature of 70°K (-203°C/-334°F), utilising a reverse-Brayton-cycle cryocooler. The NCS uses non-expendable neon gas as a coolant in a closed system with high cooling capacity using a miniature cryogenic circulator to remove heat from the NICMOS instrument and transport it to the cryocooler.

The primary coolant loop is the heart of the cryocooler and contains a compressor, a turboalternator and two heat exchangers. This loop implements a reverse-Brayton thermodynamic cycle and provides cooling power for the entire system. It produces a significant amount of heat – up to 500W – which is removed by a capillary pumped loop connecting the main components generating heat to external radiators. Heat is removed by evaporating ammonia on the hot end and recondensing it at the cold end. The small turbine turns at up to 400,000rpm and is virtually vibration-free, ideal for the HST.

The major operating functions of the NCS are contained within the equipment section, which has an 8051 microprocessor implementing control laws for the cooler functions. It also controls the capillary pumped loop reservoir temperatures and regulates the amount of heat transferred to the radiators. In addition it collects and transmits telemetry from the NCS and provides data on the general handling of the equipment, relaying commands to the NCS and sending data through the HST communications channels, down to Earth.

Power Control Unit (PCU)

The original PCU had been with the HST since it was built, and the fourth servicing mission would provide a replacement, doing just the same job of distributing electrical power from the solar arrays to the various systems, subsystems and components. The PCU also protects the Telescope from power surges and electrical spikes. The new unit weighs 160lb (72.6kg) and occupies a volume 45in x 25in x 12in (114cm x 63.5cm x 30.5cm), modified to afford better handling for the spacewalkers.

The original PCU was not designed or installed as an Orbital Replacement Unit and its removal posed challenges for the astronauts. Built by Lockheed Martin Space Systems, its installation required the HST to be completely powered down for the first time since launch, a harrowing time for flight controllers as well as scientists. Like its predecessor, the new PCU has 36 connectors that had to be removed from the old unit and reinserted in the new one. The old unit had shown signs of age, several relays that control battery charging having failed. An electrical joint had also drifted loose causing some loss of power and a risk of the battery overheating.

Solar Array III (SA-3)

Enduring problems with the original flexible solar arrays built by British Aerospace resulted in modestly redesigned arrays replacing them during the first servicing mission in 1993, but a completely new rigid array system was designed and built at Lockheed Martin and adapted by the Goddard Space Flight Center and these were installed on the Telescope in SM3B.

The wings comprise eight fixed panels and were originally designed for the commercial communications satellite Iridium. Mounted to aluminium-lithium support frames, the new wing assemblies support Gallium-Arsenide photo-voltaic cells and have one-third less surface area but produce 20% more power.

When deployed each rigid array has a length of 24.75ft (7.1m) and a width of 8ft (2.6m), compared to 39.67ft (12.1m) and 9.75ft (3.3m) respectively for the flexible SA-2 arrays. They each weigh 640lb (290kg), versus 339lb (153.8kg) for the SA-2 arrays. After six years of operation, the SA-2 array was producing 4.6kW of electrical power while the SA-3 arrays were estimated to produce 5.27kW at the same stage of life.

New Solar Array Drive Mechanisms were also loaded aboard *Columbia* for the SM3B mission and these replaced existing units. The wings and the SADMs were tested in 2000 at the European Space Agency's facility at ESTEC in the Netherlands.

Because they have less surface area the SA-3 arrays produce less drag, which, although operating in the relative vacuum of space, is subject to the cumulative effect of random molecules colliding with the spacecraft. There is also an effect from solar radiation, although this is not as great as it would be if the spacecraft were to be placed outside the magnetosphere in the solar wind of charged particles from the Sun.

Rigid Array Carrier (RAC)

This pallet is identical to the one used under the ORUC designation on earlier servicing missions and was located in *Columbia*'s forward payload bay where it held the SA-3 and diode boxes, also supporting the returned SA-2 arrays

BELOW The Rigid Array Carrier was located in the forward section of the payload bay and carried the two SA-3 solar arrays and support structure for the solar arrays they replaced, which had been on the HST since December 1993, almost nine years. *(Rockwell International)*

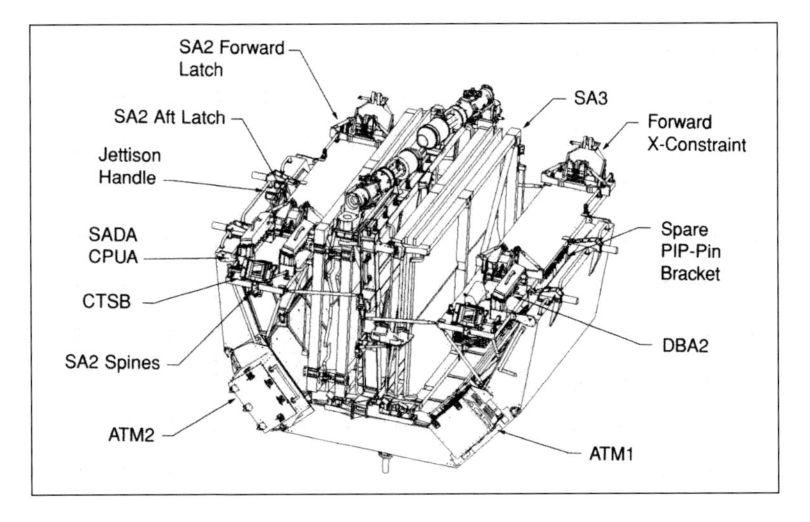

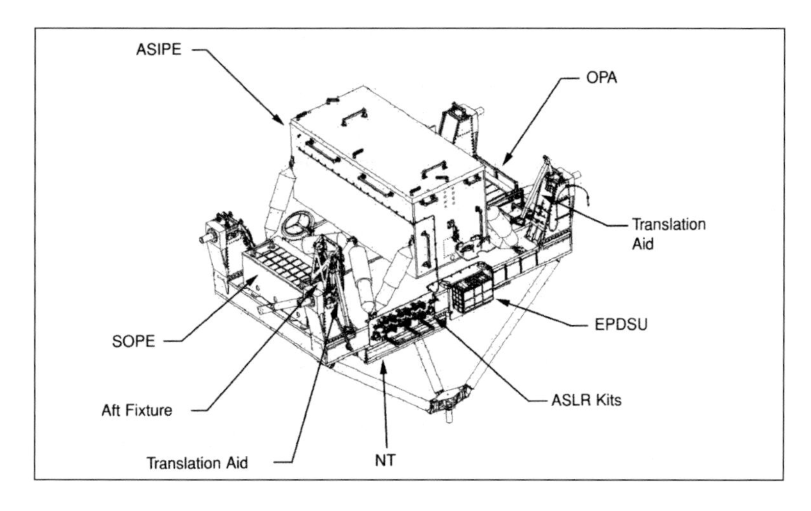

ABOVE The Second Axial Carrier (SAC) made a return to space carrying the Advanced Camera for Surveys, the Power Control Unit, the NICMOS cryocooler and new protective enclosures for the Equipment Bay doors on the SSM, in addition to sundry small items. It sat between the FSS/HST mounting and the SAC. *(Lockheed)*

retrieved from the Telescope. It also included the multi-layer insulation repair tool, two SA-2 spines, spare diode box assemblies, four solar array bistem braces and a miscellaneous array of small pieces of hardware.

Second Axial Carrier (SAC)

First used on the SM2 mission (see above), the SAC for SM3B carried the Axial Scientific Instrument Protective Enclosure (ASIPE), which contained the Advanced Camera for Surveys in a sealed environment, the Power Control Unit, the NICMOS cryocooler, the NOBL Transporter, containing protective covers for the HST equipment bay doors, and a variety of repair kits, handrail covers, safety bars, Aft Shroud harnesses and translation aids. The protective enclosure and thermal insulation controlled the temperature of the ORUs. Design features and fixtures are the same as those described for the SM2 flight. It was carried immediately forward of the Flight Support Structure to which the Telescope would be attached for servicing.

BELOW The MULE was located behind the FSS in the aft section of the payload bay, carrying the NICMOS radiator, the Electronics Support Module and two protective enclosures. *(Lockheed)*

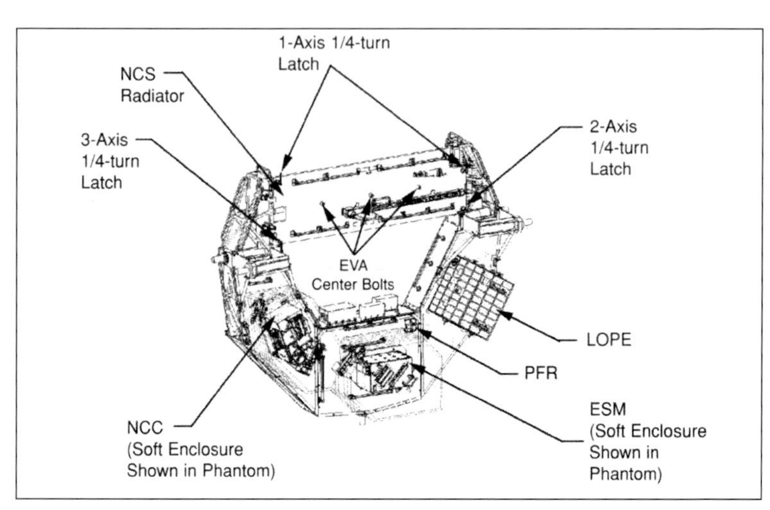

Multi-Use Lightweight Equipment Carrier (MULE)

New to HST servicing missions, the MULE is located in the aft section of the Orbiter payload bay behind the Flight Support Structure and was attached at five points: two on each longeron and one at the keel point. It carried the NICMOS cryocooler radiator, the NICMOS Electronics Support Module, a large ORU enclosure and a small protective enclosure.

STS-109 *Columbia* (SM3B)

1 March 2002
Mission duration: 10d 22hr 9min 51sec
EVA: 5
Commander: Scott D. Altman (Flt 3)
Pilot: Duane Carey (Flt 1)
MS1 (EV1): John M. Grunsfeld (Flt 4)
MS2: Nancy Currie (Flt 4)
MS3 (EV2): Richard Linnehan (Flt 3)
MS4 (EV3): James Newman (Flt 4)
MS5 (EV4): Michael Massimo (Flt 1)

The fourth servicing mission for the Hubble Space Telescope was the first use of *Columbia* in this role, the 26th flight for this vehicle. The Orbiter's most recent mission had been in July 1999 and it had been stood down for modifications that provided upgrades and improvements. It was the 108th flight of a Shuttle and was sadly the last successful flight for this venerable Orbiter, which had been the first Shuttle launched – on 12 April 1981.

Although nobody was to know it at the time, *Columbia* would be lost during re-entry after its next mission, a science mission, launched on 16 January 2003 and destroyed in the atmosphere coming back to Earth on 1 February 2003. Tragic as it was, that event would very

RIGHT Arriving at the HST less than 42 hours after launch, the Shuttle crew observe the distorted shape of the ESA solar arrays that will be replaced on this mission with a completely new rigid design. *(NASA)*

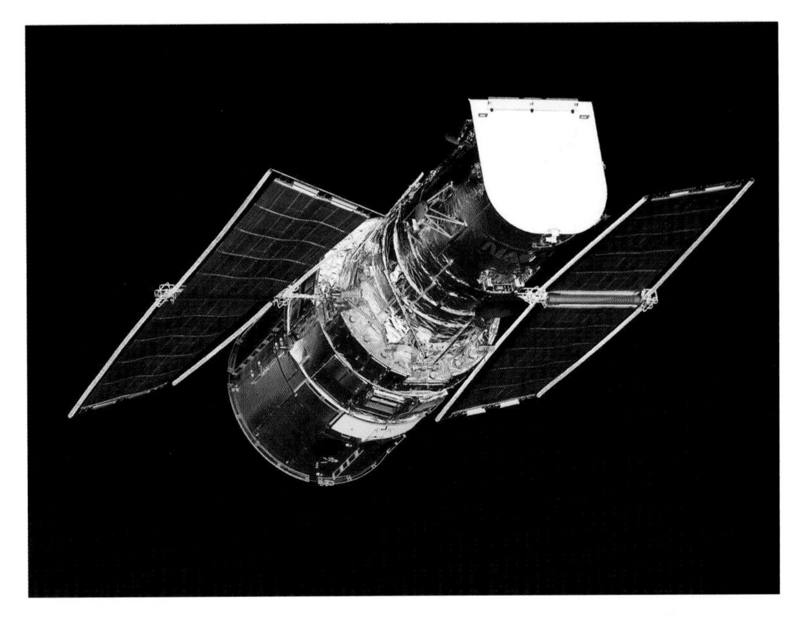

nearly deny the Hubble Space Telescope an opportunity to get one last upgrade.

The launch of *Columbia* occurred at 6:22:02am local time (11:22:02am UTC) on 1 March 2002 from LC-39A at the Kennedy Space Center. The flight had been delayed a day by unfavourable weather and began on the rescheduled time at the start of a 5min 21sec window. The usual rendezvous manoeuvres were completed as required, the RMS arm being uncradled at 1d 17hr 46min and the

RIGHT A close-up view of the attachment interface between the Flight Support Structure and the base of the HST shortly before docking. *(NASA)*

BELOW The configuration of the Flight Support Structure for the SM3B mission was almost identical to that for the previous servicing mission with only minor changes. *(Lockheed)*

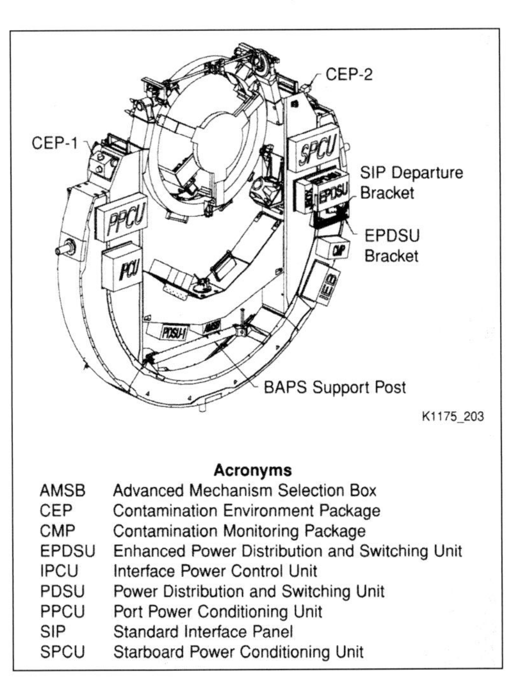

BELOW RIGHT The pre-planned EVA schedule for the five EVAs together with the task times developed in simulation and training sessions in facilities on Earth. *(Rockwell International)*

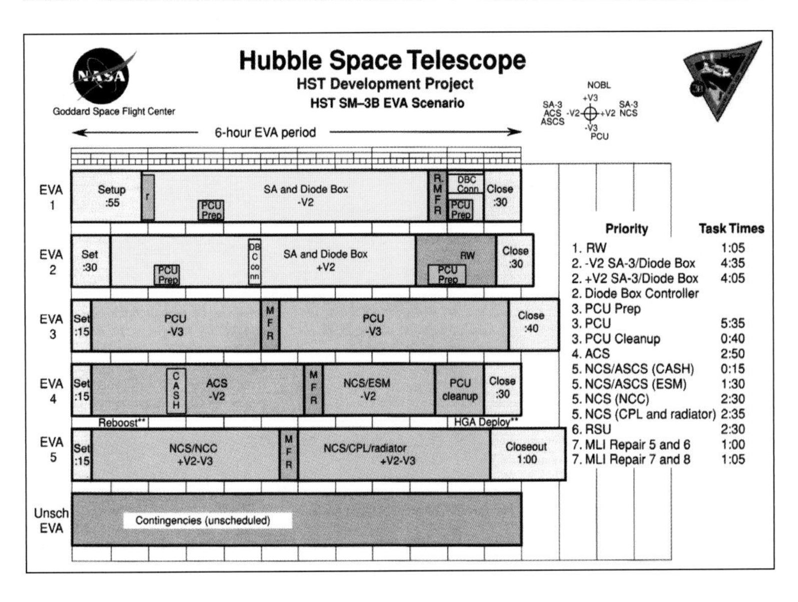

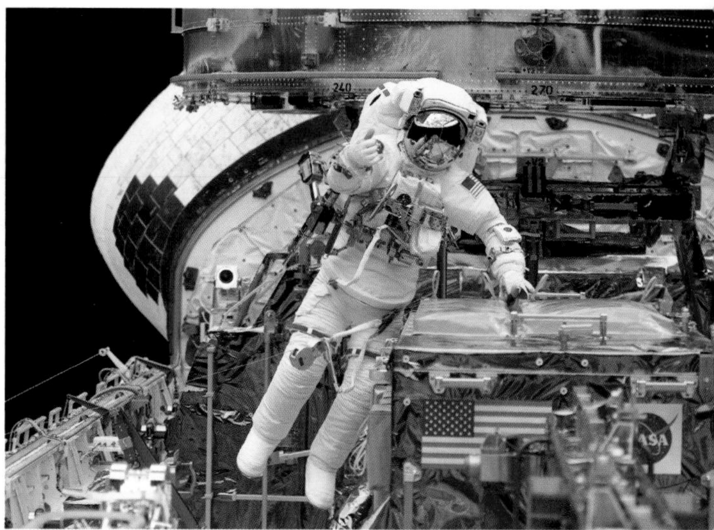

ABOVE Massimo works his way along the payload bay. *(NASA)*

ABOVE RIGHT Newman moving around the payload bay as he pauses near the Second Axial Carrier. Note the attached safety line and cable reel attached to his waist. *(NASA)*

BELOW The new solar arrays installed, attention focuses on connecting the new drive mechanisms. *(NASA)*

HST captured at 1d 22hr 9min 34sec before lock-down on the Flight Support Structure at 1d 23hr 10min. The RMS was used for a full photo-survey of the Telescope to secure a visual record of its condition.

EVA-1

The first spacewalk for Grunsfeld and Linnehan began at 2d 19hr 10min 0sec, with Nancy Currie operating the RMS arm from the operating station on the port side of the aft flight deck using TV screens and the rear-facing windows to move the astronauts around as required. The initial task was to remove the existing solar array on the -V2 side of the Telescope and replace it with one of the new ones including the diode box. As part of the installation of the NICMOS cryogenic cooler they set up the multi-layer insulation power system tent for the feed equipment. Their spacewalk ended about 30 minutes later than intended with a total duration of 7hr 1min.

EVA-2

Newman and Massimo began their EVA at 3d 19hr 14min 42sec and set about removing the +V solar array and installing the new SA-3 unit, replacing the RWA in Bay 6, and attaching a new thermal blanket for added protection from the punishing temperature variations experienced by the HST in its remorseless day/night cycles. A number of get-ahead tasks were completed, with doorstop extensions on Bay 5 together with new foot restraints for EVA-3, and tests with new bolts on the NICMOS and STIS instruments. The lower bolt required replacement and the crew used an Aft Shroud latch replacement kit to effect a repair.

During the spacewalk biomedical data from the EMU-2 suit was lost for 77 minutes, and erratic relay from this and the second suit became common on the following three spacewalks, the suspicion being that sternal harness pads separated from the crewman's chest. The stick-on pads were new, the first use of disposable electrodes of a peel-and-stick variety. Thoughts were being given to reverting to the old type for successive missions. The EVA ended at an elapsed time of 7hr 16min.

EVA-3

It was Grunsfeld and Linnehan's turn again on this spacewalk, which began at 4d 21hr 4min 45sec, starting about two hours later than

RIGHT **Adapted from a commercial communications satellite programme and built by Lockheed, the new solar arrays provide more power and the gallium-arsenide cells produced 5.68kW of electrical energy at beginning of life, more than 1.4kW better than the replaced photo-voltaic cells did when they were removed.** *(NASA)*

planned due to a water leak in the upper torso element of EMU-1 necessitating a resizing of that component of EMU-3 for Grunsfeld before fully suiting up and depressurising the airlock module.

About four hours before the EVA began flight controllers had completely powered down the Telescope for the first time in its orbital life in readiness for the arduous job of replacing the Power Control Unit. Currie was again on the RMS arm as Linnehan worked his way through 30 of the 36 bolts, removing them before moving back to the payload bay, his place taken by Grunsfeld who then removed the remaining bolts and detached the old PCU before carrying it down to the payload bay. The new PCU was installed and the connectors were mated at an EVA elapsed time of 4hr 53min. When completed the total duration of the EVA was 6hr 48min. A series of tests on the new PCU judged it to be in excellent health.

EVA-4

Newman and Massimo conducted the fourth spacewalk, which began at 5d 21hr 35min 29sec with a focus on removing the Faint Object Camera and replacing it with the Advanced Camera for Surveys. This was assigned 2hr 50min, with both men working between the Aft Shroud and the payload bay, fetching and carrying the exchanged instruments. Then the electronic module for the NICMOS cooling system was attached as a get-ahead task for the fifth EVA where the installation of the new cooler would be the focus. Some activities left over from the preceding day were finished up and the EVA ended at an elapsed time of 6hr 48min.

RIGHT **With both arrays installed, attention turns to the installation of the ACS in place of the Faint Object Camera.** *(NASA)*

ABOVE Few difficulties were encountered on this fourth servicing mission, with a near-record number of fittings and attachments completed! *(NASA)*

EVA-5

The final spacewalk of the fourth servicing mission began at 6d 21hr 20min 46sec with Grunsfeld and Linnehan starting installation of the NCS equipment. First they opened the Aft Shroud doors and generally prepared the area for the new equipment. Currie moved Linnehan down to the payload bay to bring the cryocooler up to the work area, where he and Grunsfeld fitted it to the NICMOS instrument. Cables were then attached to its equipment section, which had been set up for them the previous day. Grunsfeld rode the arm back down to the payload bay to retrieve the radiator and was returned to the outside of the Telescope to install it. Linnehan fed the wires from the radiator through the bottom of the Telescope to Grunsfeld who made the connections to NICMOS. Support equipment was configured in the payload bay for *Columbia*'s return to Earth and the EVA ended at 7hr 20min.

Recovery

Little more than an hour after the EVA ended, at a mission time of 7d 5hr 56min 1.8sec, the primary RCS thrusters began a reboost session lasting almost 36min, with thrusters turned on and off as necessary to maintain a 5° pointing deadband. The firings imparted a V of 11.8ft/sec (3.597m/sec) for a general altitude increase of 4.14 miles (6.66km) placing the spacecraft in an orbit of 362.16 x 357.44 miles (582.7 x 575.1km).

The Telescope was grappled by the RMS at 7d 19hr 46min 53sec, unberthed from the Flight Support Structure at 7d 21hr 12min and released at 7d 22hr 42min 16sec. Having been brought down to two-thirds atmospheric pressure before the rendezvous had been completed, the Orbiter crew area was brought back up to normal levels less than four hours later. With a landing back at the Kennedy Space Center, the mission had a duration of 10d 22hr 9min 51sec. All objectives had been accomplished and the HST was in an excellent state of health, with a new instrument fitted and the NICMOS back up and running.

The next servicing mission was expected around mid-2004, the last for an extended series of observations that had been approved by the various organisational bodies responsible for its operation. Although a successor to the Hubble Space Telescope had been approved and was under development, there was a keen awareness that the HST had many more productive years ahead and, as the servicing missions displayed real success in not only restoring ailing support systems but also in equipping it with new generations of science instruments, support was widespread.

The final servicing mission was to carry two new instruments, the Cosmic Origins Spectrograph and the Wide Field Camera III. These would bring a new generation of observational capabilities to the Telescope and a new wave of opportunities for astronomers and astrophysicists. As for the Shuttle, there had been extensive debate about whether to complete assembly of the International Space Station, to which it had been primarily committed since 1998, and retire it, or conduct a series of advanced upgrades to extend its life by a further two decades.

There was little enthusiasm among the political leadership in America for a more advanced form of human space flight and there was insufficient money to run Shuttle operations and invest in new deep-space exploration beyond low Earth orbit. It had to be one or the other. As the fifth HST servicing mission approached, the Shuttle looked set for a further decade or two of operations, completing the ISS and supporting it with regular logistics and crew flights. Until the events of early 2003.

On 1 February that year the Orbiter *Columbia* was destroyed during re-entry, distributing debris over a wide swathe of the mid-western and southern states in a catastrophic disaster which took the lives of its seven crewmembers. Coming 17 years after the loss of the Shuttle *Challenger*, which also took seven lives with it, the Shuttle was grounded pending an investigation and would not return to flight until July 2005 with the flight of *Discovery*. While the loss of *Challenger* on 28 January 1986 had stimulated construction of a replacement named *Endeavour*, there was to be no replacement for *Columbia*. Only three Shuttle vehicles remained in the fleet.

The investigation disclosed that a piece of the Shuttle's foam insulation on the External Tank had come off during the launch. It had struck and holed reinforced carbon-carbon thermal protection on the leading edge of the wing, opening a path that on re-entry caused hot gases to flow in. The heat melted the inside of the wing itself, eventually destroying the entire vehicle as it broke apart during re-entry at the end of its near 16-day flight. It was determined that the Shuttle programme would end after the assembly of the ISS and that no other flights would be scheduled which did not go to the ISS where cameras could examine the thermal protection to visually verify its integrity.

While *Columbia*'s mission was under way there was some suspicion that the wing leading edge had been damaged by this strike during launch but there was no way of finding that out and no way of avoiding the need to come back through the atmosphere. *Columbia*'s mission was one of a very few Shuttle flights which did not visit the ISS, where the crew could await a rescue should they discover damage to RCC insulation or tiles.

The orbit of *Columbia*'s science mission was so different to that of the ISS that the Orbiter could not reach it with the amount of propulsion capacity built into the Shuttle. A similar fate could await any other Shuttle that did not dock with the International Space Station. As part of a blanket ban on flights not supporting the ISS, on 16 January 2004 – exactly one year to the day after the launch of *Columbia* on its last flight – NASA Administrator Sean O'Keefe cancelled the fifth servicing mission for the Telescope. The orbit of the HST was such that a Shuttle flight to the Telescope could not reach the safe haven of the ISS.

This decision caused a furore, support for the pro-HST servicing mission being mobilised by none other than Barbara Mikulski, the leader of the Senate subcommittee overseeing NASA's budget, and astronaut John Grunsfeld who

LEFT NASA administrator Michael Griffin made the important decision to assign a Shuttle to fly a fifth and final servicing mission to the HST, reversing a previous decision after the loss of *Columbia* in 2003. *(NASA)*

BELOW Responding to the tragic loss of *Columbia*, the Orbiter Boom Sensor System was an extension of the RMS arm equipped with an optical device to visually inspect the exterior surface of the Orbiter to verify that the thermal protection system was intact after reaching orbit. *(NASA-JSC)*

had visited the HST twice. A large contingent from the science community, as well as some internal NASA managers, were opposed to the cancellation, and Representative Mark Udail introduced a bill requiring an independent assessment of the risk.

O'Keefe left NASA in December 2004 and was replaced by Michael Griffin who lost no time in confirming that he would reconsider the decision. After the restoration of Shuttle flights in July 2005 and implementation of a system for checking the condition of the thermal protection with a sophisticated imaging system operating at the end of an extension to the RMS arm, plus the availability of a rescue plan to recover the crew of a damaged Orbiter without them having to risk re-entry, the die was cast. On 31 October 2006 Griffin announced that the fifth servicing mission was back on and would take place in 2008.

As part of the precautions ensuring the safety of the crew, NASA would provide sufficient consumables for a powered-down Orbiter to await rescue, should it be needed, for as long as 25 days in orbit. This was sufficient time for a second Shuttle to be launched to retrieve the crew of the stranded Orbiter, a mission known as STS-400. For the final servicing flight, *Atlantis* would fly the prime mission with *Endeavour* as its backup waiting on the adjacent launch pad.

As for the Telescope, the ACS and NICMOS as well as the WF/PC II remained operational but the STIS failed in August 2004. It remained in safe mode and would be restored to functionality by SM4. Then in January 2007 the Wide Field Camera and the High Resolution Camera in the Advanced Cameras for Survey instrument became unavailable after a failure in the electronics.

BELOW The Wide Field/ Planetary Camera III would replace its predecessor as the radial science instrument on the HST. *(NASA-GSFC)*

Renewal

Objectives for the fifth servicing mission included installation of the Wide Field Camera 3 (WFC-3) and the Cosmic Origins Spectrograph (COS), replacement of the SI C&DH with a backup unit, replacement of all six nickel-hydrogen batteries, installation of Fine-Guidance Sensor-2, and installation of new multi-layer outer insulation panels (NOBLE). It was also to attempt repairs to the Advanced Camera for Surveys by installing new circuit boards and attaching a power supply module, and the Space Telescope Imaging Spectrograph by replacing the Low Voltage Power Supply-2 board in the Main Electronics Box 1. The Orbiter also carried a Soft Capture Mechanism for installing on the Telescope to facilitate any future requirement that, up to the time of writing, was not planned. Science instruments removed included the WF/PC II and the COSTAR corrective optical device.

Wide Field Camera 3 (WFC-3)

Designed to significantly improve the capabilities of the WF/PC II instrument that it replaced, like its predecessor the WFC-3 incorporated corrective optics to compensate for the aberration in the primary telescope mirror. It incorporated some components from the WF/PC originally installed before the launch of the Telescope in 1990, an instrument returned by Shuttle, and was designed to incorporate broad wavelength coverage, wide field of view and high sensitivity. In combining two optical/ultraviolet CCDs with near-infrared arrays, the instrument was capable of high resolution imaging over a wavelength range of 200–1,700nm and was equipped with a comprehensive range of wide, intermediate and narrow band filters.

The WFC-3 occupied a volume of 5ft x 3ft x 1.7ft (1.3m x 1m x 0.5m) with a radiator section of 7ft x 2.6ft (2.2m x 0.8m) and a weight of 890lb (404kg). The primary instrument scientist was Randy Kimble of the Goddard Space Flight Center, and it was developed by the Goddard Space Flight Center together with Ball Aerospace and the Space Telescope Science Institute, with consultative support and guidance from Professor Robert O'Connell at the University of Virginia, who also chaired the Science Oversight Committee.

The optical design features an ultraviolet and optical (UVIS) channel sensitive to 200–1,000nm and an infrared (IR) channel sensitive to 800–1,700nm. A channel selection mirror detects on-axis light from the Optical Telescope Assembly to the IR channel or the mirror can be turned to defect light to the UVIS channel but simultaneous observations are not possible, although both UVIS and IR observations can be made sequentially on the same orbit. The light sensitive detectors in both channels are solid-state devices, with two large CCDs similar to those used in digital cameras for the UVIS channel. The 2,000 x 4,000 pixel arrays are butted together creating an effective 4,000 x 4,000 pixel detector. The crystalline photosensitive surface in the NIR (near infrared channel) detector is composed of mercury, cadmium and tellurium (HgCdTe).

The 16 megapixel UVIS array combined with the 160 x 160 arc-sec field of view provides a 35-fold increase in the discovery power of the instrument compared to the High Resolution Camera in the Advanced Camera for Surveys (which see). The NIR channel detector is a more highly advanced one megapixel version of the 65,000-pixel detectors in the NIR instrument in NICMOS (which see) and has a 123 x 137 arc-sec field of view with a pixel size of 0.13 arc-sec. This combination of field of view, sensitivity and low detector noise results in a 15–20-fold enhancement over the NICMOS instrument.

While the fundamental components from WF/PC II were used, a new optical bench was designed and the instrument case, filter wheel assembly and radiator retained. Just as in its two precursors, the physical layout captures the image off a pick-off mirror, routing the light past a channel select mechanism (CSM) that reflects lights either to the UVIS or IR channels. In the UVIS channel the light falls on an adjustable mirror that steers the beam to another mirror that corrects for aberration, this design being a copy of the Advanced Camera for Survey WFC (which see). The beam then transits the Selectable Optical Filter Assembly (SOFA), a shutter mechanism (also copied from the WFC shutter), and finally enters the CCD detector enclosure, another copy from the WFC in the ACS.

When the CSM is in the IR channel position, the beam is directed to a fold mirror and on to a cold enclosure (-35°C) that reduces the cooling requirement of the IR detector and the internal background at IR wavelengths. Within this enclosure it then passes through a refractive corrector element that removes spherical

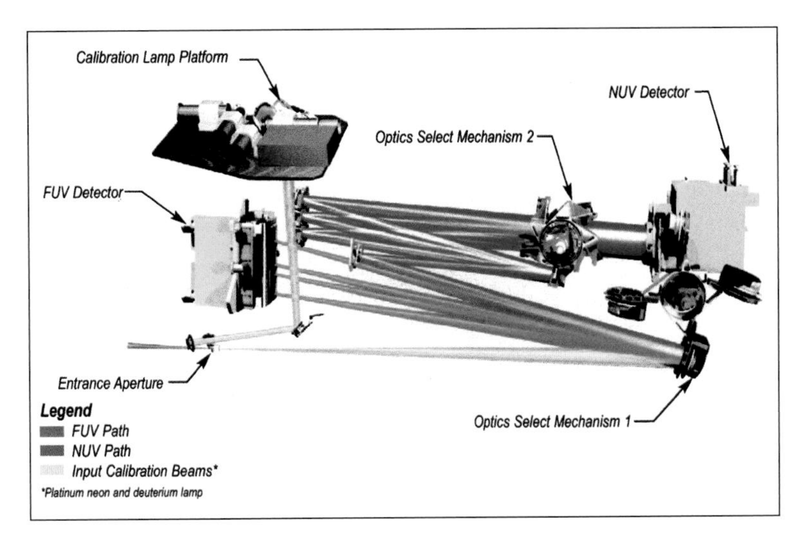

ABOVE A key design innovation with the WF/PC III is the NIR channel in which the detector is shaped so as to reject infrared light with a wavelength greater than 1,700nm. *(Lockheed)*

BELOW A view looking forward from the aft end of the payload bay showing the MULE (left foreground), the FSS which would hold the Telescope during servicing activity, the ORUC with the Cosmic Origins Spectrograph to replace COSTAR, and the Super Lightweight Interchangeable Carrier with the WF/PC III instrument. *(NASA-KSC)*

ABOVE With the FSS at the extreme left, the ORUC sits in the middle of the payload bay and behind the SLIC, which is at the forward end immediately aft of the airlock module. *(NASA-KSC)*

aberrations from the primary mirror, a cold mask and a selectable IR filter. A design innovation stems from it tailoring the detector to reject IR light longer than 1,700nm, which avoids the necessity for a cryogen to keep it cold. Instead, the detector is chilled with an electrical device known as a thermos-electric cooler.

Cosmic Origins Spectrograph (COS)

Replacing the COSTAR instrument, COS will perform spectroscopy, and with the planned repair of the STIS instrument was to restore full spectroscopic capability to the Telescope. Designed and built by Ball Aerospace, the instrument measures 7ft x 3ft x 3ft (2.2m x 0.9m x 0.9m) and weighs 850lb (385.6kg). Principal investigator was James C. Green of the University of Colorado.

The COS has two channels – far ultraviolet (FUV) covering 115–205nm and the near ultraviolet (NUV) covering 170–320nm. The light sensitive detectors of both channels were designed around thin micro-channel plates containing several thousand tiny curved glass tubes aligned in the same direction. Incoming photons of light ultimately induce showers of electrons emitted from the walls of these tubes, which are accelerated, captured and counted in electronic circuitry located behind the plates. The FUV channel has a detector array pixel size of 32,768 x 1,024 pixels with each pixel 6 x 24 microns and with three gratings. The NUV channel has a 1,024 x 1,024 pixel detector array, with each pixel 25 x 25 microns, and four gratings.

Unique to COS among the Telescope's spectrographs is the amount of throughput, the FUV channel being designed specifically to minimise the number of light bounces off the optical surface. It makes one bounce off the selectable light-dispersing grating and goes directly to the detector. The COS also has a very low level of scattered light produced at the light-dispersing gratings.

COS is complementary to STIS, the latter having the unique ability to observe the spectrum of light across spatially extended objects such as galaxies and nebulae. When operated in conjunction with the STIS, which has only 3% of the sensitivity of the COS, the new instrument is best suited for observing point sources of light, and the two spectrographs working together provide a full set of spectroscopic tools aboard the Telescope for astrophysical research.

Super Lightweight Interchangeable Carrier (SLIC)

Due to the heavy payloads carried on this mission a uniquely designed lightweight cargo carrier was developed and built to replace the Second Axial Carrier used on earlier flights. It was specifically designed using new materials to reduce the weight of the support structure and to demonstrate a new carrier for further applications. The SLIC is fabricated from carbon fibre using a cyanate ester resin and a titanium metal matrix composite and was the first all-composite cargo carrier used on the Shuttle.

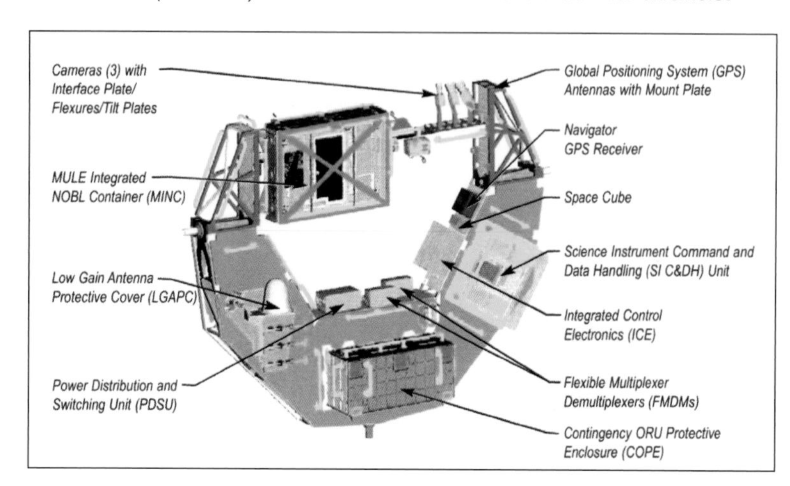

BELOW Utilised on the previous servicing mission, the Multi-Use Lightweight Carrier had GPS antennas, the replacement SI C&DH unit and a variety of subsystems for installation on the HST. *(Lockheed)*

RIGHT The ORUC for the SM4 mission carried the COS instrument and the WF/PC handhold frame. *(Lockheed)*

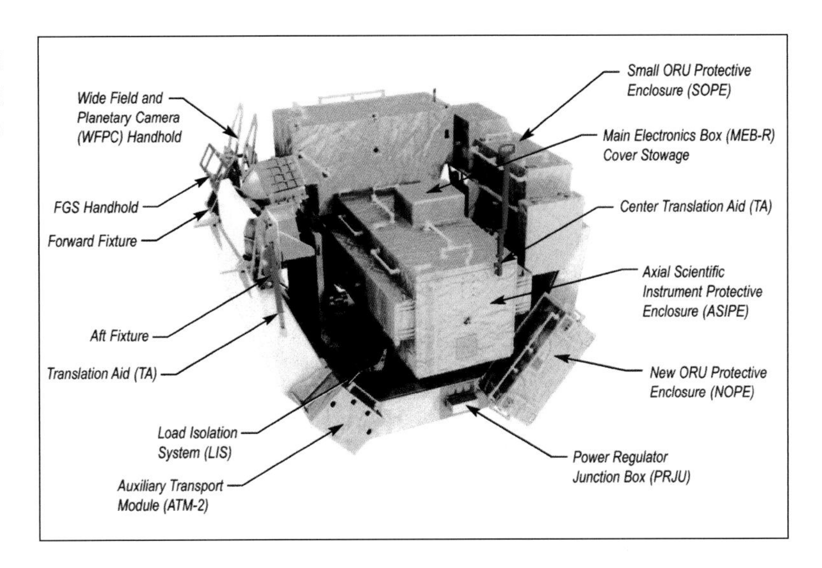

The SLIC is 15ft x 8.7ft (4.6m x 2.65m) in size and has an empty weight of 1,750lb (794kg) but with a maximum carrying capacity of 3,500lb (1,678kg). For SM4 it was required to support the 980lb (444.5kg) WFC-3, its 650lb (295kg) enclosure and two new battery modules each weighing 460lb (207kg), a total load on this final servicing flight of 2,550lb (1,157kg). The SLIC is attached to the Orbiter payload bay at two points on two longerons and by a single keel point.

ORUC

The same as that used on earlier missions, on this flight carrying the COS instrument, Fine-Guidance Sensor, three rate sensor units, and other hardware including handrails for the FGS and the WF/PC II which are stored on the port side, multi-layer insulation, translation aids, repair kits and a variety of support equipment.

Multi Use Lightweight Equipment Carrier (MULE)

The same as that used on SM3B, containing spare ORUs and tools, multi-layer insulation for the Support Systems Module's Equipment Section bay doors, and replacement SI C&DH unit.

Soft Capture and Rendezvous System (SCRS)

Because this was to be the last servicing mission for the Telescope, and to allow the next generation of human space flight vehicles to access the HST, the SCRS was carried on top of the Flight Support Structure and incorporated the Soft Capture Mechanism (SCM), and an integral Low Impact Docking System (LIDS), and the Relative Navigation System (RNS). The SCRS was to be attached to the base

BELOW The Super Lightweight Interchangeable Carrier held the replacement batteries and the original sets returned to Earth, the WF/PC II and a spare Fine Guidance Sensor, among other items. *(Lockheed)*

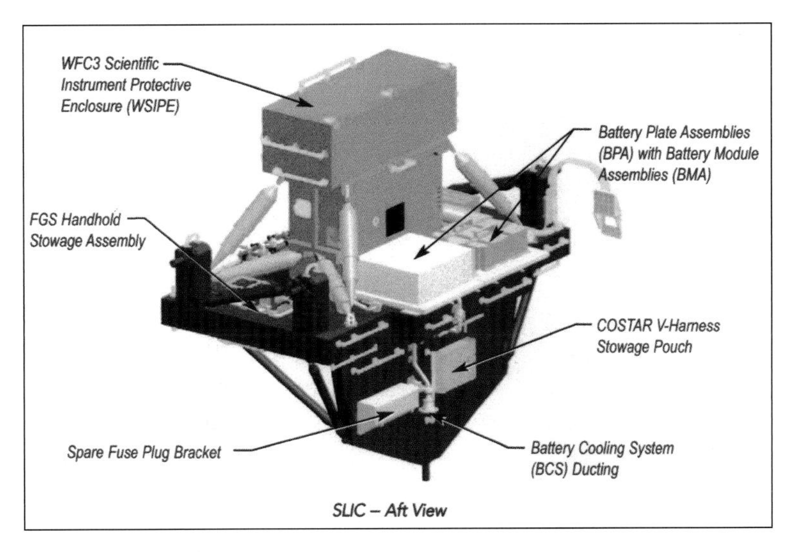

RIGHT Bay 2 nickel-hydrogen battery module for installation on the second EVA. *(Lockheed)*

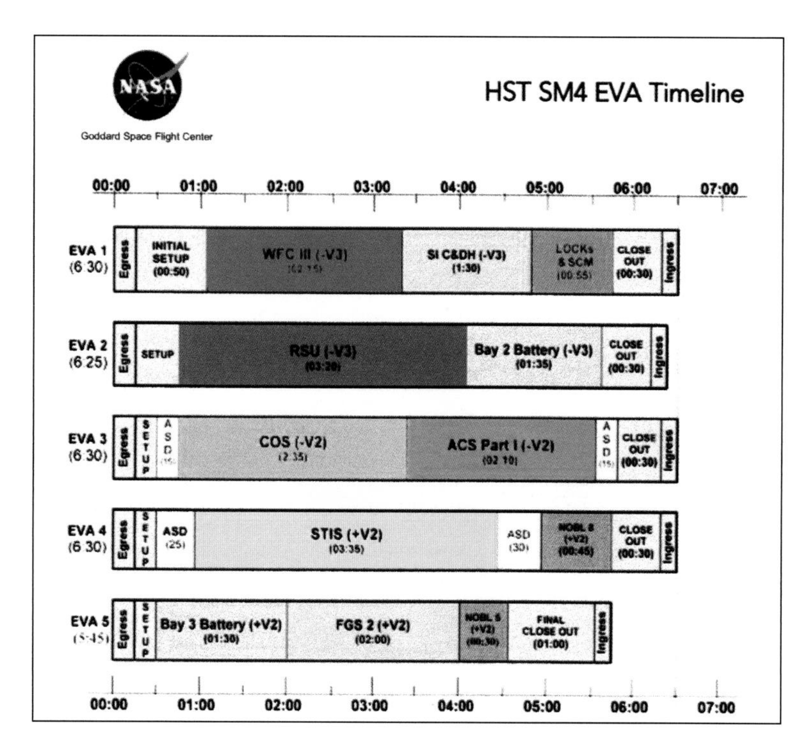

ABOVE Five periods of EVA were scheduled for the HST on this fifth and final servicing mission, the sixth Shuttle flight dedicated to the Telescope and its continued upgrade. *(Lockheed)*

of the telescope as it sat on the FSS and would remain with it when it was finally released.

With a diameter of 72in (183cm) and a height of 24in (61cm), the SCM would attach to the base of the HST by three sets of jaws that clamp on to the existing berthing pins on the aft bulkhead. The astronauts would drive a gearbox to release the SCM from the FSS when the Telescope was unberthed from the Shuttle Orbiter. The LIDS was a common docking interface that is standard on all future human space flight spacecraft whether developed by NASA or by commercial contractors. It provides a low-energy, shock absorbing interface imparting little energy into the host vehicle when attached to a smaller or more massive space vehicle.

The RNS consists of an imaging system with optical and navigation sensors with supporting avionics and was designed to collect data on the HST during capture and on deployment. It was to acquire information about the HST through video and images of the Telescope's aft bulkhead as it is released by the RMS arm back into space, an area of the spacecraft not readily visible when attached to the FSS. This information was required by NASA for determining among numerous options what to do about safely de-orbiting the spacecraft at the end of its life. It was carried on the MULE.

STS-125 *Atlantis* (SM4)

11 May 2009
Mission duration: 12d 21hr 38min 45sec
EVA: 5
Commander: Scott Altman (Flt 4)
Pilot: Gregory C. Johnson (Flt 1)
MS1: Michael Good (Flt 1)
MS2: Megan McArthur (Flt 1)
MS3: John Grunsfeld (Flt 5)
MS4: Michael Massimo (Flt 2)
MS5: Andrew Feustel (Flt 1)

The launch of the final HST servicing mission took place from LC-39A at the Kennedy Space Center with lift-off at 2:01:56pm local time (6:01:56pm UTC). The usual activities ensued, with the payload bay doors opened at 1hr 39min into the flight, but on this mission there was particular attention to the condition of the Orbiter's thermal protection system. There was no place the Orbiter could go should a defect be discovered which would threaten a safe re-entry, and *Endeavour* was waiting down on LC-39B should it be called upon to perform a rescue (known as 'Launch-On-Need'). Using a boom attached to the end of the RMS, the imaging and sensor system was able to complete a full documented survey, with the images transmitted to ground stations for detailed analysis. Nothing was found to be of concern but *Endeavour* would remain on standby until *Atlantis* got ready for re-entry at the end of the mission.

Rendezvous operations preceded arrival at the HST, with the RMS arm controlled by McArthur grasping the Telescope at 1d 23hr 22min 14sec with the Telescope berthed on the FSS 48min 52sec later. There had been some delay berthing the HST due to problems with communication between the Orbiter and the Telescope, and due to the inability to command the HST to roll as required the Shuttle flew a manual manoeuvre to secure it in the payload bay. The RMS arm was then used to conduct a full survey of the HST and the new solar arrays.

EVA-1

The first spacewalk began at 2d 18hr 36min with Grunsfeld being the free-floater and Feustel assigned to riding the arm. They accomplished

LEFT Feustel moves around the forward section of the payload bay with the SLIC carrier. The large rectangular gold-foil-covered box contains the WF/PC III. Note the airlock module attached to the forward bulkhead of the bay. *(NASA)*

removal of the WF/PC II and replaced it with the WFC-3 instrument before setting about replacement of the failed SI C&DH unit with SI C&DH-2. The A side of the unit had failed in September 2008, which had caused the delay of SM4 to allow engineers to develop hardware for the replacement and to evolve a plan to install it on the Bay 10 door.

The crew also completed transfer of the Soft Capture Mechanism from its launch mounting on the FSS to the HST and installed two of three Latch Over Centreline Kits on the –V2 door. To avoid bending the door and allowing an unacceptable light leak, they installed one Aft Shroud door latch repair kit on the lower middle latch of that door. The crew finally removed

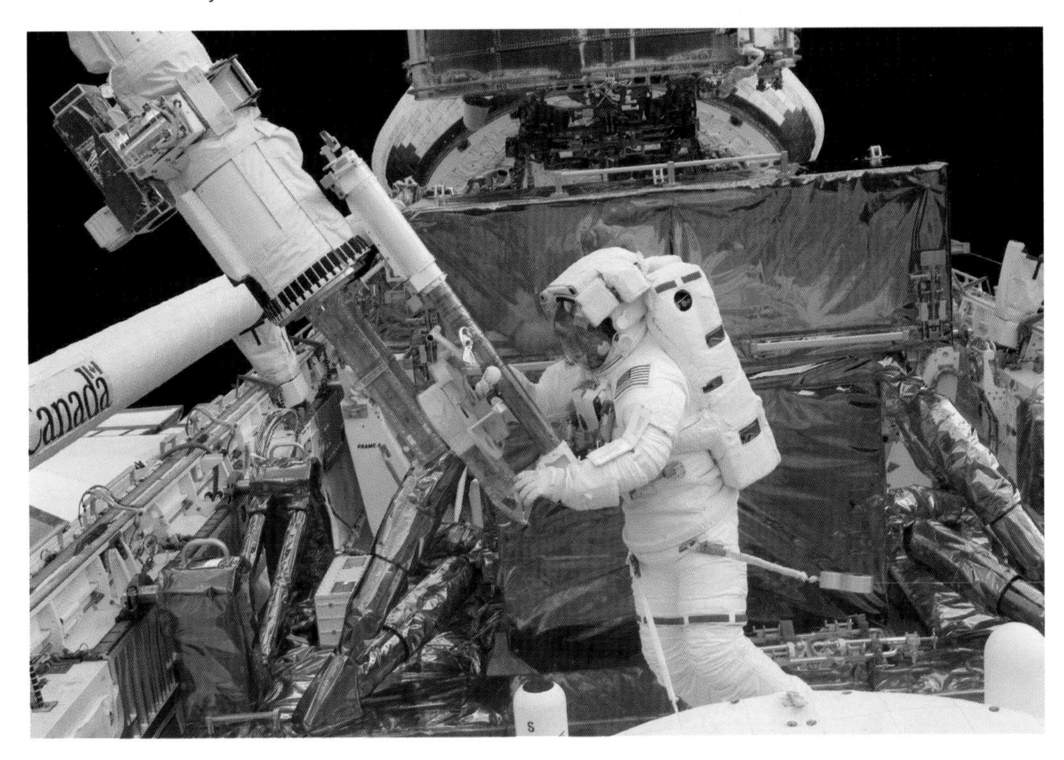

LEFT Massimo works at the forward end of the RMS arm. *(NASA)*

LEFT The Wide Field/Planetary Camera III is manoeuvred by Feustel up the side of the Orbiter as it is positioned for insertion into the radial instrument bay on the SSM. *(NASA)*

CENTRE Feustel gets a ride on the RMS arm with a handful of tools. *(NASA)*

two Bay 5 New Outer Blanket Layers (NOBL) vent plugs and applied lubricant to several door latch bolts as get-ahead tasks for the next spacewalk. The EVA lasted 7hr 20min.

EVA-2

The second spacewalk of the mission began at 3d 18hr 48min, with Massimo as the free-floater and Good on the RMS foot restraint. Priority here was installation of three RSUs each containing two new gyroscopes, Battery Module-1 being installed into Bay 2, and a get-ahead start on repairs to the Advanced Camera for Surveys. A problem arose with one of the RSUs when it was impossible to properly seat it and bolt it in securely and a spare had to be brought up from the payload bay which, when fitted, worked well. A flight software patch was uplinked to the HST to upgrade the gyroscope configuration after all the EVAs were over and the Telescope had been released. The EVA lasted 7hr 56min, longer than expected due to the RSU problem.

RIGHT Grunsfeld works on the battery module high on the HST. *(NASA)*

EVA-3

The primary activity for this spacewalk was removing COSTAR and replacing it with COS. Again, Grunsfeld was the free-floater and Feustel was on the arm. This operation was conducted swiftly, and with the two astronauts gaining on the time allotted they began early on parts 1 and 2 of the repair schedule for the Advanced Camera for Surveys. Since January 2007 an electronics failure had prevented the WFC and HRC channels from being used, only the SBC channel being effective since that date. The planned timeline had expected them to do part 1 of the repair on EVA-3 with part 2 on EVA-5, so the slot for that on the last spacewalk was revised to accommodate the Fine-Guidance Sensor-2 installation. The work on the ACS required the electronics board to be changed, and this was the first time an EVA astronaut had performed a card or board level repair on an electronics box. Before completing the EVA, which had a duration of 6hr 36min, the crew noticed a fogging in the window of the IMAX payload camera, due it was concluded to it being left with the cover open in the armed configuration.

EVA-4

The fourth spacewalk began at 5d 19hr 42min with Massimo as the free floater and Good riding the arm. Attention this day was on repairs to the STIS that consisted of replacing the failed low voltage power supply No 2 circuit card in the main electronic box No 1. To get at the device the handrail had to be removed and this proved troublesome when the lower right bolt appeared to lock up. After several unsuccessful attempts it was decided to pull on the top of the handrail and break the bolt and this was successful. When a tear was noticed in a suit glove the EVA was immediately ended, at a duration of 8hr 2min, with an unattended task of applying NOBL to Bay 8 deferred to the last spacewalk.

EVA-5

The last of 23 spacewalks to service the HST began at 6d 18hr 18min, about one hour early, with free-floater Feustel and Grunsfeld riding the RMS arm. They performed Battery Module-2 installation in Bay 3, removal and replacement of FGS-2, and replaced multi-layer insulation with NOBL on Bays 5, 7 and 8 of the Support Systems Module. During final closeout activity, EV-1 accidentally knocked the low gain antenna at the bottom of the HST, chipping off the foam tip, which required the crew to fit the Low Gain Antenna protective cover (LGAP) on to help provide some thermal protection for the exposed wiring, a repair that would not interfere with the functionality of the antenna. The EVA had lasted 7hr 2min.

Departure

With the final closeout of the Hubble Space Telescope, 19 years of HST servicing activity came to an end and the telescope was left in orbit to continue its work for astronomers and astrophysicists. The RMS arm grappled the spacecraft at 7d 16hr 43min, removing it from the FSS and presenting it at the correct attitude for release, which came at 7d 18hr 56min.

Several manoeuvres were completed to increase the separation distance between the Orbiter and the HST but the return of *Atlantis* to Earth was delayed a day due to bad weather at the Kennedy Space Center. On 24 May *Atlantis* returned to a heroic welcome and great relief at an elapsed time of 12d 21hr 38min 45sec. Never again would an Orbiter conduct a mission where it could not reach the safe haven

BELOW The Soft Capture Ring System attached to the aft end of the HST in this artist's view would allow a future spacecraft to dock with the Telescope. *(NASA)*

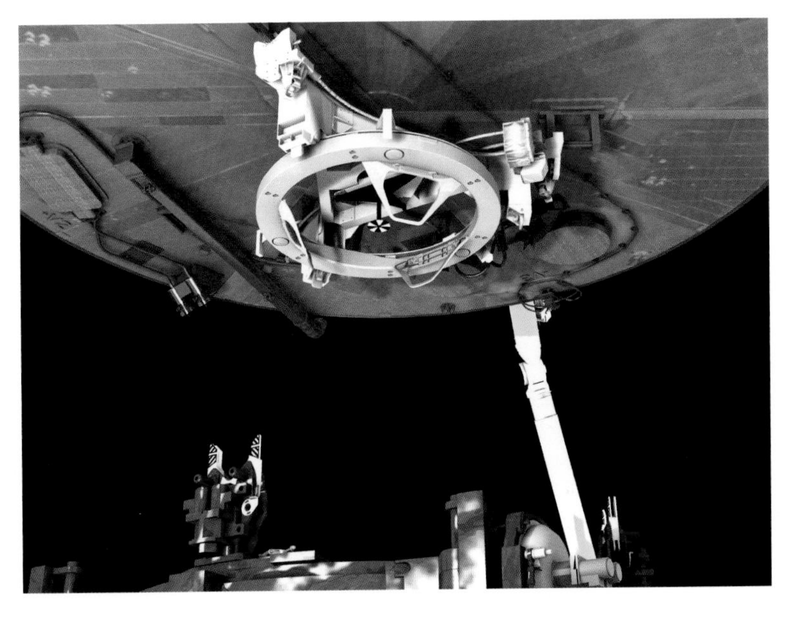

LEFT The Relative Navigation System uses an optical system to collect data during release to obtain specific data on the aft section. *(NASA)*

CENTRE The moment of release, almost 18 days after launch into the last servicing mission mounted by the Shuttle. *(NASA)*

of the International Space Station. The Shuttle programme itself was on the way out but all four Orbiters had visited the Telescope in one of the most profound examples of the value of on-orbit servicing yet demonstrated.

In a dramatic announcement, in January 2004 President George W. Bush and NASA announced that the Shuttle programme would end in 2010 and that America was heading back to the Moon in a programme embracing new technologies known as Constellation. It would involve development of a new spacecraft capable of carrying humans beyond low Earth orbit and into deep space once again. Looking like an enlarged Apollo spacecraft, the new vehicle called Orion would be launched by a new generation of rockets called Ares, but this programme would not survive the change in presidency.

In early 2010 the Obama administration killed Constellation and replaced it with a set of commercial projects to replace the Shuttle for carrying cargo and eventually people to the International Space Station. The last Shuttle mission took place in 2011 and future ISS crewmembers flew on Russia's Soyuz spacecraft. Meanwhile, the US Congress forced the hand of the White House and Orion was resurrected, together with a new rocket called the Space Launch System with a potential lifting capacity greater than the Saturn V which sent Apollo to the Moon. But there would be no capacity for visiting the Hubble Space Telescope.

The first flight of the SLS and an unmanned Orion spacecraft is planned for 2018 at the earliest. The HST is presently in an orbit from

LEFT The Telescope drifts away, separating from the Orbiter *Atlantis* which has fired its thrusters to begin moving to a safe distance prior to re-entry. *(NASA)*

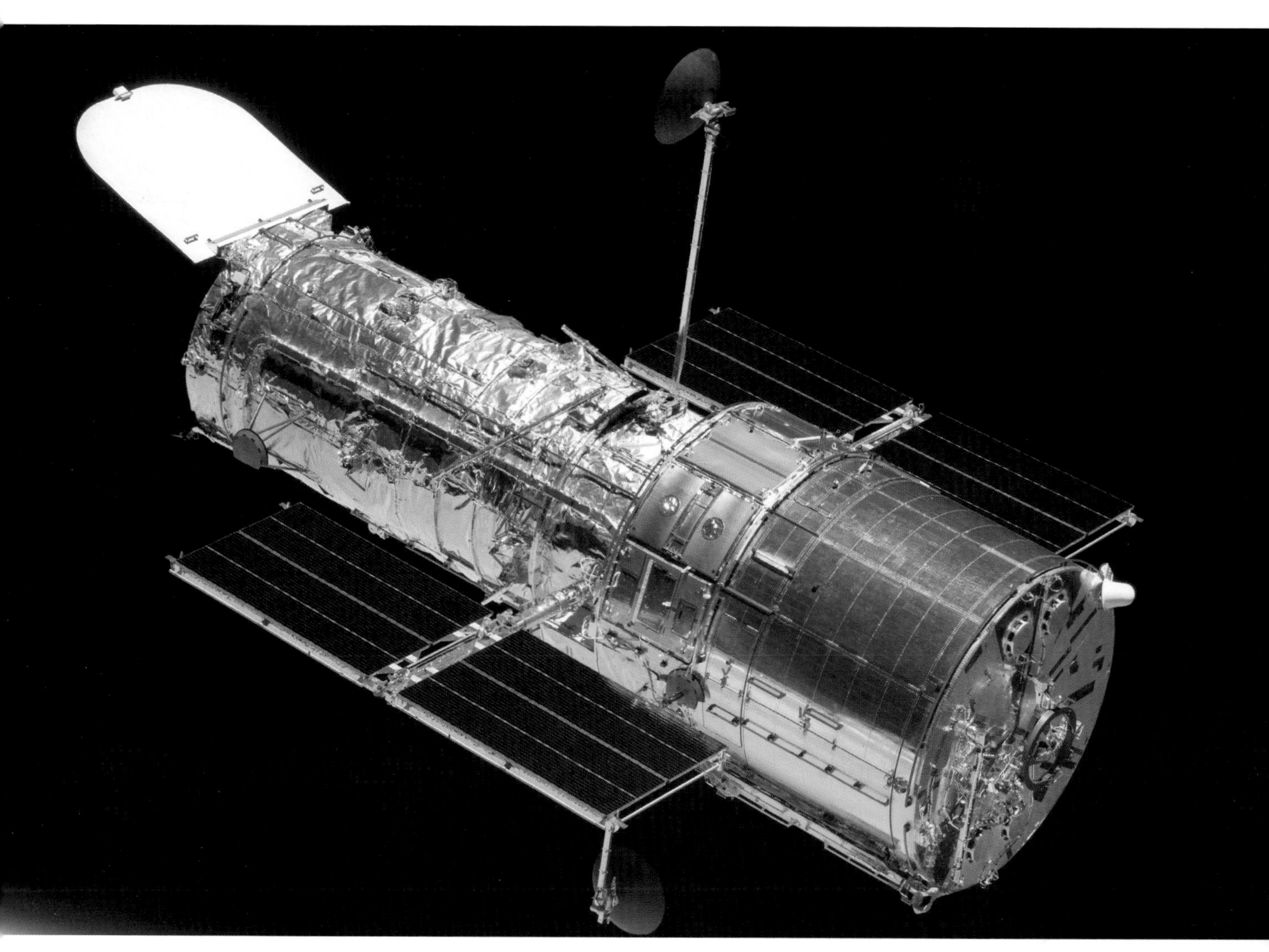

ABOVE A parting shot from a camera inside *Atlantis*. *(NASA)*

which it is likely to re-enter the atmosphere between 2022 and 2028. Its successor, the James Webb Space Telescope (JWST) is scheduled for launch in 2018 on a European Ariane V rocket, a deal struck similar to that for the HST where hardware contributions offset observing time for European astronomers. But the JWST will not be serviceable and the age when spacecraft were built for astronauts to repair, refurbish, improve and upgrade them in space is over.

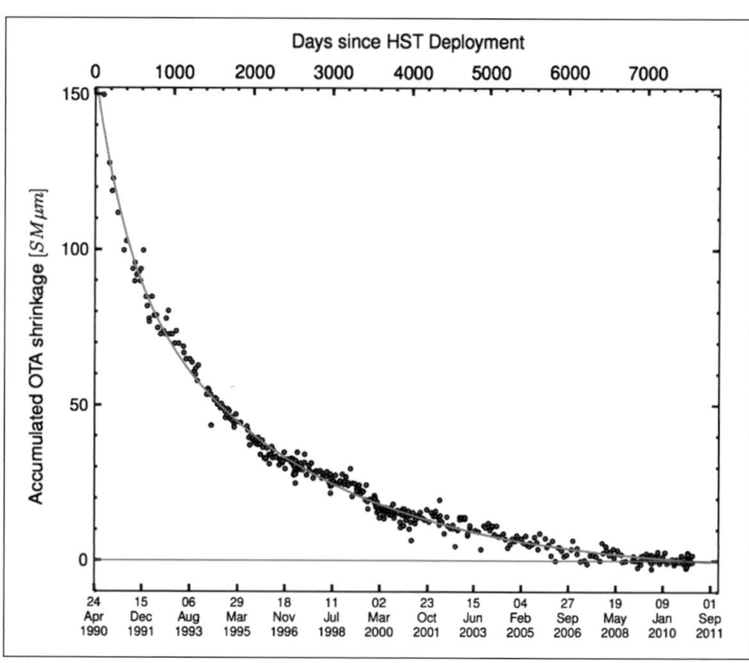

RIGHT Precise measurement of the effects of 21 years in space, the shrinkage of the Optical Telescope Assembly provides valuable data for space engineering applications. *(David Baker)*

Chapter Nine

Successor to Hubble

Originally known as the Next Generation Space Telescope, this massive successor to Hubble was named the James Webb Space Telescope in 2002 in honour of the NASA Administrator from 1961 to 1969 who saw the US civilian space agency through its most challenging years of spectacular growth and supreme achievement. It is the first space observatory not named after a scientist and has been in development since 1997.

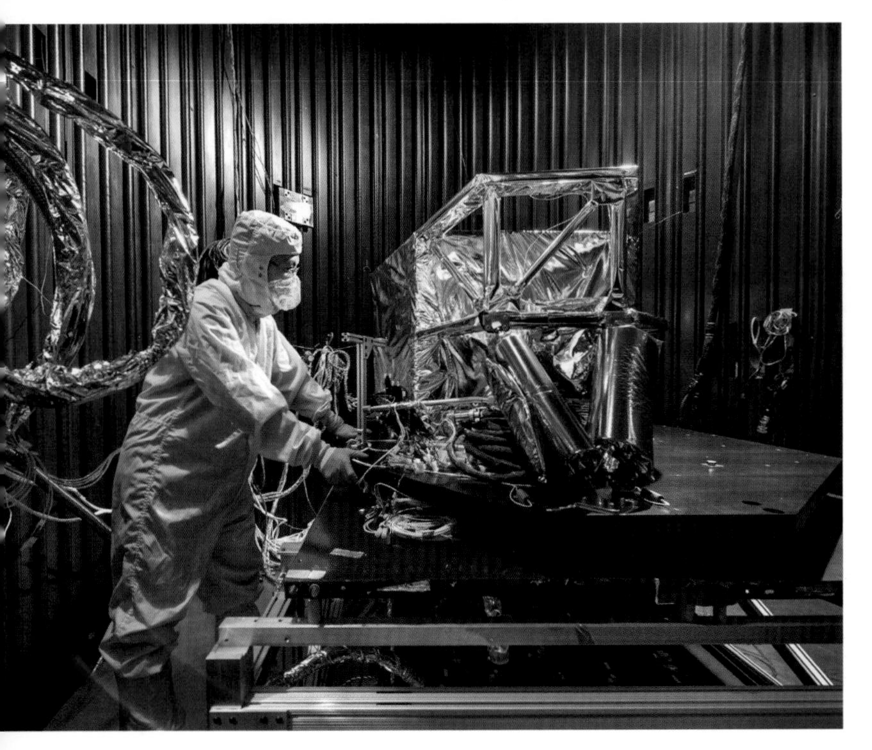

The JWST will weigh 14,330lb (9,700kg), with an effective mirror measuring 6.5m (21.3ft), compared to the HST's 2.4m mirror. Composed of 18 hexagonal segments each weighing 88lb (40kg), the mirror will have nine times the light-gathering capacity of the HST mirror that weighed more than twice as much. In addition to its large light-gathering capacity, the JWST will operate at near- and mid-infrared wavelengths, allowing it to study infrared emissions from objects formed within less than one billion years of the Big Bang, which formed the physical universe. It will have the ability to 'see' objects 400 times fainter than those currently detected by ground-based observatories or the best space-based infrared telescopes of today.

Designed for a five-year life, unlike the HST the James Webb Space Telescope will not be serviceable and it will be placed in an orbit at the Lagrangian 2 point, 932,100 miles (1.5 million km) distant on the opposite side of the Earth to the Sun. When deployed, complete with Sun shield, the JWST will span 69ft (21m). Having been in development for more than 20 years by the time it gets into space, the telescope is scheduled for launch on an Ariane 5 rocket in 2018. It is not designed for upgrade visits. However, it is not inconceivable that someday a future crewed spacecraft will visit the JWST and conduct a retrieval or attach additional science instruments, but not for a long, long time.

LEFT Tests of the thermal shield on the JWST. The 18 hexagonal-shaped mirror segments come in three shapes and unfold in sections to come together as a single monolithic mirror. To perform like that a wavefront sensor and control system will detect and correct any distortion in the optics. *(NASA)*

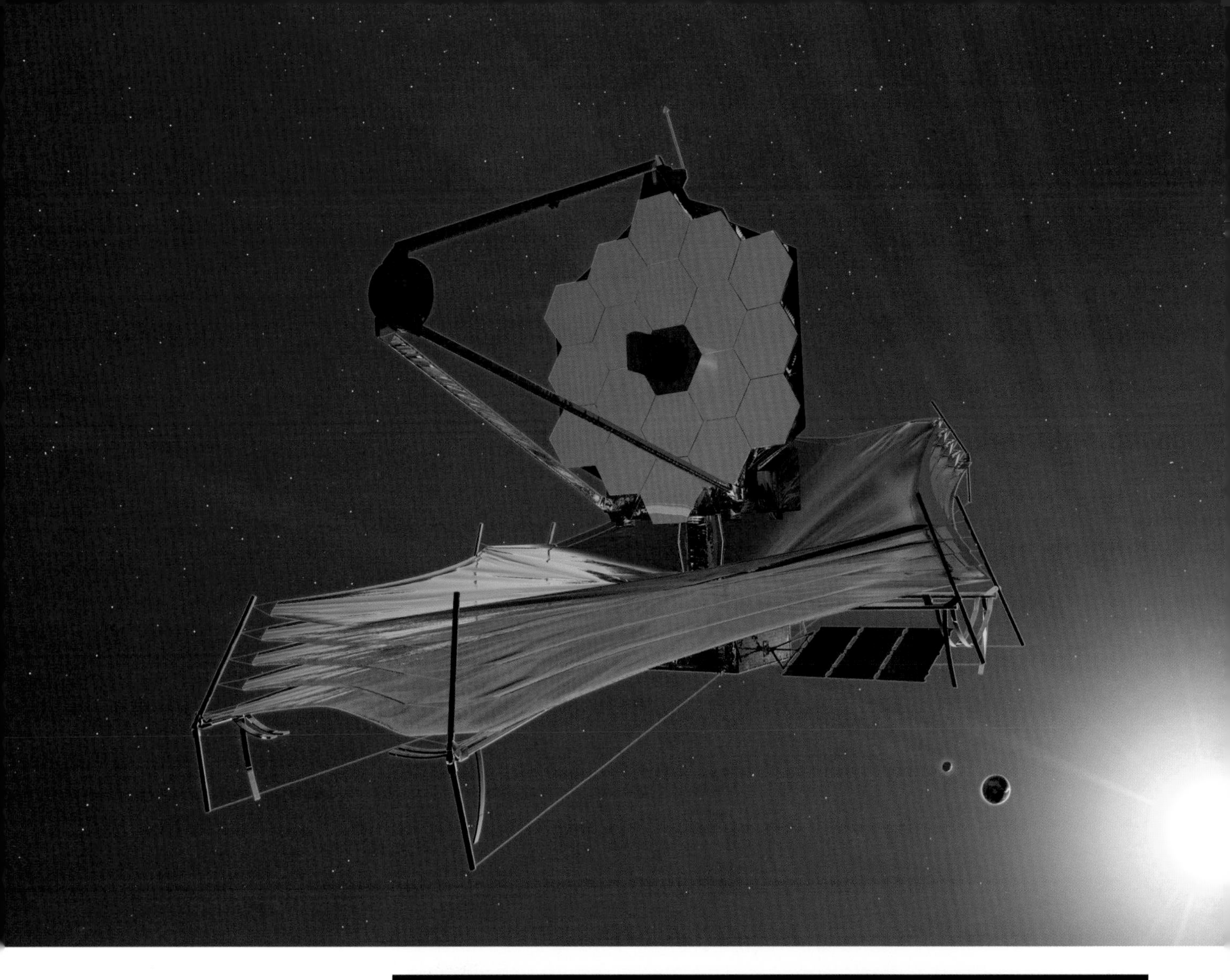

ABOVE Developed as a cooperative venture between NASA, the European Space Agency and the Canadian Space Agency, the James Webb Space Telescope will extend observations further into the infrared than is possible with the Hubble Space Telescope. Built by prime contractor Northrop Grumman, it comprises a deployable, five-layer sunshield the size of a tennis court and the telescope itself is cooled to –248°C. *(NASA)*

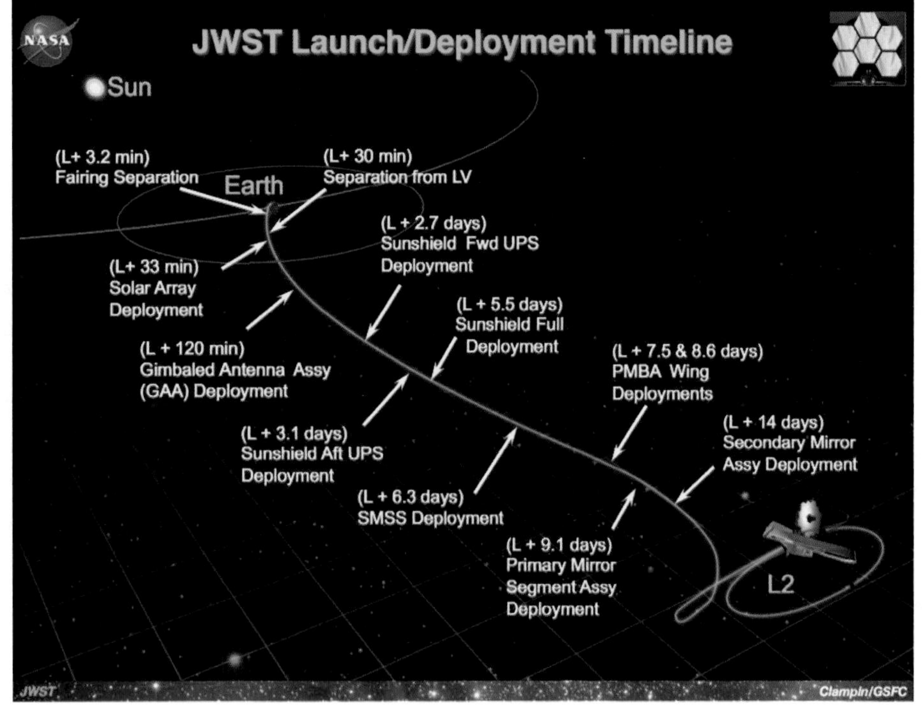

RIGHT The post-launch and deployment sequence demands a complicated series of mechanical events to 'unfold' the spacecraft from its encapsulated position on top of the Ariane 5 rocket. *(NASA)*

Appendix A

What is astrometry?

Not known by this term until relatively recently, astrometry is the oldest recorded science in the world and probably predates such written texts as have been discovered to date. In essence it is the measurement made of the actual location and the relative position of celestial objects. It forms the basis of astronomical observation and has been recorded as a scientific study since Babylonian times. It framed the context in which observations were made of the planets, the stars and the transient appearances of comets in several cultures including those of Egypt, Assyria and Greece. In modern times, astrometry is that part of astronomy that supports optical and non-optical observations of the heavens, and has become a pillar of that science.

A fundamental part of astrometry is the description of the motions observed by these bodies in the sky and to define those using very precise terms involving angles so small they are measured in degrees and seconds of arc. In terrestrial navigation the spherical world is divided into northern and southern hemispheres defined by an equator perpendicular to the axis of rotation. In this system the sphere is divided into 360°, each degree subdivided into 60 minutes of arc and each minute into 60 seconds of arc, there being 3,600 arc-seconds in one degree.

That measure of one degree at the equator translates into a radial distance of 6,076ft (1,852m) between degrees of longitude, which is used in terrestrial navigation as one nautical mile since longitude is the essential determinate from stellar measurements and time comparison. This replaced the system of 'leagues', which were equivalent to three Roman miles (5,000ft or 1,524m), although the Portuguese used four Roman miles. There has been some dispute as to the precise measure of a Roman mile, since it is based upon the

RIGHT The search for a clear view of the sky has driven astronomers to remote and high-altitude locations where the shimmering effect of the Earth's atmosphere is minimised, as with this telescope in Gran Canaria. *(David Baker)*

average length of the human foot, which was smaller in the time of the Roman Empire, and variations were introduced in different countries.

Astrometry was an important part of navigation from early times and in 1484 King John II of Portugal decreed that because the Pole Star could not be seen from the southern hemisphere, the Sun could be used instead, in conjunction with the solar tables of Zacuto of Salamanca. The nautical mile is also used today as a measure of speed for boats and ships at sea: one nautical mile per hour is defined as one knot. Most land-users refer to distance in statute miles because the British Parliament in 1592 established multiples of a furlong (660ft) as the length of a furrow an ox team could plough in a day. Since Parliament decreed that a mile should be measured as eight furlongs, the distance equated to 5,280ft (1,609.344m).

Precise measurements of angular variations are different for celestial navigation and the standard formats of degrees, minutes and seconds used in terrestrial navigation are too broad to be effective here. Astrometrists do not use radians. A radian is the standard unit of angular measure numerically equal in length to the length of a corresponding arc of a unit circle; one radian is the angle subtended at the centre of a circle by an arc equal in length to the radius of the circle. One radian (as a pure number) equals 57.2958° arc radius, there being 6.28318 radians in a complete circle. Stellar measurements in astrometry use degrees and seconds of arc and usually incorporate milliseconds, referred to as mas, or milliarc-seconds.

For astrometrists a position in the sky is determined by two spherical coordinates. The primary plane is the plane of the Earth's equator, which is also the celestial equator – not to be confused with the plane of the ecliptic (the plane in which the Earth orbits the Sun), because the Earth is tilted 23.5° on its axis of rotation; the ecliptic is, therefore, inclined 23.5° to the celestial (the Earth's) equator. Beginning at the vernal equinox where the celestial equator intersects the plane of the ecliptic, the measurement of right ascension is made counterclockwise in degrees and seconds radius. The second coordinate is the declination, which is measured from the equator, plus to the north and minus to the south, and in astronomical terms is the equivalent of a line of latitude on a celestial sphere.

Since these are hypothetical lines in the sky they are anchored by way of fiducial points, a separate set of coordinates assigned to a range of objects. These are said to constitute the

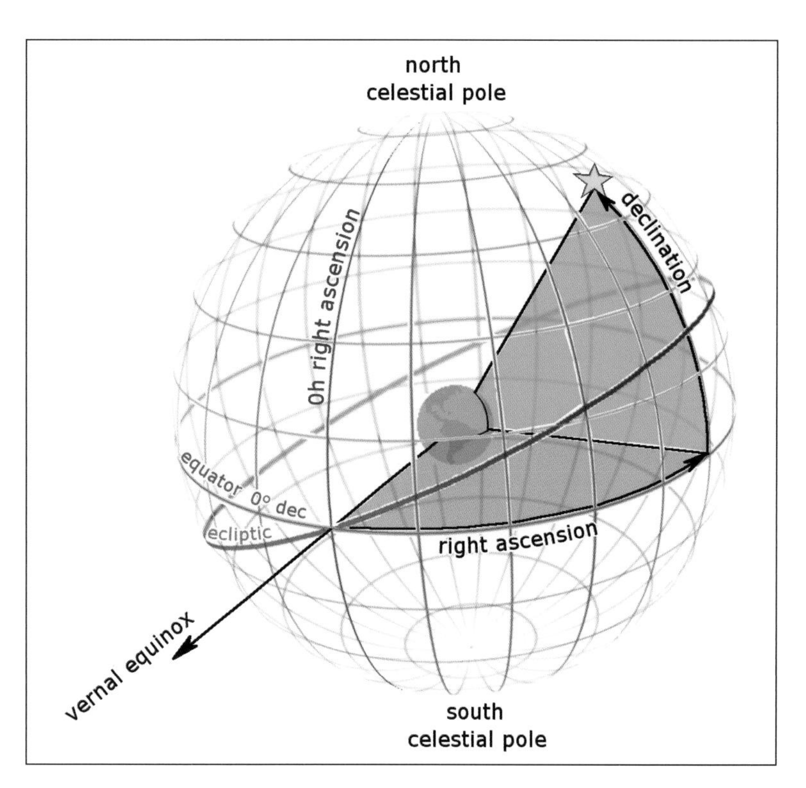

ABOVE The right ascension and declination of the celestial sphere is the coordinate system by which astronomers and both amateur and professional observers establish a reference matrix for all objects in the sky. *(David Baker)*

LEFT Undergoing thermal tests at ESTEC in the Netherlands, Europe's Hipparcos was launched a year before Hubble and provided a relatively short-range refinement on the precise position of more than 118,000 stars. *(ESA)*

reference frame, and the position of a celestial object is quoted in relation to these fiducial points. One vital component of the selected fiducial object is that it does not change position in time. Only very distant objects such as other galaxies or quasars several million light years distant which will not change as the Earth orbits the Sun, or as the solar system travels around the nucleus of the Milky Way galaxy, are appropriate. These points are integrated into a catalogue known as the International Celestial Reference Frame (ICRF), which is essentially a radio-source of fiducial points extended to the optical portion of the spectrum by data from the Hipparcos satellite (see next column).

Atmospheric refraction and aberration are problems faced by terrestrial observers but avoided for astrometrists working with the Hubble Space Telescope. For observers on Earth, light from radiating sources in space is bent progressively as it enters the atmosphere, which is composed of layers of different refractive indexes. The integrated effect depends upon the pressure, temperature and humidity of the atmosphere and the wavelength of the light. The lower the object in the sky, the larger and the more uncertain the correction to be applied. This effect is simply non-existent in the vacuum of space.

Aberration is experienced in both Earth- and space-based astrometry but the relevant conditions and reasons are different. In practice, ground-based astrometric observations are not performed below 60° zenith distance. The apparent direction of a source is a combination of the direction from which the light arrives and the velocity of the observer. In ground-based observations, the diurnal aberration is distinguished by the motion of the observer as a consequence of Earth's rotation and stellar aberration due to the motion of Earth around the Sun. In astrometry from space, the diurnal aberration is replaced by the orbital aberration due to the motion of the satellite in its orbit.

In 1989, less than a year before the Hubble Space Telescope, the European Space Agency launched its Hipparcos satellite. Named after the Greek astronomer Hipparchus, the satellite's name is acquired as an acronym from the title High Precision Parallax Collecting Satellite. Hipparcos was designed to measure the parallax of stars, the semi-angle of inclination between two sight-lines to a designated star when the Earth (and its orbiting satellite) is on opposite sides of the Sun. Measurements taken at six-month intervals will provide accurate near-term distance measurements and is an important part of astrometry. From an elliptical orbit of 22,300 x 315 miles (35,888 x 507km) the 2,513lb (1,140kg) satellite operated until the end of March 1993 and built up a catalogue of 118,200 stars.

The Hipparcos programme arose from a seemingly insoluble set of problems which threatened to limit the ability of ground-based observers to improvements in accuracy dominated by atmospheric aberration, thermal problems with mirrors in the atmosphere, gravitational effects and the absence of clear skies. The programme was proposed in the mid-1960s, astronomers seeing in the Earth-orbiting telescope a range of opportunities which, on a broader base, stimulated intense interest in developing what would eventually become the Hubble Space Telescope itself. Hipparcos helped establish a relatively short-range astronometric refinement that directly assisted with improving conditions under which the HST could do its job.

The Hubble Space Telescope is not primarily designed to perform astrometric

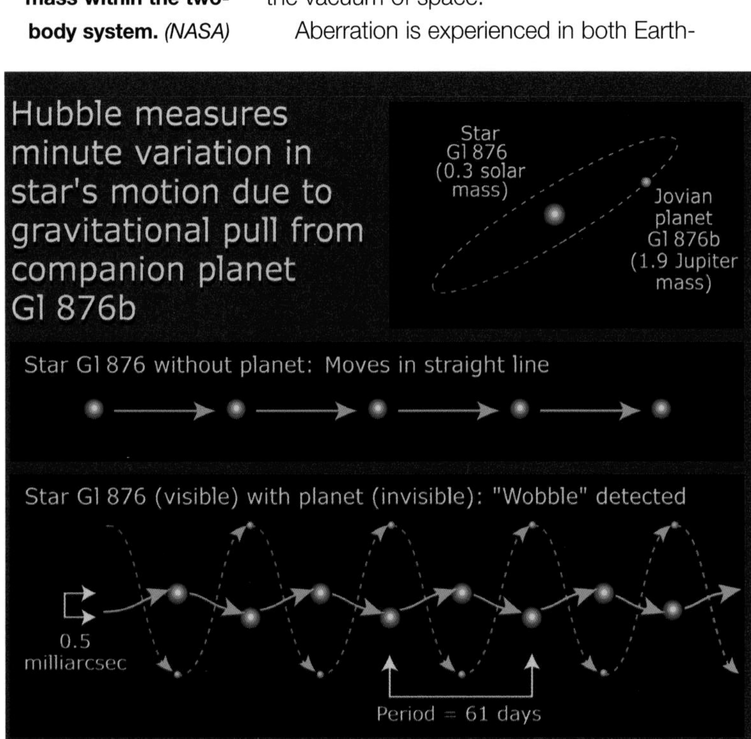

BELOW Hubble is able to precisely measure minor variations in the movement of a star as a companion pulls and tugs to perturb its own rotational motion due to an offset centre of mass within the two-body system. *(NASA)*

observations, but to support several scientific instruments whose common requirement is that the Telescope must be able to point in any given direction in the sky and stay pointed with very high stability as long as necessary. As described in Chapter 3, this is achieved by three Fine-Guidance Sensors, with two sufficient to locate a target and stay pointed at it. The third one remains free with the possibility to do astrometry within the field of view. So, in general, astrometric measurements are confined to a certain field in the vicinity of the region studied by other instruments.

However, some astrometric programmes are scheduled for their own sake, and then the choice of targets is left to the discretion of the astrometrist. The Wide-Field/Planetary Camera (WF/PC) is sometimes used to perform, despite its name, very narrow field astrometry. Although a rare occurrence, the Faint Object Camera (FOC) may also be used for this purpose.

Shortly after launch and orbital deployment it was determined that the main mirror had a severe case of spherical aberration and that the secondary mirror was slightly tilted and off-centre. In addition, there was an important jitter due to the excitation of the solar panel assemblies when the satellite passed into or out of direct sunlight. All of this significantly impaired the astrometric quality of the Telescope. The Hubble repair mission in December 1993 suppressed the jitter and the WF/PC was replaced by a new camera with modified optics to correct the defects of the Telescope. The Faint Object Camera (FOC) was corrected by additional optics provided by the multi-corrector COSTAR instrument. But the FGSs remained untouched, and the only improvement came from replacing the solar cell arrays that suppressed the jitter.

Otherwise, the situation remained characterised by the major spherical aberration added to the expected astigmatism inherent in a Ritchey-Chrétien telescope configuration, a configuration chosen to avoid an important coma that would have been a more severe penalty to astrometry from the Fine-Guidance Sensors. The point-spread function has significant features even a few seconds of arc away from the centre of an image and in addition depends strongly on the position in the field of view. Only 15% of the light is concentrated in the 0.1sec central circle instead of more than 50%, as anticipated. This corresponds to a loss of more than one magnitude in access to fainter stars and some general degradation of astrometric capabilities. However, even if it could have been better, the Fine-Guidance Sensors, with careful calibrations, make the HST a remarkable tool for astrometry.

BELOW Only by measuring precisely the relative position of stars and the distances from Earth has it been possible to extrapolate beyond the galaxy to more distant places and to construct, partially with this information, a history of the evolution of our universe. Although not designed for astrometry, the Fine Guidance Sensor played a dual role in enabling that search. *(ESA)*

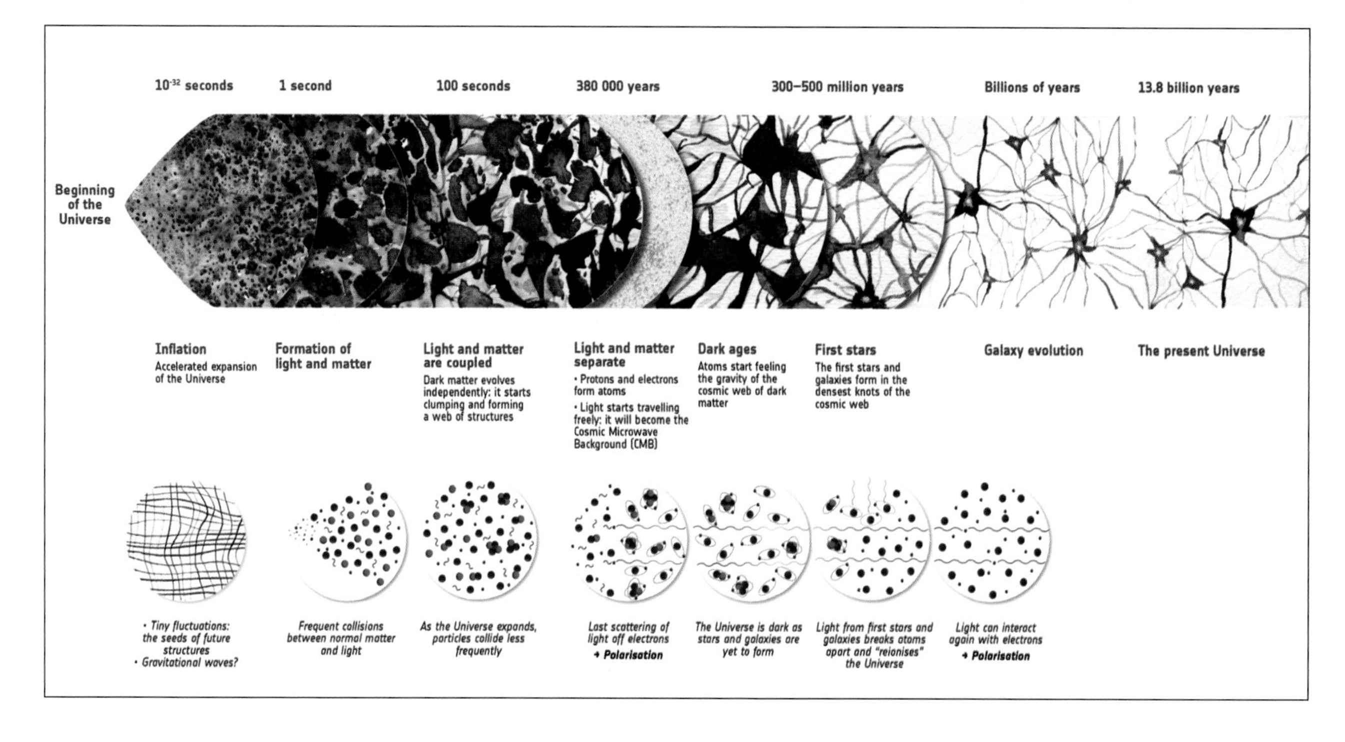

Appendix B

The physics of the telescope

As an optical instrument the telescope is more than 400 years old, with Galileo Galilei being the first person to make recorded observations of a practical nature. In 1608, the by then ancient art of spectacle-making was taken to new heights by Hans Lipperhey and Zacharias Janssen from Middelburg, the Netherlands, who applied for a patent on 'an instrument for seeing far'. They had in mind the sight-seeing telescope which is used universally today for everything from bird-spotting to observing mountain slopes and glaciers. But it was Galileo who, just a year later, built his own telescope and turned this toward the heavens, when he first resolved the rings of Saturn and discovered the four large moons of the giant planet Jupiter – given the classical names Io, Europa, Ganymede and Callisto.

Telescopes became popular instruments, refractors dominating the scene until 1688 when Isaac Newton designed the first practical reflecting telescope and gave his name to a generation of instruments that live on to this day – the Newtonian reflector. The invention of the achromatic lens in 1733 allowed partial correction for colour aberrations in the reflector telescope, allowing smaller and more capable telescopes of this type to emerge. By the end of the 19th century the majority of astronomical telescopes were reflectors and by the early 20th century, accompanied by a series of groundbreaking discoveries in physics, astronomy came of age. But it was not technology that imposed limits to what could be achieved with both reflectors and refractors: the very Earth itself posed significant problems for astronomers seeking to peer through the atmosphere.

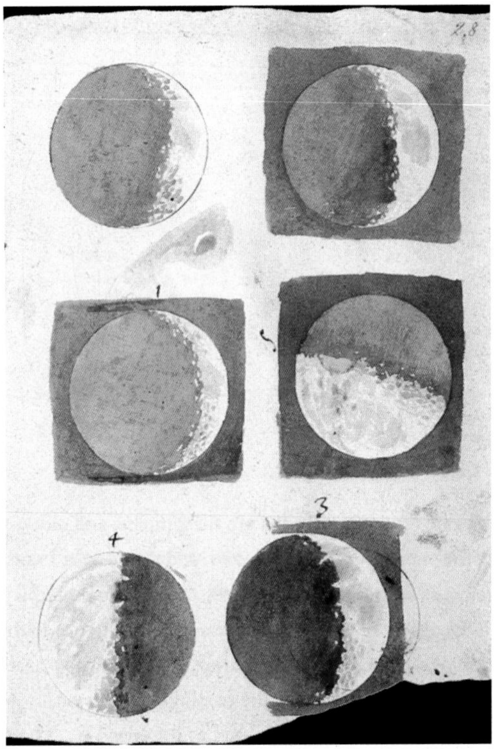

RIGHT From the early 17th century astronomers have been observing the heavens with telescopes, but Galileo Galilei was the first to observe phenomena not seen with the naked eye. This early drawing by Galileo shows his depictions of lunar phases. *(David Baker)*

The effects of the atmosphere are significant and while many of the properties of the envelope surrounding Earth limit the effectiveness of the telescope they are important for the preservation of life. Nevertheless, the presence of the ozone layer, a screen protecting life from the harshest effects of ultraviolet radiation, absorbs up to 2% of visible light. But there are other characteristics that benefit us on Earth, the Rayleigh scattering effect being the one which turns the sky blue instead of black. Named after the British physicist Lord Rayleigh, it is the electric polarisation of particles in the atmosphere that can be at the molecular or even the atomic level. They work through a very simple effect.

The oscillating excitation of the electric field of a light wave acts on the charge within the particle, causing it to move at the same frequency. This scattering causes the diffuse radiation to colour the sky blue since it is more effectively observed in the visible portion of the spectrum at this wavelength, where more than 75% of scattering is in the blue spectral bands below 460nm. The Rayleigh effect even works in moonlight, though we do not see the sky blue under that condition because at exceptionally low light levels vision comes from rods in the human eye which do not have colour response.

The atmosphere has a visually limiting effect at low angles to the horizon due to the added effect of the cumulative influence of the atmosphere, where the linear penetration is considerably greater than directly up through the least amount of atmosphere. Particles of dust, humidity and gaseous pollution all work against the observer, less than 50% of the light received by the eye looking straight up. Unfortunately, ground-based telescopes are ineffective if their viewing angle is restricted to a line-of-sight perpendicular to the surface!

Another effect forced upon ground-based observers is the shimmering caused by turbulence in the atmosphere, although this is greatly reduced with altitude, which is the reason why several observatories are located at the tops of mountains. Astronomers work to basic assumptions that there are three primary areas of distortion (turbulence). The first 330ft (100m) is where convection currents flow from thermal energy released by organic, geographic, geologic or man-made structures; the height from 330ft to approximately 6,500ft (2,000m) is where the central troposphere is disturbed by thermal upwelling, large geographic undulations or wind eddies; and from 20,000ft (6,100m) to 40,000ft (12,200m) in the high troposphere is the region where jet streams can seriously affect atmospheric turbulence.

Even the telescope itself, if sufficiently large to require a domed observatory for weather protection when not in use, can be the cause of its own debilitation. Thermal energy stored in the structure can take time to stabilise and to release a thermal soakback during the heat of the daylight hours and this affects the lowest level of perturbation. At the central-tropospheric level wind raised by cities and large industrial areas creates thermal upwelling which causes severe distortion in the atmosphere. At the highest level weather systems accumulate and change the density (pressure) of the air, and this has a negative effect on the visual clarity of the objects under observation.

Of other effects, light pollution has become progressively one of the greatest limiting effects caused by humans. Street lights, night lights, industrial illumination across open workplaces and the general glow that attends large urban conurbations, all take their toll on the observer, playing havoc with the potential of the telescope. Added to the atmospheric effects caused by the presence of the atmosphere and the particulates and aerosols which occur naturally, ground-based astronomy beguiled the first 380 years of serious optical observation until the launch of the Hubble Space Telescope liberated astronomers from the limitations of ground-based observation.

ABOVE The 36in Lick telescope was set up on Mount Hamilton, California, in the late 19th century by James Lick. This observatory did much to advance the science of stellar spectroscopy. *(USIS)*

LEFT The 50cm refractor at Nice Observatory. *(David Baker)*

RIGHT Mount Wilson Observatory was set up by George Ellery Hale as a place where the sky is clear 300 days in the year and atmospheric effects are reduced by its altitude of 5,700ft (1,737m). This reflecting telescope has a 60in mirror and variable focal length adjustments. *(Mount Wilson Observatory)*

Whether on Earth or in space, the laws of physics apply equally to ground-based astronomers or those fortunate enough to programme work time on the Hubble Space Telescope. There are certain terms and phrases used by astronomers to describe the physical parameters of their work that may seem unfamiliar or even daunting to the uninitiated. A few non-technical descriptions may help to unravel the mystery and also to show why, and how, the HST is such a breakthrough in observational astronomy. It will also add purpose to the selection of science instruments carried on board the HST.

Every object in the sky radiates energy and, for very distant objects, it is through this energy that we know it is there. Light is the part of the electromagnetic spectrum with which we see objects in the night sky and almost all those we see *emit* energy as light. Only the Moon, the planets, the comets and observable asteroids are seen by reflected light from our Sun, which is of course a star in its own right. But the energy radiated by a star also tells astronomers about the heat within the system. Light is an expression for a discrete unit of electromagnetic energy known as a photon and is measured by counting the number of electrons released by photons when they strike certain materials. To discover the properties, light detectors channel photons through chemically coated windows where the released electrons can be counted.

All optical images are made of light that is a form of electromagnetic radiation (see later). More precisely, a telescope image is made by imaging a countless number of light-emitting point sources from distant objects in the universe, just like similar observations through a terrestrial telescope. Light waves emitted

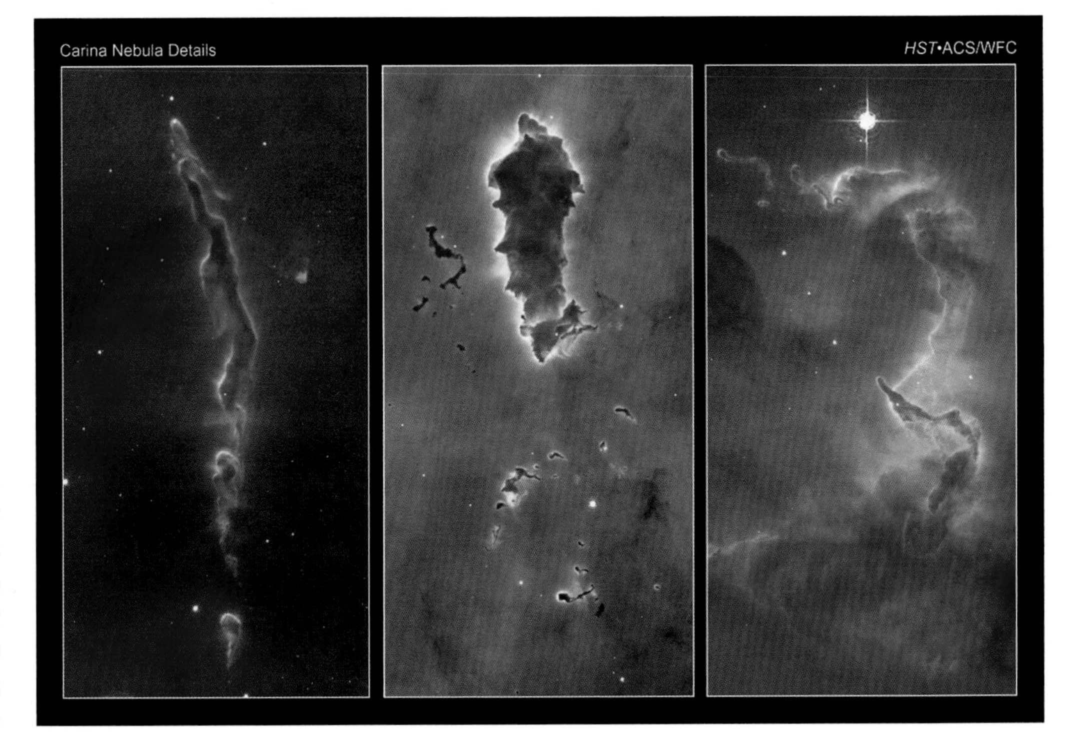

RIGHT The Carina nebula with gas and dust as viewed in a combination of images from the WF/PC and ACS instruments on the Hubble Space Telescope. *(NASA)*

by a point-source spread out in a concentric pattern, propagating as an oscillating energy field. It is convenient to present wave oscillation as a cycle, the full cycle being 360 degrees, or 2π radians. The phase of wave oscillation is defined by $o = A\sin(2\pi x/\lambda)$, where A is the wave amplitude, defined as the maximum value of wave oscillation; x is the length of the wave path from the origin or source; and λ is the wavelength of light.

Energy can therefore be manifested as light and can be regarded both as a wave and a particle. Both exist together yet each is separate. When observed as a wave the length is defined as the distance between peaks either side of the trough and the wavelength depends on the temperature; the cooler the temperature the longer the wavelength. Because different elements radiate energy at different temperatures, the pattern of the wavelength reflects the nature of the element. The light from a star, therefore, tells astronomers the temperature of the energy by the wavelength and the nature of its elemental composition by the nature of the spectrum. The spectrum is a product of the chemicals and extends from short wavelengths such as gamma rays to the longest wavelengths, radio waves.

The visible portion of the spectrum is the one to which the human eye has adapted for visual sight and extends from short violet rays at one end to longer red rays at the other end. In observing the light from a distant star the astronomer is limited by this visible part of the spectrum. In looking at a star the dominant peak energy level in its visible spectra will reveal the dominant wavelength through its colour, be it a cool star appearing red or a much hotter star appearing blue. But that will mask a more dominant energy level that may be outside the visible portion of the spectrum. For instance, the seemingly less energetic star may be emitting far higher energy levels as gamma rays, which are invisible to the human eye.

Because of this narrow window through the electromagnetic spectrum the astronomer sees with his naked eye only a very limited amount of information coming from a source that has a veritable torrent of potentially valuable measurements with which it can be characterised. But even light has more to offer the astronomer. When the electric and magnetic field components of light proliferate, they vibrate at random motion in planes perpendicular to the direction the light is travelling. These components, an effect known as polarisation, can be made to provide information about the medium through which the light is passing. When the starlight travels through magnetic clouds of dust, as it frequently does through space, the alignment of the cloud particles scatters light according to the spatial orientation on the electrical and magnetic field components of the light waves. When detected, this polarised light can indicate the existence of large magnetic fields in space.

As explained above, an important aspect of understanding the physical properties of a star lies in the measurement of its wavelength. This uses a measurement known as the angstrom (Å); 10 billion angstrom units are equal to one metre, or 10Å per nanometre. The wavelength

ABOVE Stellar graveyards in the Milky Way in images from the WF/PC II camera. *(NASA)*

RIGHT **NICMOS views of young stars 450 light years away in the constellation Taurus.** *(NASA)*

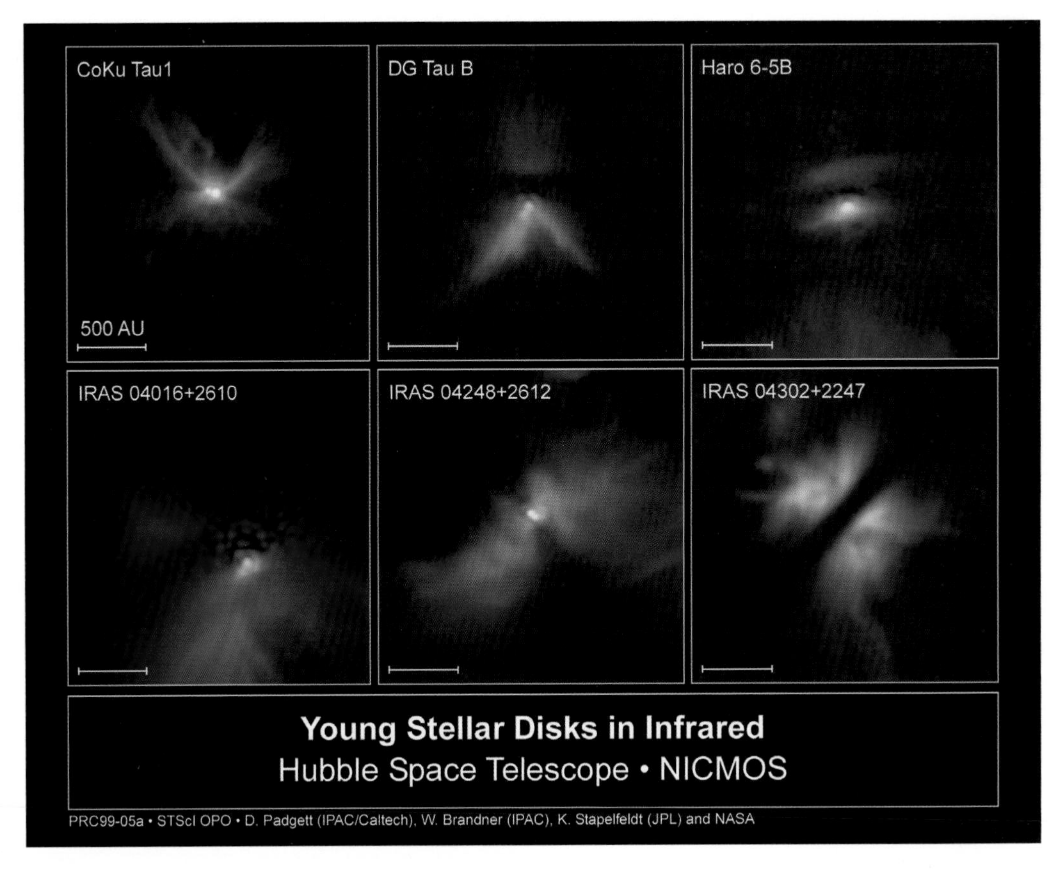

varies from a few angstroms for high-energy gamma rays to several thousand angstroms for radio waves, with visible lights covering the spectral range of 4-000–7,000Å. While most electromagnetic wavelengths are invisible to the human eye a considerable portion of the visible spectrum can be occluded by dust and gas in the Earth atmosphere. Because of this, the visible portion of light reaching Earth-based telescopes is really only a fraction of the amount of energy the star, or the intrinsically bright object observed, is actually radiating.

The Optical Telescope Assembly is capable of observing energy in wavelengths from 1,100Å in the ultraviolet to 11,000Å in the infrared, and the science instruments cover various windows within that range. But it is not enough to open a window, as the clarity – or resolution – is determined not by the size of the window into a portion of the electromagnetic spectrum, but on the size of the discriminating ability to differentiate between distances within the wavelength. Spectral features within the wavelength provide more detailed information about the target observed and some instruments on the HST are required to achieve different resolutions, as described in Chapter 3.

A conventional reflecting ground-based telescope is defined by the objective and the eyepiece. The objective captures light and uses it to form a real image; the eyepiece is a sophisticated magnifying glass that enables the eye to enlarge the projection of the image on to the retina. The objective is a single concave mirror or it can be a combination of mirrors and lenses and it is the function of the eyepiece to make all the light from the image formed by the objective to be made available to the eye – or, in the case of the HST, to the science instruments into which the reflected light is directed. In all telescopes, as in photography, the resolution is determined by the aperture, the most basic element of its structure.

With the diameter of the aperture when denoted as D and the focal length as f the relative aperture is determined by $D/f = 1/F$, with F being the focal ratio. The focal length is the distance between the objective and the point where it focuses collimated light. In other words, the point where the wavefront entering the objective is partially flat and the light rays parallel. This definition of focal length is only

LEFT About 6,500 light years distant, the famed 'Pillars of Hercules' in high definition – one of the most imposing images taken by instruments on the Hubble Space Telescope. The original 1995 image taken by Hubble is at left, and the new high-definition image taken by the WF/PC III to the right. *(NASA)*

true if it focuses all rays to a single path on the central ray, which in theory none can do because of spherical aberration, which is itself a natural property of all non-planar surfaces and is not referring to a 'spherical' characteristic in the traditional sense but rather a product of an optical surface which fails to match an incident wavefront. For the HST, spectral resolution is different according to the instrument concerned and those values can be found in Chapter 3.

The measurement of stars is a fundamental aspect of the HST and the distance is measured by observing a star's parallax, the apparent angular displacement of an object in the sky when caused by the movement of the observer's position, such as by the Earth moving halfway round the Sun or of an orbiting telescope moving around the Earth. It is only relevant when observing relatively near-field stars to a distance of about 650 light years as measured against very much more distant stars which have no apparent motion. This technique has been used by satellites to measure just that phenomenon in the universe, the European Space Agency's Hipparcos satellite being one (see Appendix A).

The star's parallax is measured as the half-angle of that angular displacement and the distance can be calculated using basic geometry. The measurement of parallax is used as the determinant for distance measurement for objects out to approximately 650 light years, one parallax second of arc being one parsec = 3.26 light years. Hence, 200 parsecs is approximately 650 light years. Astronomical distances within that radius of Earth are measured in parsecs. It is from this that the angular resolution can be determined too and is a function of how clearly an instrument forms a clear and distinct image.

As described in Chapter 3, the angular or spatial resolution is a measure of fineness; the greater the angular resolution the closer two objects can appear and still be distinguishable. This angular resolution is measured in degrees of arc and that difference is measured in arc-seconds for the science instruments. The finest angular resolution with the HST is about 0.01 arc-sec, which is about ten times better than the largest Earth-based telescope at the time the HST was launched. But the HST is capable of resolving to a far greater degree of accuracy the magnitude, or brightness, of an object in the visible portion of the electromagnetic spectrum. And here some basic observational astronomy helps for those unfamiliar with the working end of a telescope.

Since Greek philosophers (astronomers) first gazed at the night sky they realised that not all stars appear as bright as each other. Obsessed with conjecture, about the nature of the night sky, the very nature of the stars and the apparent geometry of Earth, Sun, Moon and planets, Hipparchus (190–120 BC) decided upon a classification with the brightest referred to as 1st magnitude and the faintest as 6th magnitude. All the intermediate levels of brightness, surely an arbitrary decision based on visual acuity and the age of the observer, were awarded intervening numbers 2–5.

This system was capable of broadly classifying the almost 7,000 stars visible to the naked eye in classical Greece. It is somewhat lower today due to the deterioration of the Earth's atmospheric transparency resulting from several centuries of human industrial activity. Tycho Brahe (1546–1601) attempted a degree of modification when applying a component of size to the brightness of stars but that proved fallible, yet the belief persisted that the magnitude of a star reflected its true size. With the advent of the telescope early in the 17th century (see the top of this section) some further attempt was made to fully understand the relationship with brightness to distance and to the inferred size of the star.

BELOW Hubble captures views of 'Mystic Mountain', a pillar of gas and dust three light years tall at whose peak scores of infant stars are firing off jets of gas streaming from towering peaks. Situated 7,500 light years from Earth in the southern constellation of Carina, the view was taken on 1–2 February 2010 with the WF/PC III. *(NASA)*

By the mid-19th century, with bigger and better telescopes equipped with more refined and capable optics, astronomers were busy measuring the parallax of stars and hence their distance from the Earth. And therein did the difficulty emerge – some quite distant stars were brighter than some closer ones, and the rationale for determining star type and size from brightness alone fell apart. It began to appear as though stars were of different sizes and produced energy at different rates, that there were many types of star and that brightness as viewed through the telescope was no yardstick to stellar properties.

Together with the emergence of modern physics at the end of that century the way was clear for a radical change in understanding about the universe, a transformation brought full circle by the Hubble Space Telescope. As early as 1856 Norman Pogson set out a scale of magnitudes that astronomers would use to measure the brightness of an object. With new methods of observing the heavens and using a more mathematical approach to the problem, it was determined that the brightest stars were about 100 times the magnitude of the faintest, 6th magnitude, stars. Using a logarithmic scale it was agreed that a ratio of $\sqrt[5]{100} \approx 2.512$ would be established between magnitudes. For instance, the star Arcturus becomes magnitude 0 and the still brighter star Sirius becomes -1.46, the full Moon becomes -13 and the Sun -27, the brightest object in the sky. At the other end of the scale, Saturn becomes 0, Uranus becomes magnitude 5, Neptune magnitude 8 and Charon, a moon of Pluto, magnitude 16. In terms of observation, the minimum brightness a pair of 7 x 50 binoculars will resolve is magnitude 10, while the observable limit for a standard 26ft (8m) ground-based telescope is an astonishing magnitude 27. The Hubble Space Telescope will detect objects down to magnitude 32.

But this is the *apparent* magnitude, the brightness of an object in the sky as viewed from Earth. It is not the *actual* brightness of the star as measured from a fixed reference distance. While the apparent magnitude is a useful measure of what is required of a telescope to see certain stars in the sky, the actual, or *absolute*, magnitude can only be calculated if the distance to that star is known. To do that the distance modulus needs to be applied, where the distance to the star is measured in parsecs, where m is the apparent magnitude and M is the *absolute* magnitude. A precise parallax measurement determines the distance and from that the absolute magnitude can be calculated as $m - M = 5(\log_{10} d - 1)$, so that the properties of the star can be fixed against a known constant – the brightness which would be measured by an observer at a distance of 10 parsecs (32.6 light years). From this can be determined the fundamental characteristics of the stars' actual brightness on a comparable scale.

Using wavelengths to determine the amount of energy from a star, resolving the wavelengths into discrete resolution to understand the chemistry of the star, and measuring parallax more precisely to determine the distance of nearby stars, the HST can provide refined measurement of absolute magnitude and consequently the intrinsic brightness qualities. All these are a standard task for the HST, but specific science instruments are designed to search for specific answers to complex problems that arise in modern astronomy (see Chapter 3).

Collectively, the Hubble Space Telescope and other ground-based observatories work together to wrestle problems and find new questions to ask, including the changes in our understanding of the evolution of the universe from its very beginning to its uncertain future. Adopting standard and undramatic capabilities, the HST is a very ordinary telescope, but the significance of this remarkable instrument lies in its adaptability to use new sets of instruments developed as technology advances and new capabilities arise and to do its work in a place where no other optical telescope on this scale has ever been deployed.

As an observatory, it would have benefitted from a continuing series of upgrades and refurbishment, but the Shuttle was available for only 19 years of its life, which could extend to a total lifespan of more than 30 years. By then, around 2020, its successor, the James Webb Space Telescope, could be fully operational from its unique vantage point and there may yet be another generation beyond that which could once again receive upgrades and replacement of instruments from astronauts on the next generation of human space flight vehicles.

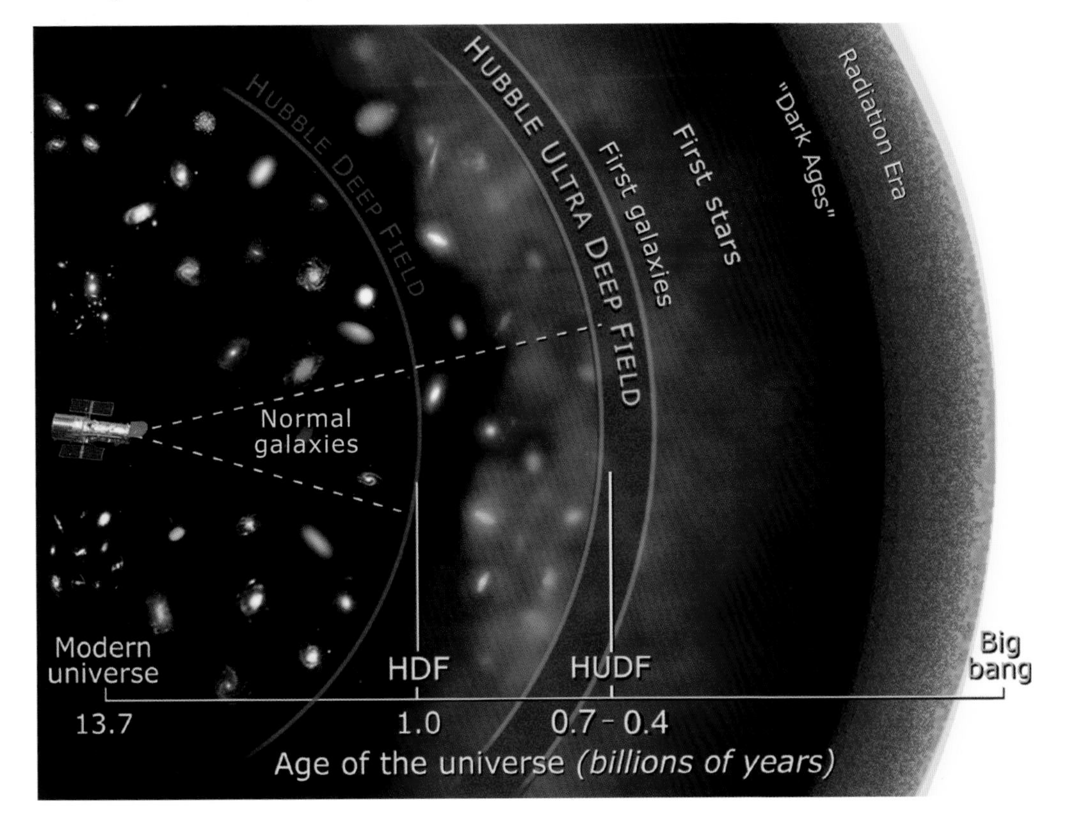

LEFT With the Hubble Space Telescope able to see structures and objects in the universe formed within the first billion years of its evolution, further advances with its successor the James Webb Space Telescope will reveal activity occurring close to the Big Bang from which all matter was formed. *(NASA)*

Abbreviations

Å Angstrom, measurement of wavelength.
AC Advanced Computer.
ACE Actuator Control Electronics.
ACS Advanced Camera for Surveys.
AD Aperture Door.
APU Auxiliary Power Unit.
arc-min A minute of arc.
arc-sec A second of arc.
ASB Aft Shroud & Bulkhead.
ASIPE Axial Scientific Instrument Protective Enclosure.
AURA Association of Universities for Research in Astronomy.
CCD Charge-couple device.
CDI Command Data Interface.
CGGM Contingency Gravity Gradient Mode.
CMG Control moment gyroscope.
CMOS Complementary metal-oxide-semiconductor.
COS Cosmic Origins Spectrograph.
COSTAR Corrective Optics Space Telescope Axial Replacement.
CPU Central processing unit.
CsI Caesium iodide.
CsTe Caesium telluride.
CU/SDF Control Unit/Science Data Formatter.
DIU Data Interface Unit.
DMS Data Management Subsystem.
DMU Data Management Unit.
DOB Deployable Optical Bench.
DOV Director of Orbital Verification.
ECU Electronic Control Unit.
EMU Extravehicular Mobility Unit.
EPS Electrical Power Subsystem.
EP/TCE Electrical Power/Thermal Control Electronics.
ES Equipment Section.
ESA European Space Agency.
ESRO European Space Research Organisation.
EST Eastern Standard Time.
ESTR Engineering Science Tape Recorder.
EVA Extravehicular activity.
FEP Fluorinated ethylene propylene.
FGS Fine Guidance Sensor.
FOB Fixed Optical Bench.
FOC Faint Object Camera.
FOS Faint Object Spectrograph.
FOSR Teflon Flexible Optical Solar Reflector.
FOV Field of view.
FPA Focal Plane Assembly.
FPS Focal Plane Structure.
FRB Failure review board.
FS Forward Shell.
FSS Flight Support Structure.
FUV Far ultraviolet.
GHRS Goddard High Resolution Spectrograph.
GSFC Goddard Space Flight Center.
HGA High-Gain Antenna.
HgCdTe Mercury cadmium telluride.
HOSC Huntsville Operations Support Center.
HOST Hubble Space Telescope Orbital Systems Test Platform.
HRC High Resolution Channel.
HSP High Speed Photometer.
HSPM Hardware Sun Point Mode.
HST Hubble Space Telescope.
I&CS Instrumentation and Communication Subsystem.
IDT Image dissector tube.
IOU Input/output unit.
IUE International Ultraviolet Explorer.
IUS Inertial Upper Stage.
JWST James Webb Space Telescope.
LAS Large Astronomical Satellite.
LDEF Long Duration Exposure Facility.
LGA Low-Gain Antenna.
LIDS Low Impact Docking System.
LMC Large Magellanic Cloud.
LOPE Large Orbital Replacement Unit Protective Enclosure.
LS Light Shield.
LST Large Space Telescope.
MAMA Multi-Anode Microchannel Plate Array.
MCU Mechanisms Control Unit.
MLI Multi-layer insulation.
MMMS Multi-Mission Modular Spacecraft.
MR Main Ring.
MSM Mode Selection Mechanism.
MULE Multi-Use Lightweight Equipment Carrier.

NASA National Aeronautics and Space Administration.
NBF Neutral Buoyancy Facility.
NCS NICMOS Cooling System.
NICMOS Near-Infrared Camera and Multi-Object Spectrometer.
NIR Near infrared channel.
nm Nanometers.
NOBL New Outer Blanket Layer.
NUV Near ultraviolet.
OCE Optical Control Electronics.
OCE-EK Optical Control Electronics Enhancement Kit.
OCS Optical Control Subsystem or Optical Control Sensors.
OMS Orbital Manoeuvring System.
OMV Orbital Maneuvering Vehicle.
OPF Orbiter Processing Facility.
ORU Orbital Replacement Unit.
ORUC Orbital Replacement Unit Carrier.
OTA Optical Telescope Assembly.
OTV Orbit Transfer Vehicle.
OV Orbital Verification.
parsec One parallax second of arc (3.26 light years).
PCR Payload Changeout Room.
PCS Pointing and Control System.
PCU Power control unit.
PDS Photo Detector System.
PDUs Power Distribution Units.
PMA Primary Mirror Assembly.
PSEA Pointing Safemode Electronics Assembly.
RAC Rigid Array Carrier.
RM Remote Module.
RMGA Retrieval Mode Gyro Assembly.
RMS Remote Manipulator System.
RNS Relative Navigation System.
RP Reaction Plate.
RSU Rate sensor unit.
RWA Reaction-wheel assembly.
SAA South Atlantic Anomaly.
SAC Solar Array Carrier (subsequently Second Axial Carrier).
SADE Solar Array Drive Electronics.
SADM Solar Array Drive Mechanism.
SAGA Solar Array Guidance Augmentation.
SAO Smithsonian Astrophysical Observatory.
SBC Solar Blind Channel; also single-board computer.
SCM Soft Capture Mechanism.

ABOVE The fifth and final servicing mission (SM4) involved the Shuttle Orbiter Atlantis on Launch Complex 39A (foreground) with the Shuttle Orbiter Endeavour on standby at LC-39B, should a rescue be necessary. *(NASA)*

SCRS Soft Capture and Rendezvous System.
SI C&DH Science Instrument Control and Data Handling.
SIHM Software Inertial Hold Mode.
SIPE Science Instrument Protective Enclosure.
SLIC Super Lightweight Interchangeable Carrier.
SMA Secondary Mirror Assembly.
SMM Solar Maximum Mission.
SOFA Selectable Optical Filter Assembly.
SOPE Small Orbital Replacement Unit Protective Enclosure.
SPVM Sun Point Vehicle Mode.
SRB Solid rocket booster.
SSB Space Studies Board.
SSM Support Systems Module.
SSR Solid State Recorder.
STAR Self-test and repair.
ST-ECF Space Telescope European Coordinating Facility.
STINT Standard Interface circuit board.
STIS Space Telescope Imaging Spectrograph.
STOCC Space Telescope Operations Control Center.
STScI Space Telescope Science Institute.
STV Space Transfer Vehicle.
SV Scientific Verification.
TDRS Tracking and Data Relay Satellites.
TDRSS Tracking and Data Relay Satellite System.
TR Tape Recorder.
UARS Upper Atmosphere Research Satellite.
UTC Universal Time Coordinated.
UVIS Ultraviolet and optical channel.
WF/PC Wide Field/Planetary Camera.
WFC Wide Field Channel.

Index